Lecture Notes in Computer Science 12514

More information about this subseries at http://www.springer.com/series/7407

Stéphane Devismes · Neeraj Mittal (Eds.)

Stabilization, Safety, and Security of Distributed Systems

22nd International Symposium, SSS 2020
Austin, TX, USA, November 18–21, 2020
Proceedings

 Springer

Editors
Stéphane Devismes 🆔
Grenoble Alpes University
Saint Martin d'Hères, France

Neeraj Mittal 🆔
The University of Texas at Dallas
Richardson, TX, USA

ISSN 0302-9743   ISSN 1611-3349   (electronic)
Lecture Notes in Computer Science
ISBN 978-3-030-64347-8   ISBN 978-3-030-64348-5   (eBook)
https://doi.org/10.1007/978-3-030-64348-5

LNCS Sublibrary: SL1 – Theoretical Computer Science and General Issues

This Springer imprint is published by the registered company Springer Nature Switzerland AG
The registered company address is: Gewerbestrasse 11, 6330 Cham, Switzerland

Preface

This volume contains the papers presented at the 22nd International Symposium on Stabilization, Safety, and Security of Distributed Systems (SSS 2020), held during November 18-21, 2020.

SSS is an international forum for researchers and practitioners in the design and development of distributed systems with a focus on systems that are able to provide guarantees on their structure, performance, and/or security in the face of an adverse operational environment.

SSS started as the Workshop on Self-Stabilizing Systems (WSS), the first two of which were held in Austin, USA, in 1989 and in Las Vegas, USA, in 1995. After 1995, the workshop was held biennially until 2005 when it became an annual event. Starting from then, it broadened its scope and attracted researchers from other communities. In 2006, the name of the conference was changed to the International Symposium on Stabilization, Safety, and Security of Distributed Systems (SSS). The last three SSS conferences were held in Boston, USA (2017), Tokyo, Japan (2018), and Pisa, Italy (2019).

This year, the conference was initially planned to be held in Austin, USA. However, due to the COVID-19 worldwide pandemic, we decided to switch to a virtual event. Anish Arora (The Ohio State University, USA) and Sandeep Kulkarni (Michigan State University, USA) served as general chairs. Stéphane Devismes (Université Grenoble Alpes, France) and Neeraj Mittal (The University of Texas at Dallas, USA) served as program chairs. The authors were asked to align their submission with one of the four tracks:

- Track A: Self-Stabilization (chaired by Sayaka Kamei, Hiroshima University, Japan)
- Track B: Foundations of Concurrent and Distributed Computing (chaired by Luìs E. T. Rodrigues, Universidade de Lisboa, Portugal)
- Track C: Mobile and Robot Computing (chaired by Gokarna Sharma, Kent-State University, USA)
- Track D: Fault tolerance, Security, and Privacy (chaired by Emmanuel Anceaume, IRISA, CNRS, France)

A total of 44 submissions were received. The submissions spanned a wide spectrum of distributed computing topics, from fundamental principles to real world applications. All the manuscripts went through a rigorous review process. Each submission was reviewed by at least three Program Committee members. We are grateful to the reviewers for their hard work and valuable feedback. Overall, the committee decided to accept 16 regular papers and 7 brief announcements. The proceedings also include

2 invited papers. We hope that you will enjoy reading these excellent and very selective papers.

October 2020
Stéphane Devismes
Neeraj Mittal

Organization

Program Committee

Yehuda Afek	Tel Aviv University, Israel
Tristan Allard	Université de Rennes 1, France
Emmanuelle Anceaume	IRISA, CNRS, France
Anish Arora	The Ohio State University, USA
James Aspnes	Yale University, USA
John Augustine	Indian Institute of Technology, Madras, India
Rida Bazzi	Arizona State University, USA
Alysson Bessani	Universidade de Lisboa, Portugal
Olivier Blazy	Université de Limoges, France
Silvia Bonomi	Sapienza University of Rome, Italy
Costas Busch	Augusta University, USA
Armando Castaneda	UNAM, Mexico
Antonella Del Pozzo	CEA LIST, France
Stéphane Devismes	Université Grenoble Alpes, France
Felicita Di Giandomenico	ISTI-CNR, Italy
Giuseppe Antonio Di Luna	Sapienza University of Rome, Italy
Shlomi Dolev	Ben-Gurion University of the Negev, Israel
Swan Dubois	Sorbonne University, France
Xavier Défago	Tokyo Institute of Technology, Japan
Paola Flocchini	University of Ottawa, Canada
Roy Friedman	Technion, Israel
Vijay Garg	The University of Texas at Austin, USA
Sukumar Ghosh	The University of Iowa, USA
Vincent Gramoli	The University of Sydney, Australia
Michel Hurfin	Inria, France
Hirotsugu Kakugawa	Ryukoku University, Japan
Sayaka Kamei	Hiroshima University, Japan
Ajay Kshemkalyani	University of Illinois at Chicago, USA
Sandeep Kulkarni	Michigan State University, USA
Anissa Lamani	Université de Strasbourg, France
Christoph Lenzen	MPI for Informatics, Germany
Romaric Ludinard	IMT Atlantique, France
Partha Sarathi Mandal	Indian Institute of Technology Guwahati, India
Toshimitsu Masuzawa	Osaka University, Japan
Neeraj Mittal	The University of Texas at Dallas, USA
Anisur Rahaman Molla	Indian Statistical Institute, Kolkata, India
Achour Mostéfaoui	Université de Nantes, France

Krishnendu Mukhopadhyaya	Indian Statistical Institute, Kolkata, India
Mikhail Nesterenko	Kent State University, USA
Fukuhito Ooshita	Nara Institute of Science and Technology, Japan
Matthieu Perrin	Université de Nantes, France
Maria Potop-Butucaru	UPMC - Sorbonne Universités, LIP6, France
Giuseppe Prencipe	Università di Pisa, Italy
Binoy Ravindran	Virginia Tech, USA
Luis Rodrigues	Universidade de Lisboa, Portugal
Stephane Rovedakis	Conservatoire National des Arts et Métiers, France
Christian Scheideler	Paderborn University, Germany
Elad Schiller	Chalmers University of Technology, Sweden
Gokarna Sharma	Kent State University, USA
Yuichi Sudo	Osaka University, Japan
Sébastien Tixeuil	Université Pierre et Marie Curie, France
Jerry Trahan	Louisiana State University, USA
Philippas Tsigas	Chalmers University of Technology, Sweden
Nitin Vaidya	Georgetown University, USA
Ramachandran Vaidyanathan	Louisiana State University, USA

Additional Reviewers

Bramas, Quentin
Cohen, Avi
Durand, Anaïs
Götte, Thorsten
Hinnenthal, Kristian
Hood, Kendric
Masetti, Giulio
Moses Jr., William K.
Muzina, Joe
Oglio, Joseph
Pattanayak, Debasish

Contents

Store-Collect in the Presence of Continuous Churn with Application to Snapshots and Lattice Agreement

Hagit Attiya[1]([⊠]), Sweta Kumari[1], Archit Somani[1], and Jennifer L. Welch[2]

[1] Department of Computer Science, Technion, Haifa, Israel
{hagit,sweta,archit}@cs.technion.ac.il
[2] Department of Computer Science and Engineering, Texas A&M University,
College Station, TX, USA
welch@cse.tamu.edu

Abstract. We present an algorithm for implementing a store-collect object in an asynchronous crash-prone message-passing dynamic system, where nodes continually enter and leave. The algorithm is very simple and efficient, requiring just one round trip for a store operation and two for a collect. We then show the versatility of the store-collect object for implementing churn-tolerant versions of useful data structures, while shielding the user from the complications of the underlying churn. In particular, we present elegant and efficient implementations of atomic snapshot and generalized lattice agreement objects that use store-collect.

Keywords: Store-collect object · Dynamic message-passing systems · Churn · Crash resilience · Atomic snapshots · Generalized lattice agreement

1 Introduction

A popular programming technique that contributes to designing provably-correct distributed applications is to use shared objects for interprocess communication, instead of more low-level techniques. Although shared objects are a convenient abstraction, they are not generally provided in large-scale distributed systems; instead, nodes keep copies of the data and communicate by sending messages to keep the copies consistent.

Dynamic distributed systems allow computing nodes to enter and leave the system at will, either due to failures and recoveries, moving in the real world, or changes to the systems' composition, a process called *churn*. Motivating applications include those in peer-to-peer, sensor, mobile, and social networks, as well as server farms. We focus on the situation when the network is always fully connected, which could be due to, say, an overlay network. A broadcast mechanism is assumed through which a node can send a message to all nodes present in the system.

The usefulness of shared memory programming abstractions has been long established for static systems (e.g., [4,5]), which have known bounds on the number of fixed computing nodes and the number of possible failures. This success has inspired work

Supported by ISF grant 380/18 and NSF grant 1816922; full paper in [9].

S. Devismes and N. Mittal (Eds.): SSS 2020, LNCS 12514, pp. 1–15, 2020.
https://doi.org/10.1007/978-3-030-64348-5_1

on providing the same for newer, dynamic, systems. However, most of this work has shown how to simulate a shared read-write register (e.g., [2,6,10,11,18]). We discuss a couple of exceptions [12,21] below.

In this paper, we promote the *store-collect* shared object [7] (defined in Sect. 2) as a primitive well-suited for dynamic message-passing systems with an ever-changing set of participants. Each node can store a value in a store-collect object with a STORE operation and can collect the latest value stored by each node with a COLLECT operation. Inherent in the specification of this object is an ability to track the set of participants and to read their latest values.

Below we elaborate on three advantageous features of the store-collect object: The store-collect semantics is well-suited to dynamic systems and can be implemented easily and efficiently in them; the widely-used atomic snapshot object can be implemented on top of a store-collect object; and a variety of other commonly-used objects can be implemented either directly on top of a store-collect or on top of an atomic snapshot object. These implementations are simple and inherit the properties of being churn-tolerant and efficient, showing that store-collect combines algorithmic power and efficiency.

A churn-tolerant store-collect object can be implemented fairly easily. We adopt essentially the same system model as in [6], which allows ongoing churn as long as not too many churn events take place during the length of time that a message is in transit. To capture this constraint, there is an assumed upper bound D on the maximum message delay, but no (positive) lower bound. Nodes do not know D and have no local clocks, causing consensus to be unsolvable [6]. The model differentiates between nodes that crash and nodes that leave; nodes that have entered but not left are considered present even if crashed. The number of nodes that can be crashed at any time is bounded by a fraction of the number of nodes present at that time. During any time interval of length D, the number of nodes entering or leaving is a fraction of the number of nodes present in the system at the beginning of the interval. (See Sect. 3 for model details.)

Our algorithm for implementing a churn-tolerant store-collect object is based on the read-write register algorithm in [6]. It is simple and efficient: once a node joins, it completes a store operation within one round-trip, and a collect operation within two round-trips. The store-collect object satisfies a variant of the "regularity" consistency condition, which is weaker than linearizability [19]. In contrast to our single-round-trip store operation, the write operation in the algorithm of [6] requires two round trips. Another difference between the algorithms is that in ours, each node keeps a local set of tuples with an entry for each known node and its value instead of a single value; when receiving new information, instead of overwriting the single value, our algorithm merges the new information with the old. One contribution of our work in this paper is a significantly revised proof of the churn management protocol that is much simpler than that in [6], consequently making it easier to build on the results. (See Sect. 4.)

Building an atomic snapshot on top of a store-collect object is easy! We present a simple algorithm with an elegant correctness proof (Sect. 5.1). One may be tempted to implement an atomic snapshot in our model by plugging churn-tolerant registers (e.g., [6]) into the original algorithm of [1]. Besides needlessly sequentializing accesses to the registers, such an implementation would have to track the current set of participants. A store-collect object which encapsulates the changing participants and collects information from them in parallel, yields a simple algorithm very similar in spirit to the original but whose round complexity is linear instead of quadratic in the number of participants. The key subtlety of the algorithm is the mechanism for knowing when to borrow a scan in spite of difficulties caused by the churn, in order to ensure termination.

Atomic snapshot objects have numerous uses in static systems, e.g., to build multi-writer registers, concurrent timestamp systems, counters, and accumulators, and to solve approximate agreement and randomized consensus (cf. [1,4]). In addition to analogous applications, we show (Sect. 5.2) how a churn-tolerant atomic snapshot object can be used to provide a churn-tolerant generalized lattice agreement object [15]. This object supports a PROPOSE operation whose argument is a value belonging to a lattice and whose response is a lattice value that is the join of some subset of all prior input values, including its own argument. Generalized lattice agreement is an extension of (single-shot) *lattice agreement*, well-studied in the static shared memory model [8]. Generalized lattice agreement has been used to implement many objects [13,15], most notably, *conflict-free replicated data types* [21,24,25].

The store-collect object specification is versatile. Our atomic snapshot and generalized lattice agreement algorithms demonstrate that layering linearizability on top of a store-collect object is easy. Yet not every application needs the costs associated with linearizability, and store-collect gives the flexibility to avoid them. Our approach to providing churn-tolerant shared objects is modular, as the underlying complications of the message-passing and the churn is hidden from higher layers by our store-collect implementation. As evidence, we observe in [9] that store-collect allows very simple implementations of max-registers, abort flags, and sets, in which an implemented operation takes at most a couple of store and collect operations. The choice of problems and the algorithms follow [21] but the algorithms inherit good efficiency and churn-tolerance properties from our store-collect implementation.

Related Work: An algorithm that directly implements an atomic snapshot object in a *static message-passing system*, bypassing the use of registers, is presented in [14]. This algorithm includes several nice optimizations to improve the message and round complexities. These include speeding up the algorithm by parallelizing the collect, as is already encapsulated in our store-collect algorithm. Our atomic snapshot algorithm works in a *dynamic* system and has a shorter and simpler proof of linearizability.

Aguilera [3] presents a specification and algorithm for atomic snapshots in a *dynamic model* in which nodes can continually enter and communicate via *shared registers*. This algorithm is then used for group membership and mutual exclusion in that model. Variations of the model were proposed in [17,22], which provided algorithms for election, mutual exclusion, consensus, collect, snapshot, and renaming. Spiegelman and Keidar [23] present atomic snapshot algorithms for a crash-prone dynamic system in which processes communicate via shared registers. Their algorithms uniquely identify each scan operation with a version number to help determine when a scan can be borrowed; we use a similar mechanism in our snapshot algorithm. However, our atomic snapshot algorithm uses a shared store-collect object which tolerates *ongoing* churn. Our use of a non-linearizable building block requires a more delicate approach to proving linearizability, as we cannot simply choose, say, a specific write to an atomic register as the linearization point of an update, as can be done in [23].

The problem of implementing shared objects in the presence of *ongoing* churn and crash failures in *message-passing systems* is studied in [10, 11], for read-write registers, and [12], for sets. Unlike our results, these papers assume the system size is restricted to a fixed window and the system is eventually synchronous. Like our algorithms, the set algorithm in [12] uses unbounded local memory at the nodes.

A popular alternative way to model churn in *message-passing systems* is as a sequence of quorum configurations, each of which consists of a set of nodes and a quorum system over that set (e.g., [2, 16, 18, 20, 21]). Explicit reconfiguration operations replace older configurations with newer ones. The assumptions made in [2, 16, 18, 20, 21] are incomparable with those in [6] and in our paper, as the former assume churn eventually stops while the latter assume the churn is bounded.

Most papers on generalized lattice agreement have assumed static systems (cf. [8, 13, 15, 24, 25]. A notable exception is [21], which considers dynamic systems subject to changes in the composition due to reconfiguration. This paper provides an implementation for a large class of shared objects, including conflict-free replicated data types, that can be modeled as a lattice. By showing how to view the state of the system as a lattice as well, the paper elegantly combines the treatment of the reconfiguration and the operations on the object. Unlike our work, the algorithms in [21] require that changes to the system composition eventually cease in order to ensure progress.

2   The Store-Collect Problem

A shared *store-collect object* [7] supports concurrent *store* and *collect* operations performed by some set of clients. Each operation has an invocation and response. For a *store* operation, the invocation is of the form $\text{STORE}_p(v)$, where v is a value drawn from some set and p indicates the invoking client, and the response is of the form ACK_p, indicating that the operation has completed. For a *collect* operation, the invocation is of the form COLLECT_p and the response is of the form $\text{RETURN}_p(V)$, where V is a *view*, that is, a set of client-value pairs without repetition of client ids. We use the notation $V(p)$ to indicate v if $\langle p, v \rangle \in V$ and $\bot$ if no pair in V has p as its first element.

Informally, the behavior required of a store-collect object is that each *collect* operation should return a view containing the latest value stored by each client. We do not require the *store* and *collect* operations to appear to occur instantaneously, that is, the object is not necessarily linearizable.

A sequence σ of invocations and responses of *store* and *collect* operations is a *schedule* if, for each client id p, the restriction of σ to invocations and responses by p consists of alternating invocations and matching responses, beginning with an invocation. Each invocation and its matching following response (if present) together make an operation. If the response of operation op comes before the invocation of operation op' in σ, then we say op *precedes* op' (in σ) and op' *follows* op. We assume that every value written in a *store* operation in a schedule is unique (a condition that can be achieved using sequence numbers and client ids).

A schedule σ satisfies *regularity for the store-collect problem* if:

- For each *collect* operation *cop* in σ that returns V and every client p, if $V(p) = \bot$, then no *store* operation by p precedes *cop* in σ. If $V(p) = v \neq \bot$, then there is a $\text{STORE}_p(v)$ invocation that occurs in σ before *cop* completes and no other *store* operation by p occurs in σ between this invocation and the invocation of *cop*.
- For every two *collect* operations in σ, cop_1 which returns V_1 and cop_2 which returns V_2, if cop_1 precedes cop_2 in σ, then for every $\langle p, v_1 \rangle \in V_1$, there exists v_2 such that $\langle p, v_2 \rangle \in V_2$ where either $v_1 = v_2$ or the $\text{STORE}_p(v_1)$ invocation occurs before the $\text{STORE}_p(v_2)$ invocation in σ. We denote this as $V_1 \preceq V_2$.

3   Overview of System Model

The events that can occur at a node p are entering the system (ENTER_p), leaving the system (LEAVE_p), crashing (CRASH_p), receiving a message m ($\text{RECEIVE}_p(m)$), and invoking an operation (COLLECT_p or $\text{STORE}_p(v)$). The occurrence of an event at node p results in changes to p's local state; optionally, a message to be broadcast; and optionally, a response, which is $\text{RETURN}_p(V)$ for a collect, ACK_p for a store, and JOINED_p for the enter. An *execution* is a collection of sequences of events, one sequence for each node, that satisfies certain conditions. The key points are the following.

A node enters, leaves, and crashes at most once. A node does nothing before it enters and after it crashes or leaves.

We assume that a nonnegative real number is associated with each event in an execution, which is the time when the event occurs. A node is *present* at time t if it entered but did not leave before time t; a crashed node is considered to be present. $N(t)$ is the number of nodes present at time t. There is a constant N_{min} such that $N(t) \geq N_{min}$ for all $t \geq 0$. A node is *active* at time t if it is present and not crashed at time t.

At time 0, the system consists of a finite nonempty set of nodes, S_0, that are considered by definition to be active. Initially nodes in S_0 have knowledge about all the nodes in S_0; every other node, which enters after time 0, has no initial knowledge about any node other than itself. A node is *joined* (or a *member*) at time t if it is present (has not left) and either is in S_0 or has experienced JOINED_p.

A broadcast service reliably delivers each message sent by node p at time t to each node q that is active throughout $[t, t + D]$, where $D > 0$ is a maximum message delay unknown to the nodes, if p's next event is not CRASH_p. If p's next event is to crash or if q enters, leaves, or crashes during the interval, there is no guarantee whether q receives the message. This is a weaker broadcast specification than in [6], which assumed broadcasts were atomic with respect to crashes. Messages from the same sender are received in the order they are sent. Every message that is received has delay in $(0, D]$.

Let $\alpha > 0$ and $0 < \Delta < 1$ be real numbers that denote the *churn rate* and *failure fraction*, respectively. The parameters α and Δ are known to the nodes. For all times $t \geq 0$, there are at most $\alpha \cdot N(t)$ enter and leave events in $[t, t + D]$ (*churn assumption*), and at most $\Delta \cdot N(t)$ nodes are crashed at time t (*failure fraction assumption*).

An algorithm is a *correct implementation of a store-collect object* if in every execution: *(1)* Every node that enters the system after time 0 and remains active eventually joins, and no joined node p experiences JOINED_p. *(2)* Every store or collect operation invoked at a node that remains active eventually completes. *(3)* The schedule resulting from restricting the execution to the *store* and *collect* invocations and responses satisfies regularity for the store-collect problem.

4   The Continuous Churn Collect (CCC) Algorithm

In our algorithm, nodes run *client* threads, which invoke *collect* and *store* operations, and *server* threads. We assume that the code segment that is executed in response to each event executes without interruption.

Our implementation adds a sequence number, *sqno*, to each value in a view, which is now a set of triples, $\{\langle p, v, sqno \rangle, \ldots\}$, without repetition of node ids. We use the notation $V(p) = v$ if there exists *sqno* such that $\langle p, v, sqno \rangle \in V$, and $\perp$ if no triple in V has p as its first element. A *merge* of two views picks the latest value in each view. That is, given two views V_1 and V_2, $merge(V_1, V_2)$ is the subset of $V_1 \cup V_2$ consisting of every triple whose node id is in one of V_1 and V_2 but not the other, and, for node ids that appear in both V_1 and V_2, it contains only the triple with the larger sequence number. Note that $V_1, V_2 \preceq merge(V_1, V_2)$.

A node p tracks the composition of the system with a set *Changes* of events concerning the nodes that have entered the system. Initially, node p's *Changes* set equals $\{enter(q)|q \in S_0\} \cup \{join(q)|q \in S_0\}$, if $p \in S_0$, and $\emptyset$ otherwise. Node p also maintains a set of nodes that it believes are present: *Present* $= \{q|enter(q) \in Changes \wedge leave(q) \notin Changes\}$, i.e., nodes that have entered, but have not left, as far as p knows. The code for managing these sets (Algorithm 1) is the same as [6] except for Line 5, which merges newly received information with current local information instead of overwriting it. Once a node has joined, its client thread handles *collect* and *store* operations (Algorithm 2) and its server thread (Algorithm 3) responds to clients. The client at node p maintains a derived variable *Members* $= \{q \mid join(q) \in Changes \wedge leave(q) \notin Changes\}$ of nodes it considers as members, i.e., nodes that have joined but not left.

Each node keeps a local copy of the current view in its *LView* variable. In a *collect operation*, a client thread requests the latest value of servers' local views using a **collect-query** message (Line 29). When a server node p receives a **collect-query** message, it responds with its local view (*LView*) through a **collect-reply** message (Line 53) if p has joined the system. When the client receives a **collect-reply** message, it merges its *LView* with the *received view (RView)*, to get the latest value corresponding to each node (Line 31). Then the client waits for sufficiently many **collect-reply** messages before broadcasting the current value of its *LView* variable in a **store** message (Line 36). When server p receives a **store** message with a view *RView*, it merges *RView* with its local *LView* (Line 50) and, if p is joined, it broadcasts **store-ack** (Line 48). The client waits for sufficiently many **store-ack** messages before returning *LView* to complete the *collect* (Line 47); this threshold is recalculated in Line 34 to reflect possible changes to the system composition that the client has observed.

In a *store operation*, a client thread updates its local variable *LView* to reflect the new value by doing a merge (Line 39) and broadcasts a **store** message (Line 42). When server p receives a **store** message with view *RView*, it merges *RView* with its local *LView* (Line 48) and, if p is joined, it broadcasts **store-ack** (Line 50). The client waits for sufficiently many **store-ack** messages before completing the *store* (Line 46).

Algorithm 1. CCC—Common code managing churn, for node p.

Local Variables:

$LView$: set of (node id, value, sequence number) triples, initially $\emptyset$ // local view

is_joined: Boolean, initially false   // true iff p has joined the system

$join\_threshold$: int, initially 0   // number of enter-echo messages needed for joining

$join\_counter$: int, initially 0   // number of enter-echo messages received so far

$Changes$: set of $enter(q)$, $leave(q)$, and $join(q)$ // active membership events known to p

    initially $\{enter(q)|q \in S_0\} \cup \{join(q)|q \in S_0\}$ if $p \in S_0$, and $\emptyset$ otherwise

Derived Variable:

$Present = \{q \mid enter(q) \in Changes \wedge leave(q) \notin Changes\}$

When ENTER$_p$ **occurs:**

1: **add** $enter(p)$ **to** $Changes$

2: **broadcast** $\langle$**enter,** $p\rangle$

When RECEIVE$_p\langle$**enter,** $q\rangle$ **occurs:**

3: **add** $enter(q)$ **to** $Changes$

4: **broadcast**$\langle$**enter-echo,**$Changes$, $LView$, is_joined, $q\rangle$

When RECEIVE$_p\langle$**enter-echo,** C, $RView$, j, $q\rangle$ **occurs:**

5: $LView$ **=** merge($LView$, $RView$)

6: $Changes = Changes \cup C$

7: **if** $\neg is\_joined \wedge (p == q)$ **then**

8:     **if** $(j ==true) \wedge (join\_threshold == 0)$ **then**

9:         $join\_threshold = \gamma \cdot |Present|$

10:        $join\_counter$++

11:        **if** $join\_counter \geq join\_threshold > 0$ **then**

12:            $is\_joined = true$

13:            **add** $join(p)$ **to** $Changes$

14:        **broadcast** $\langle$**join,** $p\rangle$

15:        **return** JOINED$_p$

When RECEIVE$_p\langle$**join,** $q\rangle$ **occurs:**

16: **add** $join(q)$ **to** $Changes$

17: **add** $enter(q)$ **to** $Changes$

18: **broadcast** $\langle$**join-echo,** $q\rangle$

When RECEIVE$_p\langle$**join-echo,** $q\rangle$ **occurs:**

19: **add** $join(q)$ **to** $Changes$

20: **add** $enter(q)$ **to** $Changes$

When LEAVE$_p$ **occurs:**

21: **broadcast** $\langle$**leave,** $p\rangle$

22: **halt**

When RECEIVE$_p\langle$**leave,** $q\rangle$ **occurs:**

23: **add** $leave(q)$ **to** $Changes$

24: **broadcast** $\langle$**leave-echo,** $q\rangle$

When RECEIVE$_p\langle$**leave-echo,** $q\rangle$ **occurs:**

25: **add** $leave(q)$ **to** $Changes$

The fraction β is used to calculate the number of messages that should be received (stored in local variable *threshold*) based on the size of the *Members* set, for the operation to terminate. Setting β is a key challenge in the algorithm as setting it too small might not return correct information from *collect* or *store*, whereas setting it too large might not guarantee termination of the *collect* and *store*.

We define a *phase* to be the execution by a client node p of one of the following intervals of its code: *(1)* lines 26 through 33, the first part of a *collect* operation, *(2)* lines 34 through 36 and 43 through 47, the second part of a *collect* operation called the "store-back", or *(3)* lines 37 through 46, the entirety of a *store* operation. The first kind of phase is called a *collect phase* while the second and third kinds are called a *store phase*. For any completed phase φ executed by node p, define $view(\varphi)$ to be the value of $LView_p^t$, where t is the time at the end of the phase. Since a *store* operation consists solely of a store phase, we also apply the notation to an entire *store* operation.

Algorithm 2. CCC—Client code, for node p.

Local Variables:

optype: string, initially $\bot$ // indicates which type of operation (*collect* or *store*) is pending

tag: int, initially 0   // counter to identify currently pending operation by p

threshold: int, initially 0   // number of replies/acks needed for current phase

counter: int, initially 0   // number of replies/acks received so far for current phase

sqno: int, initially 0   // sequence number for values stored by p

Derived Variable:

Members $= \{q|\ join(q) \in Changes \wedge leave(q) \notin Changes\}$

When COLLECT$_p$ **occurs:**

26: *optype = collect; tag++*

27: *threshold = $\beta \cdot |Members|$*

28: *counter = 0*

29: **broadcast** ⟨**collect-query**, *tag, p*⟩

When RECEIVE$_p$⟨**collect-reply**, *RView, t, q*⟩ **occurs:**

30: **if** *(t == tag)* $\wedge (q == p)$ **then**

31:     *LView = merge(LView, RView)*

32:     *counter++*

33:     **if** *(counter $\geq$ threshold)* **then**

34:         *threshold = $\beta \cdot |Members|$*

35:         *counter = 0*

36:         **broadcast** ⟨**store**, *LView, tag, p*⟩

When STORE$_p$(v) **occurs:**

37: *optype = store; tag++*

38: *sqno++*

39: *LView = merge(LView, {⟨p, v, sqno⟩})*

40: *threshold = $\beta \cdot |Members|$*

41: *counter = 0*

42: **broadcast** ⟨**store**, *LView, tag, p*⟩

When RECEIVE$_p$⟨**store-ack**, *t, q*⟩ **occurs:**

43: **if** *(t == tag)* $\wedge (q == p)$ **then**

44:     *counter++*

45:     **if** *(counter $\geq$ threshold)* **then**

46:         **if** *(optype==store)* **then return** ACK

47:         **else return** *LView*

Algorithm 3. CCC—Server code, for node p.

When RECEIVE$_p$⟨**store**, *RView,tag, q*⟩ **occurs**

48: *LView = merge(LView, RView)*

49: **if** *is_joined* **then**

50:     **broadcast** ⟨**store-ack**, *tag, q*⟩

51: **broadcast** ⟨**store-echo**, *LView*⟩

When RECEIVE$_p$⟨**collect-query**, *tag, q*⟩ **occurs:**

52: **if** *is_joined* **then**

53:     **broadcast** ⟨**collect-reply**, *LView, tag, q*⟩

When RECEIVE$_p$⟨**store-echo**, *RView*⟩ **occurs:**

54: *LView = merge(LView, RView)*

To prove the correctness of the algorithm, consider any execution of the algorithm. The correctness of the algorithm relies on the following constraints ($Z = \left[(1-\alpha)^3 - \Delta \cdot (1+\alpha)^3\right]$, which is the fraction of nodes that survive an interval of length $3D$):

$$N_{min} \geq \frac{1}{Z + \gamma - (1+\alpha)^3} \qquad (A)$$

$$\gamma \leq Z/(1+\alpha)^3 \qquad (B)$$

$$\beta \leq Z/(1+\alpha)^2 \qquad (C)$$

$$\beta > \frac{(1-Z)(1+\alpha)^5 + (1+\alpha)^6}{((1-\alpha)^3 - \Delta \cdot (1+\alpha)^2)((1+\alpha)^2 + 1)} \qquad (D)$$

Fortunately, there are values for the parameters α, Δ, γ, and β that satisfy these constraints. In the extreme case when $\alpha = 0$ (i.e., no churn), the failure fraction Δ can be as large as 0.21; in this case, it suffices to set both γ and β to 0.79 for any value of N_{min} that is at least 2. As α increases up to 0.04, Δ must decrease approximately linearly until reaching 0.01; in this case, it suffices to set γ to 0.77 and β to 0.80 for any value of N_{min} that is at least 2. The following technical claims hold:

Lemma 1. *For all $i \in \mathbb{N}$ and all $t \geq 0$, (a) at most $((1 + \alpha)^i - 1) \cdot N(t)$ nodes enter during $(t, t + i \cdot D]$; and (b) $N(t + i \cdot D) \leq (1 + \alpha)^i \cdot N(t)$.*

Lemma 2. *For any interval $[t_1, t_2]$ with $t_2 - t_1 \leq 3D$, where S is the set of nodes present at t_1, at least $Z \cdot |S|$ of the nodes in S are active at t_2.*

Below, a local variable name is subscripted with p and superscripted with t to denote its value in node p at time t; e.g., v_p^t is the value of node p's local variable v at time t.

In the analysis, we will frequently be comparing the data in nodes' *Changes* sets to the set of ENTER, JOINED, and LEAVE events that have actually occurred in a certain interval. We are especially interested in these events that trigger a broadcast invoked by a node that is not in the middle of crashing, as these broadcasts are guaranteed to be received by all nodes that are present for the requisite interval. We call these *active membership events*. Because of the assumed initialization of the nodes in S_0, we use the convention that the set of active membership events occurring in the interval $[0, 0]$ is $\{enter(p) | p \in S_0\} \cup \{join(p) | p \in S_0\}$. The next lemma holds:

Lemma 3. *For every node p and all times t such that p is joined and active at t, $Changes_p^t$ contains all the active membership events for $[0, \max\{0, t - 2D\}]$.*

We can prove that a node that is active sufficiently long eventually joins.

Theorem 1. *Every node p that enters at some time t and is active for at least $2D$ time joins by time $t + 2D$.*

Theorem 2. *A phase invoked by a client that remains active completes within $2D$ time.*

Proof sketch. Consider a phase invoked by node p at time t. Let S be the set of nodes present at time $\max\{0, t - 2D\}$. Lemma 2 and Theorem 1 imply that at least $Z \cdot |S|$ of them are joined by time t and active at time $t + D$, and thus respond to p's message. We argue that $Z \cdot |S|$ is at least as large as the value of *threshold* computed by p in Line 27 or 34 or 40 of Algorithm 2. We first show that $|S| \geq |Present_p^t|/(1 + \alpha)^2$, by Lemmas 3 and 1(a). Then we note that Constraint C implies that $Z/(1 + \alpha)^2 \geq \beta$, and so $Z \cdot |S| \geq \beta \cdot |Present_p^t|$. Since $|Present_p^t| \geq |Members_p^t|$, the definition of *threshold* gives the result. $\square$

The next lemma shows that the view of a store phase is smaller, in the partial order $\preceq$, than the view of a subsequent, non-overlapping, collect phase.

Lemma 4. *For any store phase s and any collect phase c, if s finishes before c starts and c terminates, then $view(s) \preceq view(c)$.*

Theorem 3. *The schedule resulting from the restriction of the execution to the* store *and* collect *invocations and responses satisfies regularity for the store-collect problem.*

Proof. (1) Suppose *cop* is a *collect* operation that returns view V. Let c be the collect phase of *cop*. Let p be a node. If $V(p) = \perp$ and a *store* operation by p, consisting of store phase s, precedes *cop*, then, by Lemma 4, $view(s) \preceq view(c)$. Hence, $view(s)$ contains a tuple for p with a non-$\perp$ value, which is a contradiction.

Therefore, $V(p) = v \neq \perp$. We show that a $\text{STORE}_p(v)$ invocation occurs before *cop* completes and no other *store* operation by p occurs between this invocation and the invocation of *cop*. A simple induction shows that every (non-$\perp$) value for one node in another node's *LView* variable at some time comes from a STORE invocation by the first node that has already occurred. Since V is the value of the invoking node's *LView* variable when *cop* completes, there is a previous $\text{STORE}_p(v)$ invocation.

Now suppose for the sake of contradiction that the $\text{STORE}_p(v)$ completes—call this operation *sop*—and there is another *store* operation by p, call it *sop'*, that follows *sop* and precedes *cop*. Let v' be the value of *sop'*; by the assumption of unique values, $v \neq v'$. Since *sop* and *sop'* are executed by the same node, it is easy to see from the code that $view(sop) \preceq view(sop')$. By Lemma 4, $view(sop') \preceq view(c) = V$. But then value v is superseded by value $v' \neq v$, contradicting the assumption that $V(p) = v$.

(2) Suppose cop_1 and cop_2 are two *collect* operations such that cop_1 returns V_1, cop_2 returns V_2, and cop_1 precedes cop_2. Note that cop_1 contains a store phase s which finishes before the collect phase c of cop_2 begins. By Lemma 4, $view(s) \preceq view(c)$. Regularity holds since $view(s) = V_1$ and $view(c) = V_2$, implying that $V_1 \preceq V_2$. □

By Theorem 1, every node that enters and remains active sufficiently long eventually joins. Since a *store* operation consists of a store phase and a *collect* operation consists of a collect phase followed by a store phase, by Theorem 2, every operation eventually completes as long as the invoker remains active. Finally, Theorem 3 ensures regularity.

Corollary 1. CCC *is a correct implementation of a store-collect object, in which each* Store *or* Collect *completes within a constant number of communication rounds.*

5   Implementing Distributed Objects Despite Continuous Churn

We show implementations of two objects using store-collect. Three additional objects are discussed in [9]. For all applications, we assume that the conditions for store-collect termination hold, which guarantees termination of the operations.

5.1   Atomic Snapshots

Like other atomic snapshot algorithms [1, 14, 23], our algorithm uses repeated collects to identify an atomic scan when two collects return the same collected views. Updates help scans to complete by embedding an atomic scan that can be borrowed by over-lapping scans they interfere with. The set from which the values to be stored in the snapshot object are taken is denoted Val_{AS}. A *snapshot view* is a subset of $\Pi \times Val_{AS}$, i.e., a set of (node id, value) pairs, without duplicate node ids.

An *atomic snapshot* provides two operations: SCAN(), which has no arguments and returns a snapshot view, and UPDATE(v), which takes a value $v \in Val_{AS}$ as an argument and returns ACK. Its sequential specification consists of all sequences of updates and scans in which the snapshot view returned by a SCAN contains the value of the last preceding UPDATE for each node p, if such an UPDATE exists, and no value, otherwise.

An implementation should be *linearizable* [19]. Roughly speaking, for every execution α, we should find a sequence of operations, containing all completed operations in α and some of the pending operations, which is in the sequential specification of an atomic snapshot, and preserves the real-time order of non-overlapping operations in α.

Our algorithm to implement an atomic snapshot uses a store-collect object, whose values are taken from the set ($\mathcal{P}$ indicates the power set of its argument):

$$ Val_{SC} = Val_{AS} \times \mathbb{N} \times \mathbb{N} \times \mathcal{P}(\Pi \times Val_{AS}) \times \mathcal{P}(\Pi \times \mathbb{N}) $$

The first component (*val*) holds the argument of the most recent update invoked at p. The second component (*usqno*) holds the number of updates performed by p. The third component (*ssqno*) holds the number of scans performed by p. The fourth component (*sview*) holds a snapshot view that is the result of a recent scan done by p; it is used to help other nodes complete their scans. The fifth component (*scounts*) holds a set of counts of how many scans have been done by the other nodes, as observed by p. The projection of an element v in Val_{SC} onto a component is denoted, respectively, $v.val$, $v.usqno$, $v.ssqno$, $v.sview$, $v.scounts$.

A *store-collect view* is a subset of $\Pi \times Val_{SC}$, i.e., a set of (node id, value) pairs, with no duplicate node ids. We extend the projection notation to a store-collect view V, so that $V.comp$ is the result of replacing each tuple $\langle p, v \rangle$ in V with $\langle p, v.comp \rangle$; when $v.comp = \bot$, the tuple is omitted. Recall that for any kind of view V, $V(p)$ is the second component of the pair whose first component is p ($\bot$ if there is no such pair).

To execute a SCAN, Algorithm 4 increments the scan sequence number (*ssqno*) (Line 55) and stores it in the shared store-collect object with all the other components unchanged, indicated by the $-$ notation. Then, a view is collected (Line 57). In a while loop, the last collected view is saved and a new view is collected (Line 59). If the two most recently collected views are equal (Line 60), the latest collected view is returned (Line 61). We call this a *successful double collect*, and say that this is a *direct scan*. Otherwise, the algorithm checks whether the last collected view contains a node q that has observed its own *ssqno*, by checking the *scounts* component (Line 62). If this condition holds, the snapshot view of q is returned (Line 63); we call this a *borrowed scan*.

An UPDATE first obtains all scan sequence numbers from a collected view and assigns them to a local variable *scounts* (Line 64). Next, the value of an embedded scan is saved in a local variable *sview* (Line 65). Then it sets its *val* variable to the argument value and increments its update sequence number (Lines 66 and 67). Finally the new value, update sequence number, collected view, and set of scan sequence numbers are stored; the node's own scan sequence number is unchanged (Line 68).

To prove linearizability, we consider an execution and specify an ordering of all the completed scans and all the updates whose store on Line 68 takes effect. The ordering takes into consideration the embedded scans, which are inside updates, as well as the "free-standing" scans; since scans do not change the state of the atomic snapshot object, it is permissible to do so. We first show that direct scans are comparable in the $\preceq$ order.

Algorithm 4. Atomic snapshot: code for node p.

Local Variables:
$ssqno$: int, initially 0   // counts how many scans p has invoked so far
$scounts$: set of (node id, integer) pairs with no duplicate node ids; initially $\emptyset$
val: an element of Val_{AS}, initially $\bot$ // argument of most recent update invoked by p
$usqno$: int, initially 0   // number of updates p has invoked so far
$sview$: a snapshot view, initially $\emptyset$ // the result of recent embedded scan by p
V_1, V_2: store-collect views, both initially $\emptyset$

When $\text{SCAN}_p()$ **occurs:**
55: $ssqno$++
56: $\text{STORE}_p(\langle -, -, ssqno, -, - \rangle)$
57: $V_1 = \text{COLLECT}_p()$
58: **while** $true$ **do**
59:      $V_2 = V_1; V_1 = \text{COLLECT}_p()$
60:      **if** $(V_1 == V_2)$ **then**
61:          **return** $V_1.val$ // **direct scan**
62:      **if** $\exists q$ such that
          $\langle p, ssqno \rangle \in V_1(q).scounts$ **then**

63:          **return** $V_1(q).sview$
            // **borrowed scan**

When $\text{UPDATE}_p(v)$ **occurs:**
64: $scounts = \text{COLLECT}_p().ssqno$
65: $sview = \text{SCAN}_p()$ // **embedded scan**
66: $val = v$
67: $usqno$++
68: $\text{STORE}_p(\langle val, usqno, -, sview, scounts \rangle)$
69: **return** ACK

Lemma 5. *If a direct scan by node p returns V_1 and a direct scan by node q returns V_2, then either $V_1 \preceq V_2$ or $V_2 \preceq V_1$.*

Proof. Let cop_p^1 and cop_p^2 be the last two collects of p (both returning V_1), and cop_q^1 and cop_q^2 be the last two collects of q (both returning V_2). We have that either cop_p^1 completes before cop_q^2 starts or cop_q^1 completes before cop_p^2 starts. In the former case, by the regularity of store-collect, $V_1 \preceq V_2$, while in the latter case, $V_2 \preceq V_1$. □

Consider all direct scans in the order they complete and place them by the comparability order. If a direct scan returning snapshot view V_1 precedes another direct scan returning snapshot view V_2, then the regularity of store-collect ensures $V_1 \preceq V_2$. Hence, this ordering preserves the real-time order of non-overlapping direct scans.

The next lemma helps to order borrowed scans. Its statement is based on the observation that if a scan sop_p by node p borrows the snapshot view in $V_1(q)$, then there is an update uop_q by q that writes this view (via a store).

Lemma 6. *If a scan sop_p by node p borrows from a scan sop_q by node q, then sop_q starts after sop_p starts and completes before sop_p completes.*

Proof. Let uop_q be the update in which sop_q is embedded. Since sop_p borrows the snapshot view of sop_q, its $ssqno$ appears in $scounts$ of q's value in the view collected in Line 59. The properties of store-collect imply that the collect of uop_q (Line 64) does not complete before the store of p (Line 56) starts. Hence, sop_q (called in Line 65) starts after sop_p starts. Furthermore, since the collect of p returns the snapshot view stored after sop_q completes (Line 68), sop_q completes before sop_p completes. □

For every borrowed scan sop_1, there exists a chain of scans $sop_2, sop_3, \ldots, sop_k$ such that sop_i borrows from sop_{i+1}, $1 \leq i < k$, and sop_k is a direct scan *from which*

sop_1 *borrows*. Consider all borrowed scans in the order they complete and place each borrowed scan after the direct scan it borrows from, as well as all previously linearized borrowed scans that borrow from the same direct scan. Applying Lemma 6 inductively, sop_k starts after sop_1 starts and completes before sop_1 completes, i.e., the direct scan from which a scan borrows is completely contained, in the execution, within the borrowing scan. This fact, together with the rule for ordering borrowed scans, implies that the real-time order of any two scans, at least one of which is borrowed, is preserved since direct scans have already been shown to be ordered properly.

Finally, we consider all updates in the order their stores (Line 68) start. Place each update, say *uop* by node p with argument v, immediately before the first scan whose returned view includes $\langle p, v' \rangle$, where either $v' = v$ or v' is the argument of an update by p that follows *uop*. If there is no such scan, then place *uop* at the end of the ordering. Note that all later scans return snapshot views that include $\langle p, v' \rangle$, where either $v' = v$ or v' is the argument of an update by p that follows *uop*. This rule for placing updates ensures that the ordering satisfies the sequential specification of atomic snapshots.

Note that if a scan completes before an update starts, then the scan's returned view cannot include the update's value; similarly, if an update completes before a scan starts, then the scan's returned view must includes the update's value or a later one. This shows that the ordering respects the real-time order between non-overlapping updates and scans. The next lemma deals with non-overlapping updates.

Lemma 7. *Let V be the snapshot view returned by a scan sop. If $V(p)$ is the value of an update uop_p by node p and an update uop_q by node q precedes uop_p, then $V(q)$ is the value of uop_q or a later update by q.*

Consider an update uop_p, by node p, that follows an update uop_q, by node q, in the execution. If uop_p is placed at the end of the (current) ordering because there is no scan that observes its value or a later update by p, then it is ordered after uop_q. If uop_p is placed before a scan, then the same must be true of uop_q. By construction, the next scan after uop_p in the ordering, call it *sop*, returns view V with $V(p)$ equal to the value of uop_p or a later update by p. By Lemma 7, $V(q)$ must equal the value of uop_q or a later update by q. Thus uop_q cannot be placed after *sop*, and thus it is placed before uop_p.

We now consider the termination property of the algorithm. Let V_1 and V_2 be two collect views returned by consecutive collects cop_1 and cop_2, within a scan sop_q by node q. If this double collect is not successful, then $V_1 \neq V_2$. If $V_1(p).usqno \neq V_2(p).usqno$, then it is immediate that for some update uop_p of node p, either uop_p's *scounts* includes the scan sequence number of sop_q, or uop_p starts before sop_q starts. Let t be the time that sop_q starts, and note that at most $N(t)$ updates are pending at time t. Further note that if uop_p's *scounts* includes the scan sequence number of sop_q, then sop_q can borrow the scan view of uop_p's embedded scan. This implies that sop_q has at most $N(t)$ unsuccessful double collects before it can borrow a scan view, and therefore it executes at most $O(N(t))$ collects. Hence, UPDATE executes at most $O(N(t))$ collects and stores. Putting the pieces together, we have:

Theorem 4. *Algorithm 4 is a linearizable implementation of an atomic snapshot object. The number of communication rounds in a SCAN or an UPDATE operation is at most linear in the number of nodes present in the system when the operation starts.*

5.2 Generalized Lattice Agreement

Let $\langle L, \sqsubseteq \rangle$ be a lattice, where L is the domain of lattice values, ordered by $\sqsubseteq$. We assume a join operator, $\sqcup$, that merges lattice values. A node p calls a PROPOSE operation with a lattice input value, and gets back a lattice output value. The input to p's i-th PROPOSE is denoted v_i^p and the response is w_i^p. The following conditions are required: *(1)* Every response value w_i^p is the join of some values proposed before this response, including v_i^p, and all values returned to any node before the invocation of p's i-th PROPOSE (*validity*). *(2)* Any two values w_i^p and w_j^q are comparable (*consistency*). This definition is a direct extension of *one-shot* lattice agreement [8], following [21]. The version studied in [15] is weaker and lacks real-time guarantees across nodes.

Our algorithm uses an atomic snapshot object, in which each node stores a single lattice value (*val*). A PROPOSE operation is simply an UPDATE of a lattice value which is the join of all the node's previous inputs, followed by a SCAN returning the analogous values for all nodes, whose join is the output of PROPOSE.

Validity and consistency are immediate from atomic snapshot properties. Clearly, the algorithm terminates within $O(N)$ collects and stores, where N is the maximum number of nodes concurrently active during the execution of PROPOSE. Since PROPOSE includes one UPDATE and one SCAN, it terminates if the node does not crash or leave.

6   Conclusion

We have advocated for the usefulness of the store-collect object as a powerful, flexible, and efficient primitive for implementing a variety of shared objects in dynamic systems with continuous churn. If the level of churn is too great, our store-collect algorithm is not guaranteed to preserve the safety property; that is, a collect might miss the value written by a previous store, essentially by the same counter-example as that given in [6]. This behavior is in contrast to the algorithms in [2, 18, 21], which never violate the safety property but only ensure progress once reconfigurations cease. In future work, we would like to either improve our algorithm to avoid this behavior or prove that any algorithm that tolerates ongoing churn is subject to such bad behavior.

Our correctness proof for our store-collect algorithm requires that the parameters defining the churn rate and failure fraction satisfy certain conditions. These conditions imply that even in the absence of churn the failure fraction tolerable by our algorithm is smaller than in the static case (namely, less than one-third versus less than one-half). Some degradation is unavoidable when allowing for the possibility of churn, since an argument from [6] can be adapted to show that when implementing store-collect in a system with churn rate α, the fraction of failures must be less than $1/(\alpha + 2)$. It would be nice to find less restrictive constraints on the parameters, either through a better analysis or a modified algorithm, or to show that they are necessary.

Another desirable modification to the store-collect algorithm would be reducing the size of the messages and the amount of local storage by garbage-collecting the *Changes* sets. In the same vein, we would like to know if modifying the atomic snapshot specification to remove from returned views entries of nodes that have left, as is done in [23], can lead to a more space-efficient algorithm.

References

1. Afek, Y., Attiya, H., Dolev, D., Gafni, E., Merritt, M., Shavit, N.: Atomic snapshots of shared memory. J. ACM **40**(4), 873–890 (1993)
2. Aguilera, M.K., Keidar, I., Malkhi, D., Shraer, A.: Dynamic atomic storage without consensus. J. ACM **58**(2), 7:1–7:32 (2011)
3. Aguilera, M.K.: A pleasant stroll through the land of infinitely many creatures. SIGACT News **35**(2), 36–59 (2004)
4. Aspnes, J.: Notes on theory of distributed systems. http://www.cs.yale.edu/homes/aspnes/classes/465/notes.pdf. Accessed 8 Aug 2020
5. Attiya, H., Bar-Noy, A., Dolev, D.: Sharing memory robustly in message-passing systems. J. ACM **42**(1), 124–142 (1995)
6. Attiya, H., Chung, H., Ellen, F., Kumar, S., Welch, J.: Emulating a shared register in an asynchronous system that never stops changing. TPDS **30**(3), 544–559 (2018)
7. Attiya, H., Fouren, A., Gafni, E.: An adaptive collect algorithm with applications. Dist. Comp. **15**(2), 87–96 (2002)
8. Attiya, H., Herlihy, M., Rachman, O.: Atomic snapshots using lattice agreement. Dist. Comp. **8**(3), 121–132 (1995)
9. Attiya, H., Kumari, S., Somani, A., Welch, J.L.: Store-collect in the presence of continuous churn with application to snapshots and lattice agreement. CoRR abs/2003.07787 (2020). https://arxiv.org/abs/2003.07787
10. Baldoni, R., Bonomi, S., Kermarrec, A.M., Raynal, M.: Implementing a register in a dynamic distributed system. In: ICDCS, pp. 639–647 (2009)
11. Baldoni, R., Bonomi, S., Raynal, M.: Implementing a regular register in an eventually synchronous distributed system prone to continuous churn. TPDS **23**(1), 102–109 (2012)
12. Baldoni, R., Bonomi, S., Raynal, M.: Implementing set objects in dynamic distributed systems. J. Comput. Syst. Sci. **82**(5), 654–689 (2016)
13. Chordia, S., Rajamani, S., Rajan, K., Ramalingam, G., Vaswani, K.: Asynchronous resilient linearizability. In: DISC pp. 164–178 (2013)
14. Delporte-Gallet, C., Fauconnier, H., Rajsbaum, S., Raynal, M.: Implementing snapshot objects on top of crash-prone asynchronous message-passing systems. TPDS **29**(9), 2033–2045 (2018)
15. Faleiro, J.M., Rajamani, S., Rajan, K., Ramalingam, G., Vaswani, K.: Generalized lattice agreement. In: PODC, pp. 125–134 (2012)
16. Gafni, E., Malkhi, D.: Elastic configuration maintenance via a parsimonious speculating snapshot solution. In: DISC, pp. 140–153 (2015)
17. Gafni, E., Merritt, M., Taubenfeld, G.: The concurrency hierarchy, and algorithms for unbounded concurrency. In: PODC, pp. 161–169 (2001)
18. Gilbert, S., Lynch, N.A., Shvartsman, A.A.: RAMBO: a robust, reconfigurable atomic memory service for dynamic networks. Dist. Comp. **23**(4), 225–272 (2010)
19. Herlihy, M.P., Wing, J.M.: Linearizability: a correctness condition for concurrent objects. Trans. Prog. Lang. Sys. **12**(3), 463–492 (1990)
20. Jehl, L., Vitenberg, R., Meling, H.: SmartMerge: a new approach to reconfiguration for atomic storage. In: DISC, pp. 154–169 (2015)
21. Kuznetsov, P., Rieutord, T., Tucci-Piergiovanni, S.: Reconfigurable lattice agreement and applications. In: OPODIS, pp. 31:1–31:17 (2019)
22. Merritt, M., Taubenfeld, G.: Computing with infinitely many processes. In: DISC, pp. 164–178 (2000)
23. Spiegelman, A., Keidar, I.: Dynamic atomic snapshots. In: OPODIS (2016)
24. Zheng, X., Garg, V., Kaippallimalil, J.: Linearizable replicated state machines with lattice agreement. In: OPODIS (2019)
25. Zheng, X., Hu, C., Garg, V.: Lattice agreement in message passing systems. In: DISC (2018)

Invited Paper: Homomorphic Operations Techniques Yielding Communication Efficiency

Dor Bitan[1]([✉]) and Shlomi Dolev[2]

[1] Department of Mathematics, Ben-Gurion University of the Negev,
Beer Sheva, Israel
dorbi@post.bgu.ac.il
[2] Department of Computer Science, Ben-Gurion University of the Negev,
Beer Sheva, Israel

Abstract. This paper describes our recent results in information theoretically secure homomorphic encryption. The main question that stands
in the basis of these works concerns the possibility of modifying encrypted
data obliviously. This possibility is useful for various applications, e.g.,
multiparty computation, outsourcing of computations, and quantum key
distribution (QKD).

The works presented here consider the scenario in which a user wishes
to outsource the storage and computation of confidential data to an
untrusted server. The first two works consider the approach of employing multiple servers and distributing secret shares of the data among
the servers. The first work introduces a method for evaluating quadratic
functions over *a dynamic* database, with no communication between the
servers. The second work allows communication and considers a method
for homomorphic evaluation of polynomials of arbitrary degree over non-
zero secret shares in a single round of communication. We present protocols that enable the evaluation of multivariate polynomials over shares of
a non-zero secret without requiring a secret sharing phase invoked in an
offline preprocessing phase, and deal with possibly-zero secrets in several
ways.

The third work reviewed here considers the approach of employing a
single server. That work assumes that the user and server have quantum capabilities, and attempts to enable the homomorphic evaluation
of encrypted classical data using quantum devices. The homomorphic
encryption scheme presented in that work is used to construct a QKD
scheme resilient against weak measurements. Weak measurement based
attacks over known QKD schemes are also introduced in the third work,
along with the innovative concept of *securing entanglement*.

We would like to thank the Lynne and William Frankel Center for Computer Science,
the Rita Altura Trust Chair in Computer Science. This work was also partially supported by a grant from the Ministry of Science and Technology, Israel & the Japan
Science and Technology Agency (JST), and the German Research Funding (DFG,
Grant#8767581199).

S. Devismes and N. Mittal (Eds.): SSS 2020, LNCS 12514, pp. 16–28, 2020.
https://doi.org/10.1007/978-3-030-64348-5_2

Keywords: Secret sharing · Homomorphic encryption · Multiparty computation · Quantum computation · Quantum key distribution

1   Background

Cloud services have become very common in recent years. Many computer companies offer information storage and processing services for individuals, companies, and organizations. These services allow customers to enjoy an enormous storage space, massive processing power, and cheap, convenient and efficient management facilities. Many customers all over the world enjoy such services. However, in many cases, the information that the customer wants to export to the cloud is confidential. In such cases, there is a concern that the confidentiality of the information will be compromised due to hacking or even by the cloud company's employees. One possible way of maintaining privacy is to encrypt the information by the user before sending it to the cloud (and keep the encryption keys secretly). This way is suitable in cases where the customer is only interested in storing the information. However, in many cases, the customer is also interested in processing the data by the cloud servers. In such cases, a question arises – how (if at all) can we enjoy the cloud's processing services while keeping the confidentiality of the data? This problem is hereafter referred to as *the secure delegation problem*. The works reviewed in this manuscript seek solutions for this problem.

The secure delegation problem was first raised in 1978 by Rivest, Adelman, and Dertouzos. In their seminal paper [34], they suggested to use 'privacy homomorphisms,' nowadays known as *homomorphic encryption schemes*, to encrypt the data and enable oblivious processing of it by (honest-but-curious) remote servers (possibly in the cloud). The data is typically represented by finite field elements. Encryption schemes are composed of algorithms for encryption, decryption, and key generation (typically denoted Enc, Dec, and Gen). To phrase the problem mathematically, let m_1, m_2 be elements of a finite field (or a ring), and c_1, c_2 their encryptions generated by an encryption system denoted π. Can c_1, c_2 be used to publicly generate $c_{\text{add}} = \text{Enc}_\pi(m_1 + m_2)$ or $c_{\text{mult}} = \text{Enc}_\pi(m_1 \cdot m_2)$? If it is possible to use c_1, c_2 to publicly generate $c_{\text{add}} = \text{Enc}_\pi(m_1 + m_2)$ (respectively, $c_{\text{mult}} = \text{Enc}_\pi(m_1 \cdot m_2)$), then π is *additively homomorphic* (respectively, *multiplicatively homomorphic*). If both tasks can be carried out, then π is a *fully homomorphic encryption* (FHE) system.

The search for fully homomorphic encryption schemes has been going on for many years and was tagged as 'the holy grail of cryptography'. The first significant breakthrough in the field occurred in 2009, when Craig Gentry proposed the first (computationally secure) fully homomorphic encryption scheme [23]. Gentry's scheme has been refined and additional FHE schemes have been proposed [2,13,24,25,37–39]. Unfortunately, the time complexity of the currently known FHE schemes is too high to make them practical [1,31].

While FHE schemes can provide solutions to the secure delegation problem, they can achieve at most *computational security*, but not *information-theoretical (IT-) security*. In IT-secure schemes, the security of the scheme is derived purely from information theory and depends neither on the computing power of the adversary nor on any computational hardness assumptions. The security of cryptographic schemes that have *computational security* is based on unproven assumptions regarding the non existence of efficient algorithms for solving specific mathematical problems and the computing power of the possible adversary. FHE schemes cannot achieve IT-security since existences of an IT-secure FHE scheme would contradict known theorems regarding private information retrieval.

Solutions for the secure delegation problem can be divided into two categories according to their overall approach. FHE schemes suggest solutions for the secure delegation problem based on *the centralized approach*. This approach assumes that the user delegates the data to be stored and processed by a single server. The second approach to the secure delegation problem is *the distributed approach*, in which the user distributes the secret information between several clouds. In the distributed approach, the user typically uses *a secret sharing scheme* to distribute *shares* of the data among the servers. In this case, the users' privacy is kept as long as no more than a predefined threshold of the servers collude in an adversarial attempt to reveal the data. Unlike FHE scheme, distributed solutions to the secure delegation problem often achieve IT-security. The first work reviewed in this manuscript [8] presents a distributed approach solution to the secure delegation problem that suggests an IT-secure delegation scheme. This scheme supports homomorphic evaluation of quadratic functions and 2-CNF circuits over a dynamic database of secrets with no communication between the servers.

The distributed approach to the secure delegation problem is related to *secure multiparty computation* (MPC). MPC is an active field of research in cryptography [6,19–22,27,33,40]. This field discusses the following problem. Several participants are holding secret inputs and wish to evaluate a multivariate function over their secret inputs while not revealing to each other any information regarding their secret inputs (except for what may be deduced from the output). MPC schemes differ in their security and efficiency levels and their assumptions regarding the behavior of the parties and the communication setting. One of the typically used ideas is to secret share inputs. Then, adding two secret shared values can implement logical OR gate, multiplication of two secret-shared values (and reducing the degree of the obtained polynomial) can implement a logical AND gate. Using these two gates, a general logical circuit can be blindly computed.

More often than not, MPC schemes provide distributed solutions to the secure delegation problem. In several cases, it also works the other way around. Namely, some distributed solutions to the secure delegation problem can be used to construct MPC schemes. The second work reviewed in this manuscript considers a specific secret sharing scheme – the distributed random matrix (DRM) scheme [10]. That secret sharing scheme is used to construct an efficient and

IT-secure preprocessing-MPC protocol and a distributed solution for the secure delegation problem. These schemes support homomorphic evaluation of polynomials over non-zero inputs with optimal round complexity. We present the one-time secrets (OTS) protocols that enable the evaluation of multivariate polynomials over shares of non-zero secrets without requiring a secret sharing phase invoked in an offline preprocessing phase. In addition, [10] deals with the problem of handling possibly zero secrets in several ways. By enabling the servers to communicate with each other, we manage to enable the homomorphic evaluation of polynomials of an arbitrary degree. Recall that in [8], we were able to support evaluation of polynomials of degree at most two with no communication between the servers.

So far, we have described the secure delegation problem and possible solutions for it assuming all participants are classic computers. An equally exciting problem arises when it is assumed that some (or all) of the participants are quantum computers. In the third work reviewed in this manuscript, [11], we take the centralized approach, assume that the user and server are quantum computers, and seek efficient and IT-secure solutions to the secure delegation problem. In that paper, the homomorphic encryption system presented is used to construct a quantum key distribution (QKD) protocol that is resistant to attacks based on weak measurements (WM). We present new proposed WM-based attacks against existing QKD schemes that cannot be applied against our system (Table 1).

Table 1. A comparison of the solutions presented here.

Work	Approach	Communication	Supported functions
[8]	Distributed	Servers - user only	Quadratic polynomials
[10]	Distributed	Servers - user, servers - servers	Polynomials of arbitrary degree
[11]	Centralized	Server - user	A family of quantum gates

In the rest of the paper, we review the works [8,10,11] in more detail– for each work, we provide some additional background, examine the overall concept and methods, and the main contributions.

2   Communication-Less Evaluation

In 1979, Adi Shamir [35] presented one of the two first (N,t)-secret sharing schemes (see [12] for the other suggestion). Such a scheme allows a user to split a piece of information (hereafter a secret) among a set of N participants in such a way that only subsets of size at least t are able to recover the secret, while smaller subsets will not be able to learn any information about the secret. Secret sharing is a vital building block in MPC schemes and distributed solutions to the secure delegation problem. In Shamir's secret sharing scheme, the secret s is an element of a finite field of order p, denoted by $\mathbb{F}_p$, and is shared by a user among a set of N parties (where $p > N$) in the following way. Each party P_i, $1 \le i \le N$, is assigned by the user with an arbitrary element α_i of $\mathbb{F}_p^\times$, where the α_i's are distinct. Random elements a_j of $\mathbb{F}_p$, $1 \le j \le t-1$, are picked by the user. Let f be the polynomial defined by $f(x) = s + \sum_{j=1}^{t-1} a_j x^j$ in the field $\mathbb{F}_p$. Each party P_i gets the value $f(\alpha_i)$. Shamir proved that, in this way, every group of t parties is able to reconstruct s, but no group of $t-1$ parties gains any information about s [35].

One prominent property of Shamir's scheme is that it is additively homomorphic. This property is based on the fact that the sum of polynomials of degree $\le t-1$ is again a polynomial of degree $\le t-1$. Unfortunately, since the product of two polynomials of degree $\le t-1$ is in general of a higher degree, Shamir's scheme is not multiplicatively homomorphic. As the degree of the polynomial gets larger, a larger coalition is required in order to extract the secret. One of the main results obtained in [8] is a method for making Shamir's scheme support one homomorphic multiplication of secrets while, in some sense, not increasing the degree of the polynomial that represents the secrets. This is our novel *function sieving method*. This method provides a way of choosing the non-free coefficients for two $N-1$ degree polynomials, f_1 and f_2, such that, if $f_1(0) = s_1$ and $f_2(0) = s_2$, then interpolating the N points $\big(\alpha_i, f_1(\alpha_i) \cdot f_2(\alpha_i)\big)$ (for $1 \le i \le N$) one obtains a polynomial whose value at zero is $s_1 \cdot s_2$. Our method enables to find the specific cases in which the polynomials f_1 and f_2 are such that, multiplying the shares of the corresponding secrets, one obtains N products of shares that represent a polynomial of degree $\le N-1$ that has the right value at 0. We define a set of $2(N-1)$-tuples. Each tuple contains suitable non-free coefficients for a pair of polynomials for which homorphic multiplication in Shamir's scheme works.

The Algorithm in a Nutshell. We now briefly sketch the outline of our constructions in more detail. Assume that the field $\mathbb{F}_p$, in which the secrets s_1 and s_2 reside, is such that $p \equiv 1 \pmod{N}$. In that case, since $\mathbb{F}_p^\times$ is cyclic, it contains a primitive root of unity of order N. Let α be such a root. For $1 \le j \le N$ denote $\alpha_j := \alpha^j$, and assign to each party P_j the value α_j. Let $a_i, b_i \in \mathbb{F}_p$, $1 \le i \le N-1$, and consider the polynomials $f_1(x) = s_1 + \sum_{i=1}^{N-1} a_i x^i$ and $f_2(x) = s_2 + \sum_{i=1}^{N-1} b_i x^i$, in $\mathbb{F}_p[x]$. Share the secrets s_1, s_2 among the parties using f_1, f_2. Let $y_j = f_1(\alpha_j) \cdot f_2(\alpha_j)$, $1 \le j \le N$. The pairs $(\alpha_j, y_j) \in \mathbb{F}_p^2$ are N distinct points through which the polynomial $(f_1 \cdot f_2)(x)$ passes.

Since $f := f_1 \cdot f_2$ is of degree $\le 2N-2$, it is uniquely determined by $2N-1$ points. Since there are only N points (α_j, y_j), interpolation of them will certainly not yield $(f_1 \cdot f_2)(x)$. Nevertheless, let $g(x)$ be the interpolation polynomial for the N points, (α_j, y_j). Obviously, g is of degree $\le N-1$. Since f and g agree on the roots of ψ, we have $g(x) \equiv f(x) \pmod{\psi(x)}$, where $\psi(x) = \prod_{j=1}^{N}(x - \alpha_j)$. Since the α_j's are all the roots of unity of order N, we have $\psi(x) = x^{n+1} - 1$. Hence, it is easy to compute g.

In fact, denote $f(x) = s_1 s_2 + \sum_{i=1}^{2N-2} c_i x^i$. We have $x^N \equiv 1 \pmod{\psi(x)}$, and therefore $g(x) \equiv f(x) \equiv s_1 s_2 + c_N + \sum_{i=1}^{N-1}(c_i + c_{N+i})x^i \pmod{\psi(x)}$. This in turn implies that $g(0) = s_1 s_2 + c_N$. Thus, if we take f_1 and f_2 such that $c_N = 0$, we get $g(0) = f(0)$. Now, $c_N = \sum_{i=1}^{N-1} a_i b_{N-i}$. Hence, instead of picking the coefficients of f_1 and f_2 uniformly at random, we pick them in such a way that $c_N = 0$.

This is, in essence, the function sieving method. Instead of using Shamir's secret sharing scheme with random polynomials from $\mathbb{F}_p[x]$, we use it with polynomials f_1, f_2, for which $c_N = 0$, which compels $g(0) = f(0)$. Such a pair (f_1, f_2) is a 1-*homomorphic multiplicative pair* of polynomials.

This method enables a user to securely distribute a confidential database of m elements to a set of N semi-trusted servers while enabling homomorphic evaluation of quadratic functions and 2-CNF circuits over the secrets efficiently, with no communication between servers, IT-secure against coalitions of up to $N - 2$ semi-honest servers, with $O(m^2)$ ciphertext, and *dynamically*. A secure outsourcing scheme is *dynamic* if it enables the user to add (or remove) new records to the database with no need for storing and re-sharing existing secrets by the dealer. The dynamic property is vital for a secure outsourcing scheme and may have significant benefits in many practical applications. Whenever one wishes to outsource the storage of a database to a set of semi-trusted servers, some pieces of data may not be known at the moment of construction of the database and are expected to be known in the future. A dynamic scheme resolves the need for storing a copy of the entire database on the user's computer. In [8], we review existing communication-less schemes that enable similar homomorphic properties (e.g., Beaver's multiplication technique [4], or other variants of Shamir's scheme) and show that these schemes are either non-dynamic or less secure.

3   Optimal-Round P-MPC

The search for solutions for the secure delegation problem often gives rise to MPC protocols, as exemplified in [10]. MPC is an extensively studied field in cryptography rooted in Yao's millionaire problem from the early 80's [40]. In their seminal work from 1988, Ben-Or et al. [6] showed that, in the plain model, every function of N inputs can be efficiently computed with perfect passive security by N parties if and only if one assumes that the majority of the participant are honest. One may enable multiparty computation of functions in the presence of a dishonest majority by switching to the preprocessing model, first suggested in [5]. *The preprocessing model* enables achieving perfect passive security against dishonest majority by enabling the parties to engage in an offline preprocessing phase before the secret inputs are known. At the end of that offline phase, the parties obtain *correlated randomness* (CR) – random coins to be used in the online phase of the protocol. Given a preprocessing MPC protocol (hereafter P-MPC), *the space complexity* of the scheme indicates how the amount of CR required for the scheme grows with respect to other parameters.

An important measure of efficiency of MPC schemes is their round complexity. Two rounds of communication are now known to be optimal for MPC – in the plain or preprocessing model [15,32]. Ishai et al. suggested in [30] two-round P-MPC protocols with perfect passive security against dishonest majority, followed by several improvements [14,16]. There already exist MPC schemes with optimal round complexity and dishonest majority. Nevertheless, all these schemes require amounts of either time, memory or communication exponential in some of the parameters: depth or size of the circuit, size of the domain, or number of parties. The space complexity of known solutions is (believed to be inherently) exponential in the size of the input and N.

In [10], we construct efficient N-party P-MPC schemes for polynomials over non-zero inputs. There already exist schemes for efficient evaluation of polynomials over non-zero inputs [26]. However, our schemes are the first not to require an additional secret sharing round during the preprocessing stage. We also suggest several ways of handling possibly-zero inputs. Each of these ways best suits different families of functions. These schemes are based on the *DRM secret sharing scheme*, a novel homomorphic secret sharing scheme established in [10]. These results were established based on our work [9], where we constructed efficient schemes for secure outsourcing of stream computations.

The DRM secret sharing scheme, presented in [10], supports homomorphic multiplications of secrets and, after a single round of communication, supports homomorphic additions of secrets. We use the DRM secret sharing scheme to construct the one-time secrets (OTS) protocols. These protocols enable the evaluation of multivariate polynomials over shares of non-zero secrets with the following properties: communication and space complexities linear in the number of monomials, optimal round complexity, perfect security against dishonest majority. The main advantage of our scheme is that we achieve all these properties without requiring a secret sharing phase invoked in an offline preprocessing phase. In addition, our paper suggests new techniques for handling possibly-zero secrets in several ways.

The Algorithm in a Nutshell. We now review the main ideas behind our method. First, we construct the Distributed Random Matrix (DRM) secret-sharing scheme. In this scheme, a secret is randomly split to a sum of field elements, and each of the addends is randomly split to a product of field elements. The factors of these products are put in the rows of a matrix, and each column of that matrix is considered a share of the secret. Namely, given an element x of a finite field $\mathbb{F}_p$, we split x to a sum of N random $\mathbb{F}_p$ elements γ_i, $1 \leq i \leq N$. Then, each of the γ_i's is split to a random product of $\mathbb{F}_p$ elements $m_{i,j}$, $1 \leq j \leq N$. Denote by C the square matrix of order N whose entries are the multiplicative shares $m_{i,j}$ of the additive shares γ_i. The $m_{i,j}$ are randomly picked under the condition that C contains zeroes only on its main diagonal, if any. The N columns of C are N DRM-shares of x. The double splitting of each secret (additively splitting the secret and multiplicatively splitting each addend) enables supporting both homomorphic multiplications and additions. In [10] we prove that, the DRM secret sharing scheme supports homomorphic multiplications with multiplicatively secret-shared $\mathbb{F}_p^\times$ elements. Furthermore, a single round of communication enables the parties to switch to additive shares of x.

Next, the DRM secret sharing scheme is used to construct a P-MPC scheme. The outline of the scheme is as follows. In the preprocessing phase, each party is supplied with a sufficient amount of CR in the form of DRM shares of $1 \in \mathbb{F}_p$ – one share for each monomial in the target polynomial. Recall that these DRM shares support homomorphic multiplications. To evaluate the polynomial over the secret inputs, for each monomial, each party multiplies the corresponding DRM-share (a column vector) with a power of the secret as required by that monomial, and obtains a new column vector. In the first communication round, the entries of this column are split among the other servers. Next, each server computes the products of the values obtained in the previous round (one product for each monomial) and adds these products to obtain an additive share of the output. Lastly, the parties distribute the additive shares of the output to each other, and each party locally adds them to obtain the output. The main advantage of our scheme is that it requires no secret sharing round in the preprocessing phase.

Our results are also extended to the client-server model, providing an IT-secure solution to the secure delegation problem. The *DRM single-round client-server scheme* enables a set of users to securely outsource the storage of their private inputs to a set of servers and have the servers evaluate polynomials over the entire collection of users-inputs (non-zero). The users obtain the result after a single round of communication between the servers.

To securely delegate non-zero secrets to the servers, the user distributes multiplicative shares of each secret among the servers. Then, to enable homomorphic evaluation of polynomials of arbitrary degree over the secrets, the user sends a query to the server containing a description of the polynomial and DRM-shares of $1 \in \mathbb{F}_p$, one for each monomial, to be used as CR. Next, the servers use the CR to evaluate the polynomial in a single round of communication and send the shares of the result to the user.

The DRM client-server scheme is perfectly secure against coalitions of up to $N-1$ honest-but-curious servers. The users do not communicate with each other during the execution of the scheme. Each user distributes secret-shares of the inputs to the servers and receives the output from the servers.

The innovative approach of the scheme enables handling high degree polynomials without being concerned with the depth of the arithmetic circuit, which is one of the main complexity bottlenecks in MPC. The communication and space complexities of our schemes are independent of the degree of the polynomial, and the required CR is independent of the function.

To emphasize the importance of round-efficiency, we note that, while processing information becomes faster as technology improves, the time it takes to transmit information between two distant places is strictly limited by the speed of light. One may consider a future need to perform MPC over inputs held by parties residing in distant places, perhaps in different continents or even in space. Denote by T the time it takes to process the computations needed for the evaluation of some function f using our schemes. If sending a message between parties takes more than T, then optimal-round schemes outperform any scheme with non-optimal round complexity.

4   Quantum HE and Applications

Quantum computers may allow feasible solutions to problems that are currently considered impractical to solve [7,18,28,36]. In view of this fact, it is natural to wonder if quantum computers can be used to achieve an IT-secure FHE scheme. In 2014, [41] showed that it is impossible to construct an efficient IT-secure quantum FHE (QFHE) scheme. Efficient IT-secure (quantum or classical) encryption schemes can support homomorphic evaluation of only a subset of all possible functions.

In a search for a quantum encryption scheme of classical data, [11] suggested the random basis encryption scheme – an efficient, IT-secure, perfectly correct, non-interactive, and fully compact encryption scheme that supports homomorphic evaluation of several quantum gates. The scheme presented in [11] shares some resemblance with the quantum one-time pad (QOTP) based encryption scheme. In QOTP based schemes, Pauli gates are randomly applied to plaintext qubits to obtain IT-secure encryption, while supporting homomorphic evaluation of Pauli gates. QOTP was suggested by Ambainis et al. in [3].

The main difference between the random basis encryption scheme and the QOTP-based schemes is that in [11], a plaintext bit is encrypted by a rotation of the corresponding qubit in an angle chosen from an immense number of possibilities, while in [3] there are only four possible different encodings for a plaintext qubit. The random basis encryption scheme essentially implements a continuous version of the (discrete) QOTP scheme. The difference between the continuous and discrete versions becomes significant in several scenarios when considering attacks based on *weak measurements* (WM).

Another advantage of our random basis encryption scheme over QOTP-based schemes is that in contrast to the legacy quantum one-time pad based HE scheme, that requires modifications of the keys by the user, our scheme is *computation agnostic*. Namely, when delegating computations, the user is not required to carry out such computations and key-adjustments and can remain utterly oblivious to the implementation method chosen by the server/cloud.

Weak measurements enable accumulating information regarding the state of a qubit while not collapsing the state, but only biasing it a little. In [11], it is shown how WM can be used to attack quantum key distribution (QKD) schemes that are based on QOTP. Namely, we demonstrate a WM attack on the [7] and [17] schemes that enables an adversary to obtain a non-negligible advantage at guessing a key-bit while reducing the risk of being caught.

Our WM attack works as follows. First, we weakly interact the subject qubit with an ancillary qubit. Then, we (strongly) measure the ancillary qubit. The outcome of the (strong) measurement of the ancillary qubit is the outcome of the weak measurement of the subject qubit. This process enables imprecisely measuring quantum states, outsmarting the uncertainty principle.

To this end, we construct a two-qubit quantum gate that is very close to the identity operator (not doing anything), but slightly tends towards the CNOT quantum gate. The CNOT quantum gate enables copying computational basis qubits ($\{|0\rangle, |1\rangle\}$) without disturbing them. If the qubits are not in the computational basis, the CNOT gate disturbs them[1]. Our two-qubit quantum gate can be taken to be arbitrarily close to the identity operator, hence enabling a tradeoff between information gain and state disturbance.

Explicitly, given $\varepsilon > 0$, let $W_\varepsilon = \sqrt{\varepsilon} \cdot i \cdot CNOT + \sqrt{1-\varepsilon} \cdot I$ a two-qubit gate (I is the identity operator and i is the square root of -1). In our WM attacks, W_ε is used to weakly interact a target qubit with an ancillary qubit. If the target qubit is in the computational basis, then measuring the ancillary qubit provides some information regarding the target qubit. If the target qubit is in the Hadamard basis, we obtain no information, but only slightly disturb the state.

In addition, [11] presents the random basis CNOT QKD scheme – an IT-secure QKD scheme that is resilient against weak measurement based attacks. Another advantage of our QKD scheme compared to other schemes is that only one side measures, and the other side can decide to blindly negate the state without knowing the chosen random base.

The random basis encryption scheme is shown to be useful in another setting – *securing entanglement*. Entanglement is an essential resource in many quantum settings – teleportation, private communication, and distinguishing quantum states [29]. The utilization of entanglement in communication, computation, and other scenarios is a very active area of research. In practice, entanglement is typically generated by direct interactions between subatomic particles. The generation of entangled systems requires efforts and expenditures. In [11] it is suggested that, once entanglement was generated, it should be secured in the sense that only its rightful owners will be able to use it. We demonstrate a process of securing entanglement using the random basis encryption scheme. Moreover, we show that our method of securing entanglement provides safer implications in the face of weak measurements compared to possible straightforward QOTP based methods for the same task.

5 Conclusions

We believe that distributed computing can benefit much from using the techniques reviewed above and in particular secure multiparty computation. The classical methods of secure multiparty computation imply high communication overhead. The reviewed works' scope is to advance the research for reducing the communication overhead in the scope of dynamic database, streaming computation, and quantum computers.

---

[1] It is not possible to copy general qubits due to the no-cloning theorem.

References

1. Acar, A., Aksu, H., Uluagac, A.S., Conti, M.: A survey on homomorphic encryption schemes: theory and implementation. ACM Comput. Surv. (CSUR) **51**(4), 1–35 (2018)
2. Akavia, A., Gentry, C., Halevi, S., Leibovich, M.: Setup-free secure search on encrypted data: Faster and post-processing free. Technical report, Cryptology ePrint Archive Report (2018)
3. Ambainis, A., Mosca, M., Tapp, A., de Wolf, R.: Private quantum channels. In: 41st Annual Symposium on Foundations of Computer Science, FOCS 2000, pp. 547–553 (2000)
4. Beaver, D.: Efficient multiparty protocols using circuit randomization. In: Feigenbaum, J. (ed.) CRYPTO 1991. LNCS, vol. 576, pp. 420–432. Springer, Heidelberg (1992). https://doi.org/10.1007/3-540-46766-1_34
5. Beaver, D.: Commodity-based cryptography. In: Proceedings of the Twenty-Ninth Annual ACM Symposium on Theory of Computing, pp. 446–455. ACM (1997)
6. Ben-Or, M., Goldwasser, S., Wigderson, A.: Completeness theorems for non-cryptographic fault-tolerant distributed computation. In: Proceedings of the Twentieth Annual ACM Symposium on Theory of Computing, pp. 1–10. ACM (1988)
7. Bennett, C. H., Brassard, G.: Quantum cryptography: public key distribution and coin tossing. IEEE, New York (2020)
8. Berend, D., Bitan, D., Dolev, S.: Polynomials whose secret shares multiplication preserves degree for 2-CNF circuits over a dynamic set of secrets. IACR Cryptol. ePrint Arch. (2019)
9. Bitan, D., Dolev, S.: One-round secure multiparty computation of arithmetic streams and functions. In: Dinur, I., Dolev, S., Lodha, S. (eds.) CSCML 2018. LNCS, vol. 10879, pp. 255–273. Springer, Cham (2018). https://doi.org/10.1007/978-3-319-94147-9_20
10. Bitan, D., Dolev, S.: Optimal-round preprocessing-mpc via polynomial representation and distributed random matrix (extended abstract). IACR Cryptol. ePrint Arch. (2019)
11. Bitan, D., Dolev, S.: Randomly choose an angle from immense number of angles to rotate qubits, compute and reverse. IACR Cryptol. ePrint Arch. (2019)
12. Blakley, G.R.: Safeguarding cryptographic keys. In: 1979 International Workshop on Managing Requirements Knowledge (MARK), pp. 313–318. IEEE (1979)
13. Brakerski, Z., Perlman, R.: Lattice-based fully dynamic multi-key FHE with short ciphertexts. In: Robshaw, M., Katz, J. (eds.) CRYPTO 2016. LNCS, vol. 9814, pp. 190–213. Springer, Heidelberg (2016). https://doi.org/10.1007/978-3-662-53018-4_8
14. Couteau, G.: A note on the communication complexity of multiparty computation in the correlated randomness model. In: Ishai, Y., Rijmen, V. (eds.) EUROCRYPT 2019. LNCS, vol. 11477, pp. 473–503. Springer, Cham (2019). https://doi.org/10.1007/978-3-030-17656-3_17
15. Damgård, I., Larsen, K.G., Nielsen, J.B.: Communication lower bounds for statistically secure MPC, with or without preprocessing. IACR Cryptol. ePrint Arch. **2019**, 220 (2019)
16. Damgård, I., Nielsen, J.B., Nielsen, M., Ranellucci, S.: The TinyTable protocol for 2-party secure computation, or: gate-scrambling revisited. In: Katz, J., Shacham, H. (eds.) CRYPTO 2017. LNCS, vol. 10401, pp. 167–187. Springer, Cham (2017). https://doi.org/10.1007/978-3-319-63688-7_6

17. Deng, F.-G., Long, G.L.: Secure direct communication with a quantum one-time pad. Phys. Rev. A **69**(5), 052319 (2004)
18. Deutsch, D., Jozsa, R.: Rapid solution of problems by quantum computation. Proc. R. Soc. Lond. A **439**(1907), 553–558 (1992)
19. Dolev, S., Garay, J., Gilboa, N., Kolesnikov, V., Yuditsky, Y.: Towards efficient private distributed computation on unbounded input streams. J. Math. Cryptol. **9**(2), 79–94 (2015)
20. Dolev, S., Garay, J.A., Gilboa, N., Kolesnikov, V., Kumaramangalam, M.V.: Perennial secure multi-party computation of universal turing machine. Theor. Comput. Sci. **769**, 43–62 (2019)
21. Dolev, S., Gilboa, N., Li, X.: Accumulating automata and cascaded equations automata for communicationless information theoretically secure multi-party computation. In: Proceedings of the 3rd International Workshop on Security in Cloud Computing, pp. 21–29. ACM (2015)
22. Dolev, S., Li, Y.: Secret shared random access machine. In: Karydis, I., Sioutas, S., Triantafillou, P., Tsoumakos, D. (eds.) ALGOCLOUD 2015. LNCS, vol. 9511, pp. 19–34. Springer, Cham (2016). https://doi.org/10.1007/978-3-319-29919-8_2
23. Gentry, C.: A fully homomorphic encryption scheme. Stanford University (2009)
24. Gentry, C., Halevi, S., Smart, N.P.: Fully homomorphic encryption with polylog overhead. In: Pointcheval, D., Johansson, T. (eds.) EUROCRYPT 2012. LNCS, vol. 7237, pp. 465–482. Springer, Heidelberg (2012). https://doi.org/10.1007/978-3-642-29011-4_28
25. Gentry, C.B., Halevi, S., Smart, N.P.: Homomorphic evaluation including key switching, modulus switching, and dynamic noise management. US Patent 9,281,941 (2016)
26. Ghodosi, H., Pieprzyk, J., Steinfeld, R.: Multi-party computation with conversion of secret sharing. Des. Codes Cryptogr. **62**(3), 259–272 (2012)
27. Goldreich, O., Micali, S., Wigderson, A.: How to play any mental game. In: Proceedings of the Nineteenth Annual ACM Symposium on Theory of Computing, pp. 218–229. ACM (1987)
28. Grover, L.K.: A fast quantum mechanical algorithm for database search. In: Proceedings of the Twenty-Eighth Annual ACM Symposium on Theory of Computing, pp. 212–219. ACM (1996)
29. Horodecki, R., Horodecki, P., Horodecki, M., Horodecki, K.: Quantum entanglement. Rev. Mod. Phys. **81**(2), 865 (2009)
30. Ishai, Y., Kushilevitz, E., Meldgaard, S., Orlandi, C., Paskin-Cherniavsky, A.: On the power of correlated randomness in secure computation. In: Sahai, A. (ed.) TCC 2013. LNCS, vol. 7785, pp. 600–620. Springer, Heidelberg (2013). https://doi.org/10.1007/978-3-642-36594-2_34
31. Naehrig, M., Lauter, K., Vaikuntanathan, V.: Can homomorphic encryption be practical? In: Proceedings of the 3rd ACM Workshop on Cloud Computing Security Workshop, pp. 113–124 (2011)
32. Patra, A., Ravi, D.: On the exact round complexity of secure three-party computation. In: Shacham, H., Boldyreva, A. (eds.) CRYPTO 2018. LNCS, vol. 10992, pp. 425–458. Springer, Cham (2018). https://doi.org/10.1007/978-3-319-96881-0_15
33. Rivest, R.: Unconditionally secure commitment and oblivious transfer schemes using private channels and a trusted initializer. Unpublished manuscript (1999)
34. Rivest, R.L., Adleman, L., Dertouzos, M.L.: On data banks and privacy homomorphisms. Found. Secure Comput. **4**(11), 169–180 (1978)
35. Shamir, A.: How to share a secret. Commun. ACM **22**(11), 612–613 (1979)

36. Shor, P.W.: Algorithms for quantum computation: discrete logarithms and factoring. In: 35th Annual Symposium on Foundations of Computer Science, 1994 Proceedings, pp. 124–134. IEEE (1994)
37. Smart, N.P., Vercauteren, F.: Fully homomorphic encryption with relatively small key and ciphertext sizes. In: Nguyen, P.Q., Pointcheval, D. (eds.) PKC 2010. LNCS, vol. 6056, pp. 420–443. Springer, Heidelberg (2010). https://doi.org/10.1007/978-3-642-13013-7_25
38. van Dijk, M., Gentry, C., Halevi, S., Vaikuntanathan, V.: Fully homomorphic encryption over the integers. In: Gilbert, H. (ed.) EUROCRYPT 2010. LNCS, vol. 6110, pp. 24–43. Springer, Heidelberg (2010). https://doi.org/10.1007/978-3-642-13190-5_2
39. Xu, J., Wei, L., Zhang, Y., Wang, A., Zhou, F., Gao, C.-Z.: Dynamic fully homomorphic encryption-based merkle tree for lightweight streaming authenticated data structures. J. Netw. Comput. Appl. **107**, 113–124 (2018)
40. Yao, A.C.-C.: Protocols for secure computations. In: FOCS, vol. 82, pp.160–164 (1982)
41. Yu, L., Pérez-Delgado, C.A., Fitzsimons, J.F.: Limitations on information-theoretically-secure quantum homomorphic encryption. Phys. Rev. A **90**(5), 050303 (2014)

Boosting the Efficiency of Byzantine-Tolerant Reliable Communication

Silvia Bonomi[1], Giovanni Farina[1,2(✉)], and Sébastien Tixeuil[2]

[1] Sapienza Università di Roma, Rome, Italy
bonomi@diag.uniroma1.it
[2] Sorbonne Université, CNRS, LIP6, 75005 Paris, France
{giovanni.farina,sebastien.tixeuil}@lip6.fr

Abstract. Reliable communication is a fundamental primitive in distributed systems prone to Byzantine (*i.e.* arbitrary, and possibly malicious) failures to guarantee integrity, delivery and authorship of messages exchanged between processes. Its practical adoption strongly depends on the system assumptions. One of the most general (and hence versatile) such hypothesis assumes a set of processes interconnected through an unknown communication network of reliable and authenticated links, and an upper bound on the number of Byzantine faulty processes that may be present in the system, known to all participants.

To this date, implementing a reliable communication service in such an environment may be expensive, both in terms of message complexity and computational complexity, unless the topology of the network is known. The target of this work is to combine the Byzantine fault-tolerant topology reconstruction with a reliable communication primitive, aiming to boost the efficiency of the reliable communication service component after an initial (expensive) phase where the topology is partially reconstructed. We characterize the sets of assumptions that make our objective achievable, and we propose a solution that, after an initialization phase, guarantees reliable communication with optimal message complexity and optimal delivery complexity.

Keywords: Reliable communication · Byzantine fault tolerance · Topology reconstruction

1   Introduction

Reliable communication primitives are fundamental building blocks for a distributed system, guaranteeing the eventual delivery of all messages sent by

This work was performed within Project ESTATE (Ref. ANR-16-CE25-0009-03), supported by French state funds managed by the ANR (Agence Nationale de la Recherche) and it has been partially supported by the INOCS Sapienza Ateneo 2017 Project (protocol number RM11715C816CE4CB). Giovanni Farina wishes to thank *Université Franco-Italienne/Universitá Italo-Francese* (UFI/UIF) for supporting his mobility through the Vinci grant 2018.

© Springer Nature Switzerland AG 2020
S. Devismes and N. Mittal (Eds.): SSS 2020, LNCS 12514, pp. 29–44, 2020.
https://doi.org/10.1007/978-3-030-64348-5_3

correct processes to their intended receivers. Their employment is particularly relevant when a fraction of processes may suffer arbitrary failures i.e., they are Byzantine and may deviate from the protocol by dropping messages, altering their content, or generating spurious messages.

The availability of a reliable communication primitive strongly depends on the system behavior and on its capability to match the set of assumptions required to ensure the correctness of the reliable communication specification. In particular, it has been shown that such a primitive can be efficiently implemented when every process can directly exchange messages with every other [5], also in presence of a bounded and known number of Byzantine processes. However, full connectivity is a strong assumption in large networks and it results impractical whenever scalability is envisioned.

When considering multi-hop networks i.e., systems where every process can communicate directly only with a subset of the others, several results exist to build a Byzantine-tolerant reliable communication primitive. In this paper, we are interested in the solutions designed for multi-hop networks where the topology is not known to the participants. In this context, [8] defined a solution working under the assumption that processes are sufficiently connected. However, providing a reliable communication service in such a general environment may mandate a huge amount of messages and may require very high computational power. Those complexity issues can somewhat be reduced to a tractable problem when either the entire network topology is known to all the processes [8] or it satisfies specific topological requirements [17]. Thus, a naive approach to build an efficient reliable communication primitive is to act in two steps: (i) run a topology reconstruction algorithm to infer the network graph and (ii) use an efficient reliable communication protocol for known network on the reconstruction just inferred. Unfortunately, Byzantine fault-tolerant topology reconstruction has been proved difficult [16], and the final topology inferred does not perfectly match the real one. Besides, correct peers may end the topology reconstruction algorithm by obtaining different network graphs.

Given a network topology G unknown to processes, our goal in this paper is to detail how properly combine the two steps of the described naive approach and to study the set of conditions that G must satisfy to correctly support it. The rationale of this work is that the high topology reconstruction overhead only needs to be paid once, afterwards, reliable communication can be achieved efficiently (otherwise, it would have remained always extremely expensive). The main difficulty is to ensure that discrepancies in the topology reconstructions do not hinder the proper functioning of the reliable communication system. Our work builds upon two reliable communication protocols (DolevR and DolevU [8]), and a topology reconstruction one (Explorer [16]). In more detail, we characterize the sets of assumptions that make our objective achievable, and we propose a solution that, after an initialization phase, guarantees reliable communication with optimal message complexity and delivery complexity.

Due to space constraints, minor proofs of Properties, Lemmas and Theorems have been reported in the technical report version of this paper [3].

2   Related Works

Several solutions have been proposed in the literature to build Byzantine-tolerant reliable communication primitives. A seminal contribution was provided by Dolev [8], assuming (i) processes interconnected through a possibly multi-hop communication network (ii) and an upper bound f on the number of Byzantine faulty processes present in the system (*globally bounded failure model*). Dolev proved that a $(2f + 1)$-connected network is required to build a reliable communication primitive in presence of f Byzantine participants i.e., the node connectivity of the communication network must be greater than twice the maximum estimated number of faulty processes. Dolev proposed two protocols working with different assumptions on the knowledge of the network topology by participating processes. More precisely, the lack of topology knowledge impacts both the message complexity (*i.e.*, the number of messages exchanged in the system during a reliable communication instance) and the delivery complexity (*i.e.*, the computational complexity of the procedure used to validate a message) of the protocol. The Dolev's protocol for unknown networks was recently revised to reduce its message complexity [4]. To the best of our knowledge, no other contribution addressed the reliable communication problem in the globally bounded failure model without considering stronger assumptions. When moving to the *locally bounded failure model* (where every process is linked to at most f Byzantine peers) other approaches have been defined [18]. The *Certified Propagation Algorithm* (CPA) was proposed as a reliable communication protocol by Pelc and Peleg, and it has been proven optimal, for the number of faulty processes that can be simultaneously tolerated [17]. Let us note that, either assuming a globally or locally bounded failure model, a dense communication network is required to enable reliable communication in a distributed system. For this reason, weaker primitives have been defined, allowing a (small) part of correct processes to either deliver spurious messages (i.e. messages not generated by their claimed author) or to never deliver a valid message [12–14]. These weaker versions enable almost reliable communication also on sparse communication networks.

All the aforementioned solutions do not necessarily rely on digital signatures or other cryptographic primitives. Indeed, the goal of Byzantine-tolerant algorithms is to withstand (computationally) unbounded adversaries that are potentially able to solve (computationally hard) problems on which cryptographic primitives are based upon. Nevertheless, links are assumed to be authenticated and reliable, so if u and v are linked, every message v received from u has been previously sent by u. Notice that cryptographic primitives are not necessary to implement authenticated links [19].

The Byzantine fault-tolerant topology reconstruction problem has been analyzed by Nesterenko and Tixeuil [16] assuming the globally bounded failure model. Then, temporary arbitrary faults have been considered by Dolev et al., defining a self-stabilizing Byzantine-tolerant solution [9].

3   Preliminaries

3.1   System Model

We consider an asynchronous distributed system [5] composed by a set P of n processes, each associated with a unique identifier i.e., $P = \{p_1, p_2, \ldots, p_n\}$.

Failure Model. We consider the *globally bounded Byzantine failure model*, namely we assume that inside the system there might be at most f Byzantine faulty processes. All other peers are assumed *correct*. Let us note that the identity of Byzantine processes is not known to correct ones.

Messages and Communication. Processes communicate by exchanging messages on top of a communication network made of *reliable* and *authenticated* links [5]. It means that messages cannot be lost or altered during their transmission and that the identity of their sender cannot be forged. Such communication network is abstracted by an undirected graph $G = (P, E)$ where the set of nodes V corresponds the set of processes participating in the system and the set of edges E contains an element $e_{i,j}$ for each existing link connecting two processes p_i and p_j. We assume the node connectivity k of G greater than twice the number of the potentially faulty processes i.e., $k > 2f$[1].

On top of the communication network, two alternative primitives are available: *unicast (UL)* and *local broadcast (LBL)* links [1,2,11]: the former interconnect single pairs of processes p_i, p_j; the latter attach a process p_i to many others, such that if a message is sent by p_i then it is received by all of its neighbors, thus preventing a faulty process to *equivocate* (i.e., to transmit conflicting messages to different neighbors).

We assume that processes are unaware about the topology of the communication network, namely the graph G: they either know the identifier of the peers they have a link with (*known neighborhood* i.e., *KN* assumption) or they have no knowledge about (*unknown neighborhood* i.e., *UN* assumption).

We refer with *source* to the advertised author of a message, and with *sender* to the process that is sending a message through a link.

3.2   Problem Specification: Reliable Communication

We investigate the *reliable communication* problem between a *source* process p_s and a *target* process p_t. Informally, when addressing this problem, the goal is to define a distributed protocol able to deliver only the messages generated by a correct source to every correct process in the system.

Let us note that, in the literature, the term *message* is commonly used instead of *content* when formalizing a problem specification based on message exchanges. However, several messages carrying the same payload can be diffused in a system to solve the reliable communication problem. Therefore, for ease of presentation,

---

[1] It is not possible otherwise to achieve reliable communication in the system model we are considering [8].

we will refer with *content* to the payload diffused by a process and with *message* to union of a content and the protocol specific overhead.

More formally, we will say that a protocol solves the reliable communication problem if, for every pair of processes p_s and p_t in the system, both the following conditions are satisfied:

- (**Safety**) if p_t is correct and it delivers a content m from p_s, then p_s previously sent m;
- (**Liveness**) if p_s and p_t are both correct and p_s sends a content m to p_t, then p_t eventually delivers m from p_s.

We refer with *spurious* content to one not sent by its claimed source (i.e. a content initially diffused by some Byzantine processes).

3.3   Evaluation Metrics

We will evaluate the solutions to the reliable communication problem in terms of (i) *message complexity* i.e., the number of messages that the protocol generates to solve the problem and (ii) *delivery complexity* i.e., the computational complexity of the procedure that allows a target process p_t to decide if a content can be delivered or not.

3.4   The Topology Reconstruction Problem

Given an unknown network G of correct and Byzantine faulty processes, the aim of a distributed protocol addressing the topology reconstruction problem is to enable every correct process p_i to reconstruct a subset of the topology of the communication network G. Such a reconstruction G_i is expected to be as complete as possible. The nodes of the communication network G can be partitioned in *correct* and *faulty*, and its edges in *correct*, *one-faulty* and *two-faulty*, respectively interconnecting two correct processes, a correct process and a faulty one, and two Byzantine processes. Likewise, the nodes and edges of a topology reconstruction G_i can be either *real* or *spurious*, respectively mapping or not nodes and edges in G.

3.5   Basic Definitions

For the sake of presentation, this section recalls some definitions and results coming from graph theory [6] that will be employed in this work.

Let us consider an undirected graph $G = (V, E)$. A *path* $\mathcal{P}$ is a sequence of nodes with no repetition i.e., $\mathcal{P} = [v_1, v_2, \ldots, v_m]$ (with $v_i \in V$) such that for each pair of adjacent elements v_i, v_{i+1} there exists an edge $e_{i,i+1} \in E$. The first and last elements of a path are referred with *endpoints*.

A pair of nodes $v_i, v_j \in V$ is *connected* if there exists at least one path $\mathcal{P}_{i,j}$ between them in G, it is *disconnected* otherwise. Given two nodes v_i and v_j, many paths between them may exist. Given a set of paths $\mathcal{P}_{i,j}^1, \mathcal{P}_{i,j}^2, \ldots, \mathcal{P}_{i,j}^x$

between two nodes v_i and v_j they are said *node disjoint* if they share no vertex except for their endpoints.

We refer with $\Pi_{i,j}$ to a *disjoint paths solution* between nodes v_i and v_j, i.e. a set of node disjoint paths having v_i and v_j as endpoints. The *local node connectivity* $\kappa_{i,j}$ between two nodes v_i, v_j is the minimum number of nodes that has to be removed from G to disconnect v_i from v_j. The *node connectivity* of a graph is the minimum value k for the local node connectivity $\kappa_{i,j}$ (i.e., $k = \min(\kappa_{i,j}), \forall v_i, v_j \in V$). A graph having node connectivity greater or equal than k is said *k-connected* graph. The local node connectivity between two nodes is equal to the maximum number of disjoint paths that exist between them (Menger theorem [15]). It is possible to compute a disjoint paths solution $\Pi_{i,j}$ between two nodes v_i, v_j of maximum size (namely $\kappa_{i,j}$) with a deterministic algorithm with computational complexity polynomial in the size of the graph [7,10]. In the following, we will consider every disjoint paths solution $\Pi_{i,j}$ always of maximum size $\kappa_{i,j}$.

4   Dolev Protocols

Dolev [8] identified the necessary and sufficient conditions to solve reliable communication in the system model we consider.

Remark 1. The reliable communication problem can be solved if and only if the node connectivity of the communication graph is greater than $2f$ i.e., $k > 2f$.

Dolev provided two protocols that work under different assumptions on the (partial) knowledge that processes have about the network topology.

4.1   Dolev's Routed Protocol (DolevR)

DolevR is a protocol solving reliable communication in routed-networks [8], i.e. systems where all messages are relayed over (and only) fixed and known paths. Specifically, processes employing DolevR route contents between each pair of process p_i, p_j over $2f + 1$ disjoint paths $\Pi_{i,j}$. The reliable and authenticated links restrict the capabilities of faulty processes, allowing them to diffuse spurious contents through at most f paths of any $\Pi_{i,j}$. The assumption of a $(2f + 1)$-connected network guarantees that at least $f + 1$ paths of any $\Pi_{i,j}$ are fault-free (i.e. they do not pass through any faulty processes). A process p_j employing DolevR delivers a content m from a process p_i if it is received through at least $f + 1$ routes of $\Pi_{i,j}$.

Protocol Complexity. The message complexity of DolevR is linear in the size of the network, whereas its delivery complexity is linear in the number of maximum assumed faults, as detailed in the following Lemmas.

Lemma 1. *DolevR solves the reliable communication problem with $O(n)$ message complexity.*

Lemma 2. *DolevR solves the reliable communication problem with $O(f)$ delivery complexity.*

The delivery complexity and message complexity of `DolevR` are optimal solving the reliable communication problem in the system model we consider.

Theorem 1. *DolevR solves the reliable communication problem in routed networks assuming the globally bounded Byzantine failure model with optimal message complexity and optimal delivery complexity.*

Proof. Given Lemmas 1 and 2, we show that no algorithm can solve the reliable communication problem, in the settings considered in this paper, with an asymptotically lower complexity without considering additional assumptions.

Let us consider two processes p_s and p_t, not connected by a link, respectively as source and target of a reliable communication instance.

The target process relies on the messages it receives from its neighbors to deliver a content. Nevertheless, up to f of its neighbors could be Byzantine faulty and process p_t cannot identify them. Thus, a $O(f)$ procedure is required.

Given that p_s and p_t are not linked, a content must be relayed over fault-free paths (i.e. not including any faulty process) to achieve liveness of reliable communication. In the worst-case scenario the length of the longest fault-free path is $n - k$. □

4.2   Dolev's Topology Unaware Protocol (`DolevU`)

`DolevU` protocol solves the reliable communication problem in unknown networks [8], where contents are flooded in the system. Specifically, `DolevU` spreads messages $\langle m, path \rangle$, in which m is the content and $path$ is a list data structure collecting the identifier of processes that are traversed by m. The source process starts the communication multicasting to all of its neighbors the content m with an empty $path$. Then, every process p_i that receives a message $\langle m, path \rangle$ from a neighbor p_j adds the identifier of p_j to $path$, it stores $\langle m, path + \{j\} \rangle$ and it relays such a message to all of its neighbors not yet included in $path + \{j\}$. Every process that succeeds identifying $f + 1$ disjoint $path$ among the ones it received with a content m delivers m.

Protocol Complexity. The message complexity of `DolevU` is factorial in the size of the network, whereas a NP-Complete problem has to be solved verifying every content, as detailed in the following Lemmas.

Lemma 3. *DolevU solves the reliable communication problem with a message complexity factorial in the number of processes.*

Lemma 4. *DolevU solves the reliable communication problem with a NP delivery complexity.*

The DolevU protocol has been recently reviewed to reduce its message complexity [4]. It has been proven that modifications can be adopted in the protocol preventing some messages to be generated. Nonetheless, it is still an open problem whether it is always possible to solve reliable communication in unknown networks, under the weakest assumptions identified by Dolev (Remark 1), with a protocol having polynomial message complexity and/or polynomial delivery complexity. For sake of simplicity, we do not employ the reliable communication protocol defined in [4], given that its worst-case delivery complexity and message complexity is unchanged with respect DolevU.

The DolevU protocol provides the following additional guarantee in case local broadcast links are assumed.

Theorem 2. *Let DolevU solve reliable communication in a network G with local broadcast links. Then, a content m is delivered by every correct process if it is delivered by any correct one.*

Proof. When the reliable communication necessary correctness condition is met (Remark 1), the DolevU protocol guarantees that if the source p_s of a content m is correct, then any correct target eventually delivers m. This is not guaranteed in case of a faulty source: it may diverge from the protocol and it may prevent some targets from delivering its contents. The local broadcast links provide an additional guarantee: every message a process sends is received by all its neighbors. A correct source in DolevU multicast message $\langle m, \emptyset \rangle$ to all of its neighbors. It follows that if a correct process delivered m, then message $\langle m, \emptyset \rangle$ has been sent to all neighbors of p_s, given the local broadcast links, and the claim follows.   □

5 Explorer

Nesterenko and Tixeuil analyzed the Byzantine fault-tolerant topology reconstruction problem [16]. Among the results they provided, two impossibilities have been identified.

Remark 2. No algorithm can decide whether a two-faulty edge exists [16].

Remark 3. No algorithm can compute a reconstruction of only real nodes and edges while including both all correct and all one-faulty edges [16].

They also defined Explorer, an algorithm that enables processes to partially reconstruct the topology of G in the globally bounded failure model assuming KN. It is specified only by the following two procedures: every process p_i *1)* broadcasts its neighborhood $\Gamma(i)$ (namely it broadcasts the identifier of processes it has a link with) and *2)* it stores all neighborhoods $\Gamma(j)$ delivered with a reliable communication primitive in a dictionary data structure $cTop_i := \bigcup \langle j, \Gamma(j) \rangle$.

We introduce a simple neighborhood discovery procedure to cope with the unknown neighborhood scenario, defined by the following actions: *1)* every process multicasts a *HELLO* message (basically a message with no payload), and *2)* every process that receives a *HELLO* message adds the identifier of the sender to its neighborhood.

Then, every process p_i broadcasts with a reliable communication primitive its neighborhood $\Gamma(i)$ every time that it changes, and it updates the entry $\langle j, \Gamma(j) \rangle \in cTop_i$ if $\langle j, \Gamma(j)' \rangle$, such that $\Gamma(j) \subset \Gamma(j)'$, is delivered.

Additionally, if local broadcast links are assumed, every process p_i that delivers two neighborhood $\Gamma(j)$ and $\Gamma(j)'$ from p_j, such that $\Gamma(j)' \not\subset (\Gamma(j) \in cTop_i)$ and $(\Gamma(j) \in cTop_i) \not\subset \Gamma(j)'$, do not consider j for the reconstruction.

Every process p_i computes the reconstruction $G_i(P_i, E_i)$ from $cTop_i$ as follows:

- $\forall \langle u, \Gamma(u) \rangle \in cTop_i \Rightarrow \exists u \in P_i$;
- $\forall \langle v, \Gamma(v) \rangle, \langle u, \Gamma(u) \rangle \in cTop_i, \ u \in \Gamma(v) \Rightarrow \exists (v, u) \in E_i$.
- $\forall v \in \Gamma(u), \langle u, \Gamma(u) \rangle \in cTop : X \leftarrow \bigcup u, |X| > f \Rightarrow \exists v \in P_i$.

We report some properties of any reconstructed topology G_i computed with the defined protocol.

Property 1. (From [16]) $j \notin P \Rightarrow j \notin P_i$ (no G_i contains non-existent nodes).

Property 2. Assuming the *unknown neighborhood* assumption (UN), some reconstruction G_i may never include some Byzantine processes.

Property 3. (From [16]) Assuming the *known neighborhood* assumption (KN), the reconstruction G_i eventually guarantees the following property: $j \in P_i \Leftrightarrow p_j \in P$ (Property 1 + all real nodes are eventually detected).

Property 4. $\forall \langle u, v \rangle \in E, \ u, v \in Correct \Rightarrow \exists \langle u, v \rangle \in E_i$ (all correct edges are eventually contained in G_i).

Property 5. $\forall \langle u, v \rangle \in E_i, \ \langle u, v \rangle \notin E \Rightarrow u \in Byzantine$ (every spurious edge contains at least one Byzantine process).

Property 6. (From [16]) Assuming the *known neighborhood* assumption (KN): $\forall \langle u, v \rangle \in E, \ u \in Correct, \ v \in Byzantine \Rightarrow \exists \langle u, v \rangle \in E_i$ (all one-faulty edges will eventually be present in any G_i).

Property 7. Assuming local broadcast links (LBL), all one-faulty edges between a Byzantine process and all of its correct neighbors are eventually either all or none present in every G_i.

Property 8. Assuming local broadcast links (LBL), all correct processes eventually share the same topology reconstruction.

Property 9. No reconstruction G_i computed assuming local broadcast links (LBL) will ever contain more real edges than one obtained assuming the known neighborhood assumption (KN).

5.1   Protocol Complexity Analysis

All correct processes p_i in `Explorer` broadcast their neighborhood $\Gamma(i)$. Supposing the know neighborhood assumption (KN), every process broadcasts such information only once. It follows that `Explorer` requires $\mathcal{O}(n)$ reliable communication executions to enable all correct processes to compute G_i. Considering the unknown neighborhood (UN) assumption, every process has to perform the neighborhood discovery and then to broadcast its $\Gamma(i)$. Unfortunately, no process p_i knows how many nodes have to be detected before diffusing $\Gamma(i)$, and thus, they may broadcast their neighborhood many times, $n - f - 1$ in the worst-case scenario. It follows that `Explorer` with neighborhood discovery executes $\mathcal{O}(n^2)$ reliable communication instances to enable all correct processes to compute G_i.

5.2   Fault-Free Disjoint Path Solution

The `Explorer` protocol enables processes to partially reconstruct the topology of G. We showed that different sets of assumptions provide more or less accurate reconstructions (Properties 1–9). We reported the `DolevR` protocol, that it leverages disjoint routes defined between all pairs of processes to achieve reliable communication. We highlighted how f Byzantine faulty processes may compromise at most f paths of any disjoint path solution $\Pi_{i,j}$ in `DolevR`, and that the liveness of such a protocol is guaranteed by the existence of disjoint path solutions of size greater than $2f$ between all pairs of processes, where at least $f + 1$ paths cannot be compromised. It follows that, if every pair of correct processes is able to identify a disjoint path solution interconnecting them where at least $f + 1$ paths are *faults-free* (i.e. they do not include any Byzantine faulty process), *real* and *disjoint* (*FF_R_D*), then they are able to achieve reliable communication.

   We analyze several sets of assumptions that enable all pairs of correct processes p_i, p_j to compute a disjoint path solutions $\Pi_{i,j}$ in G_i containing at least $f + 1$ FF_R_D paths.

Theorem 3. *The set of assumptions a)* **k > 3f**, *b)* **unicast links** *and c)* **unknown neighborhood enables** *every correct process p_i to compute a disjoint paths solution $\Pi_{i,j}$ toward any correct process p_j that contains at least $f + 1$ faults-free, real and disjoint paths.*

Proof. Let us assume processes employing `Explorer` and that all messages it generates have been already delivered by the peers. The unknown neighborhood assumption and unicast links allow Byzantine faulty processes to decide which one-faulty and two-faulty edges to declare (Remarks 2, 3), thus the local connectivity between any two processes p_i, p_j in the reconstructed topology may be reduced by at most f. It follows that any disjoint paths solution $\Pi_{i,j}$ will contain more than $2f$ paths (Property 4). Given that at most f paths of any $\Pi_{i,j}$ may contain faults the claim follows.   □

Theorem 4. *The set of assumptions a)* **k ≤ 3f**, *b)* **unicast link** *and c)* **unknown neighborhood is not sufficient** *to enable every correct process p_i to* *compute a disjoint paths solution $\Pi_{i,j}$ toward every correct process p_j containing at least $f + 1$ faults-free, real and disjoint paths with any protocol.*

Proof. The unknown neighborhood assumption and the unicast links allow faulty processes to decide which one-faulty and two-faulty edges are detectable by correct processes (Remarks 2, 3). It follows that the faulty processes may potentially be able to reduce the local connectivity between some pairs of correct processes p_i, p_j by f: the local connectivity $\kappa_{i,j}$ in G_i may be lower than $2f$ and at most $2f - 1$ disjoint path $\Pi_{i,j}$ will be identifiable between p_i and p_j, whatever algorithm is envisioned for the reconstruction. Then, up to f paths in $\Pi_{i,j}$ may include faulty processes and the claim follows.   □

Theorem 5. *The set of assumptions a)* $\mathbf{k > 2f + \lfloor f/2 \rfloor}$, *b)* **local broadcast** *links and c)* **unknown neighborhood enables** *every correct process p_i to compute a disjoint paths solution $\Pi_{i,j}$ toward any correct process p_j containing at least $f + 1$ faults-free, real and disjoint paths.*

Proof. Given Property 7, let us suppose that $f_d \leq f$ Byzantine processes decide to be detected by their neighbors and they send the `HELLO` message, whereas $f - f_d$ ones do not. Let us assume that all messages exchanged by `Explorer` have been already delivered and let us consider $\Pi_{i,j}$ as the disjoint path solution computed on G_i between a pair of correct processes p_i and p_j. The assumption on the node connectivity of G guarantees that at least $2f + \lfloor f/2 \rfloor + 1$ disjoint paths exist between p_i and p_j in the communication network. The undeclared Byzantine processes may reduce the local connectivity between p_i and p_j by $f - f_d$ in G_i. Let us temporarily assume, for the purpose of the proof, that the declared Byzantine processes behave as correct ones. It follows, from Property 4 and 7 of `Explorer`, that the size of $\Pi_{i,j}$ would be at least equal to:

$$2f + \lfloor f/2 \rfloor + 1 \ - (\mathbf{f} - \mathbf{f_d}) = f + \lfloor f/2 \rfloor + 1 + f_d$$

Specifically, all paths between p_i and p_j that contain only correct or declared Byzantine processes existing in G are present in G_i.

Let us now consider the declared Byzantine processes not reporting the edges existing between them (i.e. the two-faulty edges). It follows that the paths in G containing two-faulty edges may not be present in G_i (Remark 2). Therefore, pairs of Byzantine processes may potentially cause a reduction to the maximum size of $\Pi_{i,j}$: every couple may decrease the number of available disjoint paths in G_i between p_i and p_j by one. It follows that the size of $\Pi_{i,j}$ would be at most reduced to:

$$f + \lfloor f/2 \rfloor + 1 + f_d - \lfloor \mathbf{f_d/2} \rfloor$$

namely, f_d declared Byzantine faulty processes may reduce the local connectivity between p and q in G_i by at most $\lfloor f_d/2 \rfloor$. The f_d declared Byzantine processes may also be selected in the paths $\Pi_{i,j}$. Specifically, in the worst case scenario f_d paths in $\Pi_{i,j}$ may contain Byzantine processes. It follows that at most f_d paths would not be fault-free, and thus the remaining fault-free ones in $\Pi_{i,j}$ would be:

$$f + \lfloor f/2 \rfloor + 1 + f_d - \lfloor f_d/2 \rfloor - \mathbf{f_d} = f + 1 + \lfloor f/2 \rfloor - \lfloor f_d/2 \rfloor$$

Thus, at least $f + 1$ paths in $\Pi_{i,j}$ are faults-free, real and disjoint. $\square$

Notice that, given Property 9, the Theorem 5 extends substituting local broadcast links with the unicast ones and assuming the known neighborhood assumption.

6   CombinedRC, Reliable Communication Protocol

We combine `Explorer`, `DolevU` and `DolevR` protocols to design a new reliable communication primitive. We call such a protocol `CombinedRC`, that aims to set up an efficient reliable communication service.

The `Explorer` protocol is used to partially reconstruct the network topology, and then to enable processes to compute disjoint paths solutions through which relay contents. The `DolevU` protocol is adopted as reliable communication subprimitive by `Explorer` and `CombinedRC` during the initialization. Lastly, the `DolevR` protocol is employed as actual reliable communication primitive in `CombinedRC`, leveraging the routes computed and communicated using `Explorer` and `DolevU`.

We showed in Sect. 5.2 that `Explorer`, under certain conditions, enables every correct process p_i to identify a disjoint paths solution $\Pi_{i,j}$ interconnecting it with any other correct process p_j, such that at least $f + 1$ paths in the solution are faults-free, real and disjoint. Once that the solution $\Pi_{i,j}$ is known to both p_i and p_j, they can efficiently communicate. We claimed in Property 8 that all correct processes eventually obtain the same topology reconstruction in case local broadcast links are employed. Thus, under such an assumption, processes p_i and p_j eventually compute the same solution $\Pi_{i,j}$. Under the weaker condition of unicast links, the reconstructed topologies may differ on distinct processes, thus a source process p_i has additionally to communicate the computed solution $\Pi_{i,j}$ to a target process p_j using `DolevU`.

Any source process p_i routes contents through the computed $\Pi_{i,j}$ and any target process p_j waits for messages over $f + 1$ paths among the ones in $\Pi_{i,j}$.

The pseudo-code of `CombinedRC` is presented in Algorithm 1.

Every process relays its contents over the computed routes if available, otherwise, they are queued for subsequent transmission (lines 1–5).

Every process p_i attempts to compute a solution $\Pi_{i,j}$ toward every other process p_j of the system. In the case of local broadcast links, the reconstructed topology G_i is eventually the same in every process. Therefore, a source process has to relay its contents over the computed disjoint routes every time they change (a finite number of times). In case of unicast links, once that the local connectivity toward a target p_j reaches a value greater than $2f$, the source process p_i communicates the computed solution $\Pi_{i,j}$ via `DolevU` (lines 6–21).

Every process relays contents or computed disjoint solution following the path attached to messages (lines 22–23).

Every process that delivers a disjoint paths solution with `DolevU` adopts it to verify contents (lines 34–35) using `DolevR` (lines 36–37).

Algorithm 1. CombinedRC

1: **upon** RC_send*(m, target)* **do**
2:    $Sent \leftarrow Sent \cup \langle m, target \rangle$
3:    **if** $\Pi_{i,target} \neq \emptyset$ **then**
4:       **for** $path \in \Pi_{i,target}$ **do**
5:          **send**($\langle CNT, i, target, m, path \rangle, path[1]$)

6: **upon** G_i changes **do**
7:    **for** $j \in G_i$ such that $i \neq j$ **do**
8:       **if** LB **then**
9:          **if** local_conn$(G_i, i, j) > f + \lfloor f/2 \rfloor$ and disj_paths$(G_i, i, j) \neq |\Pi_{i,j}|$ **then**
10:             $\Pi_{i,j} \leftarrow$ disj_paths(G_i, i, j)
11:             **for** $path \in \Pi_{i,j}$ **do**
12:                **for** $\langle m, target \rangle \in Sent$ such that $j = target$ **do**
13:                   **send**($\langle CNT, i, j, m, path \rangle, path[1]$)
14:             $\Pi_{j,i} \leftarrow$ disj_paths(G_i, j, i)
15:       **else if** UC **then**
16:          **if** $\Pi_{i,j} = \emptyset$ and local_conn$(G_i, i, j) > 2f$ **then**
17:             $\Pi_{i,j} \leftarrow$ disj_paths(G_i, i, j)
18:             **for** $path \in \Pi_{i,j}$ **do**
19:                **send**($\langle ROU, i, j, \Pi_{i,j}, path \rangle, path[1]$)
20:                **for** $\langle m, target \rangle \in Sent$ such that $j = target$ **do**
21:                   **send**($\langle CNT, i, j, m, path \rangle, path[1]$)

22: **upon** **receive**($\langle CNT, s, t, m, path \rangle, j$) **do**
23:    **if** predecessor$(path, i) = j$ **then**
24:       **if** $t = i$ **then**
25:          $Paths_{cnt}[\langle m, s \rangle] \leftarrow Paths_{cnt}[\langle m, s \rangle] \cup \{path\}$
26:       **else**
27:          **send**($\langle CNT, s, t, m, path \rangle,$ successor$(path, i)$)

28: **upon** **receive**($\langle ROU, s, t, \Pi, path \rangle, j$) **do**
29:    **if** predecessor$(path, i) = j$ **then**
30:       **if** $t = i$ **then**
31:          $Paths_{rou}[\langle s, \Pi \rangle] \leftarrow Paths\_uRts[\langle s, \Pi \rangle] \cup \{path\}$
32:       **else**
33:          **send**($\langle ROU, s, t, \Pi, path \rangle,$ successor$(path, i)$)

34: **upon** DolevU_deliver($Paths_{rou}[\langle s, \Pi \rangle], s$) **do**
35:    $\Pi_{s,i} \leftarrow \Pi$

36: **upon** DolevR_deliver($Paths_{cnt}[\langle m, s \rangle], s$) **do**
37:    RC_deliver(m, s)

6.1   CombinedRC Correctness

Theorem 6. *CombinedRC provides safety of reliable communication.*

Theorem 7. *CombinedRC provides liveness of reliable communication in all cases where Explorer succeeds in identifying a disjoint path solution between two processes i, j that contains $f + 1$ FF_R_D paths.*

Proof. Let us assume that all messages exchanged by Explorer have been already delivered and that a process p_i aims to reliably communicate with a correct process p_j. In case of local broadcast links, processes p_i and p_j eventually share the same topology reconstruction G_i, thus also the disjoint path solution $\Pi_{i,j}$ will eventually be the same both on p_i and p_j. Process p_i relays the contents through $\Pi_{i,j}$ every time such a solution changes. The assumption of $f + 1$ FF_R_D paths in $\Pi_{i,j}$ guarantees reliable communication. In case of unicast channels, the solution $\Pi_{i,j}$ is diffused via DolevU and contents are routed over $\Pi_{i,j}$. The assumption of $f + 1$ FF_R_D paths in $\Pi_{i,j}$ guarantees reliable communication.   □

6.2   Protocol Complexity Analysis

CombinedRC provides reliable communication with optimal message complexity and delivery complexity (Theorem 1). Specifically, it routes contents over computed disjoint routes as DolevR, thus $\mathcal{O}(n)$ messages per content are exchanged, and an $\mathcal{O}(f)$ procedure is executed to verify any content.

CombinedRC requires an initialization phase where the network topology is partially reconstructed and the solutions containing $f + 1$ FF_R_D paths are computed between every pair of correct processes. We showed in Sect. 5 that Explorer requires at most $\mathcal{O}(n^2)$ reliable communication instances to partially reconstruct the network topology. The same solution $\Pi_{i,j}$ is eventually computed by both p_i and p_j, assuming local broadcast channels, without additional message exchanges, because the topology reconstruction will eventually be the same on every process and the disjoint paths solutions can be computed through a deterministic algorithm. On the other hand, employing unicast links, every couple of processes has to agree on a solution $\Pi_{i,j}$. Thus, an additional content exchange (with payload $\Pi_{i,j}$) using a reliable communication primitive has to be performed for each pair of correct processes. It follows that the initialization phase of CombinedRC requires the execution of $\mathcal{O}(n^2)$ DolevU instances. Notice that, in the case of known neighborhood and local broadcast links, the cost of the initialization phase reduces to $\mathcal{O}(n)$ DolevU instances, indeed each process diffuses its neighborhood only once and all correct processes eventually share the same reconstruction.

7   Conclusion

We demonstrated how to boost the efficiency of reliable communication despite some of the participants being Byzantine faulty, when the network topology is unknown to the participants, assuming reliable authenticated links. Our solution combines a costly topology reconstruction process, that is executed once, and an efficient reliable communication scheme that is optimal both in terms of exchanged messages and of local computation complexity. Without leveraging the topology reconstruction, the cost of every reliable communication instance in the same scenario would have been factorial in message complexity and NP in delivery complexity.

An interesting path for future research is to decrease the adversary capabilities. A noteworthy candidate is the computationally bounded adversary, that enables solutions based on cryptography.

References

1. Bhandari, V., Vaidya, N.H.: Implementing a reliable local broadcast primitive in wireless ad hoc networks. https://disc.georgetown.domains/publications/rbcast-tech.pdf
2. Bhandari, V., Vaidya, N.H.: On reliable broadcast in a radio network. In: Aguilera, M.K., Aspnes, J. (eds.) Proceedings of the Twenty-Fourth Annual ACM Symposium on Principles of Distributed Computing, PODC 2005, Las Vegas, NV, USA, 17–20 July 2005, pp. 138–147. ACM (2005)
3. Bonomi, S., Farina, G., Tixeuil, S.: Boosting the efficiency of byzantine-tolerant reliable communication. Tech. rep. https://hal.archives-ouvertes.fr/hal-02960087
4. Bonomi, S., Farina, G., Tixeuil, S.: Multi-hop byzantine reliable broadcast with honest dealer made practical. J. Braz. Comp. Soc. **25**(1), 9:1–9:23 (2019)
5. Cachin, C., Guerraoui, R., Rodrigues, L.E.T.: Introduction to Reliable and Secure Distributed Programming, 2nd edn. Springer, Heidelberg (2011). https://doi.org/10.1007/978-3-642-15260-3
6. Diestel, R.: Graph Theory. Springer, Heidelberg (2017). https://doi.org/10.1007/978-3-662-53622-3
7. Dinic, E.A.: Algorithm for solution of a problem of maximum flow in networks with power estimation. Soviet Math. Doklady. **11**, 1277–1280 (1970)
8. Dolev, D.: Unanimity in an unknown and unreliable environment. In: 22nd Annual Symposium on Foundations of Computer Science, Nashville, Tennessee, USA, 28–30 October 1981, pp. 159–168 (1981)
9. Dolev, S., Liba, O., Schiller, E.M.: Self-stabilizing byzantine resilient topology discovery and message delivery - (extended abstract). In: Networked Systems - First International Conference, NETYS 2013, Marrakech, Morocco, 2–4 May 2013, Revised Selected Papers, pp. 42–57 (2013)
10. Edmonds, J., Karp, R.M.: Theoretical improvements in algorithmic efficiency for network flow problems. J. ACM **19**(2), 248–264 (1972)
11. Khan, M.S., Naqvi, S.S., Vaidya, N.H.: Exact byzantine consensus on undirected graphs under local broadcast model. In: Robinson, P., Ellen, F. (eds.) Proceedings of the 2019 ACM Symposium on Principles of Distributed Computing, PODC 2019, Toronto, ON, Canada, 29 July–2 August 2019, pp. 327–336. ACM (2019)
12. Maurer, A., Tixeuil, S.: Byzantine broadcast with fixed disjoint paths. J. Parallel Distrib. Comput. **74**(11), 3153–3160 (2014)
13. Maurer, A., Tixeuil, S.: Containing byzantine failures with control zones. IEEE Trans. Parallel Distrib. Syst. **26**(2), 362–370 (2015)

14. Maurer, A., Tixeuil, S.: Tolerating random byzantine failures in an unbounded network. Parall. Process. Lett. **26**(1), 1650003 (2016)
15. Menger, K.: Zur allgemeinen kurventheorie. Fundamenta Mathematicae **10**(1), 96–115 (1927)
16. Nesterenko, M., Tixeuil, S.: Discovering network topology in the presence of byzantine faults. IEEE Trans. Parallel Distrib. Syst. **20**(12), 1777–1789 (2009)
17. Pagourtzis, A., Panagiotakos, G., Sakavalas, D.: Reliable broadcast with respect to topology knowledge. Distrib. Comput. **30**(2), 87–102 (2017)
18. Pelc, A., Peleg, D.: Broadcasting with locally bounded byzantine faults. Inf. Process. Lett. **93**(3), 109–115 (2005)
19. Zeng, K., Govindan, K., Mohapatra, P.: Non-cryptographic authentication and identification in wireless networks. IEEE Wirel. Commun. **17**(5), 56–62 (2010)

Stand Up Indulgent Rendezvous

Quentin Bramas[1]([✉]), Anissa Lamani[1], and Sébastien Tixeuil[2]

[1] ICUBE, Strasbourg University, CNRS, Strasbourg, France
`bramas@unistra.fr`
[2] Sorbonne University, CNRS, LIP6, Paris, France

Abstract. We consider two mobile oblivious robots that evolve in a continuous Euclidean space. We require the two robots to solve the rendezvous problem (meeting in finite time at the same location, not known beforehand) despite the possibility that one of those robots crashes unpredictably. The rendezvous is stand up indulgent in the sense that when a crash occurs, the correct robot must still meet the crashed robot on its last position.

We characterize the system assumptions that enable problem solvability, and present a series of algorithms that solve the problem for the possible cases.

1 Introduction

The study of swarm robotics in Distributed Computing has focused on the computational power of a set of autonomous robots evolving in a bidimensional Euclidean space. In this setting, a robot is modeled as a point in a two dimensional plane and has its own coordinate system and unit distance. Robots are usually assumed to be very weak: they are *(i)* anonymous (they can not be distinguished), *(ii)* uniform (they execute the same algorithm) and, *(iii)* oblivious (they cannot remember past actions). Robots operate in cycles that comprise three phases: Look, Compute and Move. During the first phase (Look), a robot takes a snapshot to see the position of the other robots. During the second phase (Compute), a robot decides to move or stay idle. In the case in which it decides to move, it computes a target destination. In the last phase (Move), a robot moves to the computed destination (if any). Depending on how robots are activated and the synchronization level, three models have been introduced: Fully synchronous model (FFSYNC) in which robots are activated simultaneously and execute cycles synchronously. Semi synchronous model (SSYNC) in which a subset of robots are activated simultaneously. The activated robots execute a cycle synchronously. The asynchronous model (ASYNC) in which there is no global clock. The duration of each phase is finite but unbounded. That is, one robot can decide to move according to an outdated view.

Among the various problems considered under such weak assumptions there is the gathering problem which is one of the benchmarking tasks in mobile robot

This work was partially funded by the ANR project SAPPORO, ref. 2019-CE25-0005-1.

S. Devismes and N. Mittal (Eds.): SSS 2020, LNCS 12514, pp. 45–59, 2020.
https://doi.org/10.1007/978-3-030-64348-5_4

networks. The gathering task consists in all robots reaching a single point, not known beforehand, in finite time. The particular case of gathering two robots is called *rendezvous* in the literature. In this paper, we consider the Stand Up Indulgent Rendezvous (SUIR) problem: in the case one of the two robots crashes, they still have to gather (obviously, at the position of the crashed robot); if no robot crashes, the robots are expected to gather in finite time. The SUIR problem is at least as difficult as the rendezvous problem, so classical impossibility results still apply.

Related Works. A foundational result [14] shows that when robots operate in a fully synchronous manner, the rendezvous can be solved deterministically, while if robots are allowed to wait for a while (this is the case *e.g.* in the SSYNC model), the problem becomes impossible without additional assumptions. Such additional assumptions include the robots executing a probabilistic protocol [8,9] (but the rendezvous only occurs probabilistically), the robots sharing a common $x - y$ coordinate system [14] or an approximation of a common coordinate system [12], or the robots being endowed with persistent memory [7,11,13–15]. Recent work considered the minimum amount of persistent memory that is necessary to solve rendezvous [7,11,13,15]. It turns out that exactly one bit of persistent memory is necessary and sufficient [11] even when robots operate asynchronously and are disoriented.

When robots can crash unexpectedly, two variants of the gathering problem can be defined [1]: *weak gathering* requires correct robots to gather, regardless of the position of crashed robots; *strong gathering* requires all robots to gather at the same point. Obviously, strong gathering is only feasible if all crashed robots crash at the same location. Early solutions to weak gathering in SSYNC for groups of at least three robots make use of extra hypotheses: *(i)* starting from a distinct configuration (that is, a configurations where at most one robot occupies a particular position), at most one robot may crash [1], *(ii)* robots are activated one at a time [8], *(iii)* robots may exhibit probabilistic behavior [9], *(iv)* robots share a common chirality (that is, the same notion of handedness) [4], *(v)* robots agree on a common direction [3]. It turns out that these hypotheses are *not* necessary to solve deterministic weak gathering in SSYNC, when up to $n - 1$ robots may crash [5].

The case of strong gathering mostly yielded impossibility results: with at most a single crash, strong gathering $n \geq 3$ robots deterministically in SSYNC is impossible even if robots are executed one at a time, and probabilistic strong gathering $n \geq 3$ robots is impossible with a fair scheduler [8,9]. However, probabilistic strong gathering $n \geq 3$ robots becomes possible in SSYNC if the relative speed of the robots is upper bounded by a constant [8,9].

For the special case of two robots, the strong gathering problem boils down to stand up indulgent rendezvous (SUIR), as presented above. Only few results are known:

1. The algorithm "with probability $\frac{1}{2}$, go to the other robot position" is a probabilistic solution to SUIR in SSYNC [8,9],

2. The algorithm "go to the other robot position" is a deterministic solution to SUIR in SSYNC when exactly one robot is activated at any time [8,9].

To this paper, it is unknown whether additional assumptions (*e.g.* a common coordinate system in SSYNC, or FSYNC scheduling) enable deterministic SUIR solvability.

Our Contribution. In this paper, we consider the SUIR problem and concentrate at characterizing its deterministic solvability. When robots share a common $x - y$ coordinate system, rendezvous is deterministically solvable in SSYNC [14]: the two robots simply meet at the position of the Northernmost, Easternmost position. Our main impossibility result shows that SUIR cannot be solved deterministically in this setting. Furthermore, it remains impossible if robots have *both* infinite persistent memory (this is a stronger assumption than the classical luminous robot model that permits to solve classical rendezvous in ASYNC [7,11,13,15]) and a common $x-y$ coordinate system. This motivates our focus on the FSYNC setting, where both robots always operate synchronously. Our main positive result is that SUIR is deterministically solvable in FSYNC by oblivious disoriented robots. Our approach is constructive: we first present a simple algorithm for the case the robots share a common coordinate system, and then a more involved solution for the case of disoriented robots.

An interesting byproduct of our work is an oblivious deterministic rendezvous protocol (so, assuming no faults) for the case where robots only agree on a single axis. This complements previous results where robots agree on both axes [14] or approximately agree on both axes [12].

A summary of our results is presented in the following table.

	Rendezvous	SUIR
SSYNC oblivious, disoriented	Impossible [14]	Impossible (Theorem 1)
SSYNC oblivious, common x axis	Possible (Algorithm 1)	Impossible (Theorem 1)
SSYNC oblivious, common $x - y$ axes	Possible [10,14]	Impossible (Theorem 1)
SSYNC luminous, disoriented	Possible [7,11,13–15]	Impossible (Theorem 1)
FSYNC oblivious, disoriented	Possible [2,6,14]	Possible (Algorithm 3)

2   Model

We consider two robots, evolving in a Euclidean two-dimensional space. Robots are anonymous and oblivious. The time is discrete, and at each time instant, called round, a subset of the robots is activated and executes a Look-Compute-Move cycle. Each activated robot first Looks at its surroundings to retrieve the position of the other robot in its ego-centered coordinate system. Then, it Computes a target destination, based only on the current position of the other robot. Finally, it moves towards the destination following a straight path.

If the movements are *rigid*, each robot always reaches its destination before the next Look-Compute-Move cycle. Otherwise, movements are *non-rigid*, and an adversary can stop the robot anywhere along the path to its destination, but only after the robot traveled at least a fixed positive distance δ. The value of δ is not known by the robots, and can be arbitrary small, but it does not change during the execution.

In the fully-synchronous model (FSYNC), all correct robots are activated at each round. In the *Semi-synchronous* model (SSYNC), only a non-empty subset of the correct robots may be activated at each round. In this case, we consider only *fair* schedules *i.e.*, schedules where each correct robot is activated infinitely often.

Configurations and Local Views. We consider different settings that impact how a robot retrieves the position of the other robot. Robots might agree on one or both axes of their ego-centered coordinate systems. In other words, robots may have a common North (and possibly a common East direction). They may also have different unit distance.

For the analysis, we assume a global coordinate system Z that is not accessible to the robots. A *configuration* is a set $\{r_1, r_2\}$ containing the positions of both robots in Z. Notice that r_i, $i = 1, 2$, denotes at the same time a robot and its position in $\mathbb{R}^2$ in the coordinate system Z.

To model the agreement of the robots about their coordinate system, we define the set $\mathcal{T}$ of *indistinguishable transformations*, that modify how robots see the current configuration. If robots agree on both axes and on the unit distance, then $\mathcal{T}$ only contains the identity. If robots do not agree on the unit distance, then $\mathcal{T}$ contains all the (uniform) scaling transformations. If robots do not agree on the x-axis, then $\mathcal{T}$ also contains the reflection along the y-axis. If robots does not agree on any axis, then $\mathcal{T}$ also contains all the rotations. Finally, $\mathcal{T}$ is closed by composition.

We say robots are *disoriented* if robots do not agree on any axis, nor on a common unit distance *i.e.*, $\mathcal{T}$ contains the rotations, scaling, reflection, and their compositions.

In a configuration $\{r_1, r_2\}$ the *local view* V_1 of robot r_1 is obtained by translating the global configuration by $-r_1$ (so that r_1 is seen at position $(0,0)$ and r_2 is at position $r_2 - r_1$) from which we apply the transformation function h_1 of r_1. Formally, $V_1 = \{(0,0), h_1(r_2 - r_1)\}$. Similarly the local view of r_2 is $V_2 = \{(0,0), h_2(r_1 - r_2)\}$, where $h_2 \in \mathcal{T}$ corresponds to the transformation function of robot r_2. Notice that the transformation function of a robot r is chosen by an adversary but it does not change over time.

A configuration C is said to be *distinct* if $|C| = 2$.

Configurations and Local Views in One Dimension. The evolving space of the robots can be naturally restricted to a one-dimensional space *i.e.*, robots that evolve on a line. In this case, robots in a configuration C correspond to points in $\mathbb{R}$ instead of $\mathbb{R}^2$. Similarly, the set of transformation functions $\mathcal{T}$ contains scaling if robots do not agree on the unit distance, and contains the reflection (or equivalently the π-rotation) if robots do not agree on the orientation of the single axis (*i.e.*, are disoriented).

Algorithms and Executions. An algorithm A is a function mapping local views to destinations. The local view of a robot r is centered and transformed by a function h_r, and when activated, algorithm A outputs r's destination d in its local view. So to obtain the destination of a robot in the global coordinate system Z, one should apply the inverse transformation h_r^{-1} *i.e.*, the global destination is $r + h_r^{-1}(d)$.

When a non-empty subset of robots S executes an algorithm A in a given configuration C, the obtained configuration C' is the smallest set satisfying[1]:

$$\forall r \in C \backslash S \Rightarrow r \in C'$$
$$\forall r \in C \cap S \Rightarrow r + h_r^{-1}\left(A\left(\{(0,0), h_r(r'-r)\}\right)\right) \in C', \text{with } r' \in C \backslash \{r\}.$$

In this case, we write $C \xrightarrow{A} C'$, and say C' is obtained from C by applying A. We say a robot *crashes* at time t if it is not activated at time t, and never activated after time t *i.e.*, a crashed robot stops executing its algorithm and remains at the same position.

An *execution* of algorithm A is an infinite sequence of configurations $C_0, C_1, \ldots$ such that $C_i \xrightarrow{A} C_{i+1}$ for all $i \in \mathbb{N}$. We say an execution contains one crashed robot, if one robot crashes at some round t.

The Stand Up Indulgent Rendezvous Problem. An algorithm solves the Stand Up Indulgent Rendezvous (SUIR) problem if, for any initial configuration C_0 and for any execution $C_0, C_1, \ldots$ with up to one crashed robot, there exists a round t and a point p such that $C_{t'} = \{p\}$ for all $t' \geq t$.

Informally, if one robot crashes, the correct robot goes to the crashed robot; if no robot crashes, both robots gather in a finite number of rounds.

Since we allow arbitrary initial configuration and the robots are oblivious, we can consider without loss of generality that the crash, if any, occurs at the start of the execution.

3   Impossibility Results

In this section we prove that the SUIR problem is not solvable in SSYNC, even if robots share a full coordinate system (the transformation function is the identity), and have access to infinite persistent memory that is readable by the other robot. In the literature [7], the persistent memory aspect is called the Full-light model with an infinite number of colors. We now prove that having such capabilities does not help solving the problem. The next lemma is very simple but is a key point to prove our main impossibility result.

---

[1] This definition works when $|C| = 2$ but can be easily generalized to larger configurations.

Lemma 1. *Consider the SSYNC model, with rigid movements, robots endowed with full-lights with infinitely many colors, and a common coordinate system. Assuming algorithm A solves the SUIR problem, then, in every execution suffix starting from a distinct configuration where only robot r is activated (e.g. because the other robot has crashed), there must exist a configuration where algorithm A commands that r moves to the other robot's position.*

Proof. Any move of robot r that does not go to the other robot location does not yield gathering. If this repeats infinitely, no rendezvous is achieved.   □

Theorem 1. *The SUIR problem is not solvable in SSYNC, even with rigid movements, robots endowed with full-lights with infinitely many colors, and sharing a common coordinate system.*

Proof. Assume for the purpose of contradiction that such an algorithm exists. Let r be one of the robots, and r' be the other robot. We construct a fair infinite execution where rendezvous is never achieved. At some round t, we either activate only r, only r', or both, depending on what the (deterministic) output of the algorithm in the current configuration is:

- If r is dictated to stay idle: activate only r
- If r is dictated to move to $p \neq r'$: activate only r
- If r is dictated to move to r', and r' is dictated to move: activate both robots.
- Otherwise (r' is dictated to stay idle): activate only r'

We now show that the execution is fair. Suppose for the purpose of contradiction that the execution is unfair, so there exists a round t after which only r is executed, or only r' is executed. In the first case, it implies there exists an execution suffix where r is never dictated to move to the other robot, which contradicts Lemma 1. Now, if only r' is activated after some round t, then there exists a suffix where r' is always dictated to stay idle, which also contradicts Lemma 1.

The schedule we choose guarantees the following. When only r is activated, rendezvous is not achieved as r is not moving to r'. When only r' is activated, rendezvous is not achieved as r' is idle. If both robots are activated rendezvous is not achieved as r is moving to r' while r' is moving.

Overall, there exists an infinite fair execution where robots never meet, a contradiction with the initial assumption that the algorithm solves SUIR.   □

4   Reduction to One-Dimensional Space

In this section, we prove that having an algorithm solving the SUIR problem in a one-dimensional space implies the existence of an algorithm solving the same problem in a two-dimensional space. This theorem is important as our algorithms are defined on the one-dimensional space. However, since we do not prove the converse, the impossibility result we present in the previous Section works in the most general settings, assuming a two-dimensional space. Indeed, we present after the theorem, an example of a problem (the fault-free rendezvous with one common full axis) that is solvable in a two dimensional space, but that cannot be reduced to the one-dimensional space. Despite the results being intuitive, the formal proof is not trivial.

Theorem 2. *Suppose there exists an algorithm solving the SUIR problem where robots are restricted to a one-dimensional space. Then, there exists an algorithm solving the SUIR problem in a two-dimensional space.*

Proof. Let A_1 be an algorithm solving the SUIR problem where robots are restricted to a one-dimensional space. We provide a constructive proof of the theorem by giving a new algorithm A_2 executed by robots in the two-dimensional space.

First, for a configuration $C = \{r_{min}, r_{max}\}$, with $r_{min} < r_{max}$ (using the lexicographical order on their coordinates), we define the function $\mathbf{v}$ as follows:

$$\mathbf{v}(\{r_{min}, r_{max}\}) = \frac{r_{\max} - r_{\min}}{\|r_{\max} - r_{\min}\|}$$

Let r_1 and r_2 denote the two robots, having transformation functions h_1 and h_2, respectively. Let $V_i = \{(0,0), h_i(r_j - r_i)\}$, $i = 1, 2$, $j = 3 - i$, be the local view of robot r_i. Each robot r_i can compute its own orientation vector $v_i = \mathbf{v}(V_i)$ of the line joining the two robots. Notice that, if robots remains on the same line, then v_i remains invariant during the whole execution as long as robots do not gather.

We define algorithm A_2, executed by robots in the two-dimensional space as follow. First, if the local view of a robot r_i is $\{(0,0)\}$, then A_2 outputs $(0,0)$.

Otherwise, r_i can map its local view V_i in a one-dimensional space to obtain $\overline{V_i} = \{0, b_i\}$ with b_i such that $h_i(r_j - r_i) = b_i v_i$, and execute A_1 on $\overline{V_i}$. The obtained destination $\overline{p} = A_1(\overline{V_i})$, is then converted back to the two-dimensional space to obtain the destination $p = \overline{p} v_i$. Doing so, the robots remain on the same line, and v_i remains invariant while the robots are not gathered (when they are gathered, both algorithms stop).

Let $E = C_0, C_1, \ldots$ be an arbitrary execution of A_2. We want to construct from E an execution $\overline{E}$ of A_1 such that the rendezvous is achieved in $\overline{E}$ if and only if the rendezvous is achieved in E.

Recall that we analyze each configuration C_i, $i \in \mathbb{N}$, using Z, the global coordinate system we use for the analysis. Let $v = \mathbf{v}(C_0)$. Again, since robots remain on the same line, then $v = \mathbf{v}(C_i)$ for any C_i while robots are not yet gathered.

Let O be any point of the line L joining the two robots. We define as follow the bijection m mapping points of L (in Z), to the global one-dimensional coordinate system (O, v):

$$\forall a \in \mathbb{R}, m(O + av) = a$$

We can extend the function m to configurations as follows:
$m(C) = \{m(r) \mid r \in C\}$.

Now, let $\overline{E} = \overline{C_0}, \overline{C_1}, \dots$ be the execution of A_1, in (O, v), of two robots having transformation function $\overline{h}_1$ and $\overline{h}_2$ respectively, with $\overline{h}_i(a) = b$ if and only if $h_i(av) = bv_i$.

We now show that, if $C \xrightarrow{A_2} C'$ then $m(C) \xrightarrow{A_1} m(C')$. To do so we show that the result of executing A_1 on $m(C)$ coincides with $m(C')$. Let $C = \{O + a_1 v, O + a_2 v\}$, i be an activated robot and j be the other robot. On one hand, we have $m(C) = \{a_1, a_2\}$ and, by construction, the view $\overline{V}_i$ of robot i in $m(C)$ is $\{0, b_i\}$ with $b_i = \overline{h}_i(a_j - a_i)$, so that the global destination of r_i in $m(C)$ is then $a_i + \overline{h}_i^{-1}(\overline{p})$ (with $\overline{p} = A_1(\overline{V}_i)$). On the other hand, the view V_i of robot i is $\{(0,0), h_i((a_j - a_i)v)\} = \{(0,0), b_i v_i\}$, so that the global destination of r_i in Z is $O + a_i v + h_i^{-1}(\overline{p} v_i) = O + a_i v + \overline{h}_i^{-1}(\overline{p})v$. Since $m(O + a_i v + \overline{h}_i^{-1}(\overline{p})v) = a_i + \overline{h}_i^{-1}(\overline{p})$, we obtain that $m(C) \xrightarrow{A_1} m(C')$ (assuming the same robots are activated in C and in $m(C)$.

Hence, $\overline{C}_i = m(C_i)$ for all $i \in \mathbb{N}$. Since A_1 solves the SUIR problem, there exists a point $p \in \mathbb{R}$ and a round t such that, for all $t' \geq t$, $m(C_{t'}) = \{p\}$. This implies that $C_{t'} = \{O + pv\}$, so that A_2 solves the SUIR problem. $\square$

Rendezvous Without Faults with One Full Axis. Now, we show that the converse of Theorem 2 is not true in the fault-free model. This observation justifies that, for the results to be more general, we defined our model and gave our impossibility results for the two-dimensional space.

We present an algorithm solving the (fault-free) rendezvous problem in a two-dimensional space, assuming robots agree on one full axis (that is, they agree on the direction and the orientation of the axis). Under this assumption, it is possible that the robots do not agree on the orientation of the line joining them, so that assuming the converse of Theorem 2 would imply the existence of an algorithm in the one-dimensional space with disoriented robots (which does not exists, using a similar proof as the one given in [14]).

The idea is that, if the configuration is symmetric (robots may have the same view), then robots move to the point that forms, with the two robots, an equilateral triangle. Since two such points exist, the robots choose the northernmost one (the robots agree on the y-axis, which provides a common North). Otherwise, the configuration is not symmetric, and there is a unique northernmost robot r. This robot does not move and the other robot moves to r.

Algorithm 1: Fault-free rendezvous. Robots agree on one full axis, may not have common unit distance, and movements are non-rigid

---

Data: r : robot executing the algorithm
Let $\{(0,0), (x,y)\}$ be r's local view.
if $y = 0$ **then**
1 | move to the point $\left(x/2, \left|\frac{x\sqrt{3}}{2}\right|\right)$
else if $y > 0$ **then**
2 | move to the other robot, at (x, y).

Theorem 3. *Algorithm 1 solves the (fault-free) rendezvous problem with non-rigid movements and robots having only one common full axis, and different unit distance.*

Proof. If the configuration is not symmetric, then, after executing Line 2, either the rendezvous is achieved, or the moving robots remains on the same line joining the robots, so the obtained configuration is still asymmetric, and the same robot is dictated to move towards the same destination, so it reaches it in a finite number of rounds.

Consider now that the initial configuration C is symmetric. If both robots reaches their destination, the rendezvous is completed in one round. If robots are stopped before reaching their destinations, two cases can occur. Either they are stopped after traveling different distances, or they are stopped at the same y-coordinate. In the former case, the obtained configuration is asymmetric and we retrieve the first case. In the latter, the configuration remains symmetric, but the distance between the two robots decreases by at least $\frac{2\delta}{\sqrt{2}}$. Hence, at each round either robots complete the rendezvous, reach an asymmetric configuration, or come closer by a fixed distance. Since the latter case cannot occurs infinitely, one of the other case occurs at least once, and the rendezvous is completed in a finite amount of rounds.   □

5   SUIR Algorithm for FSYNC Robots with a Common Coordinate System

Since it is impossible to solve the SUIR problem in SSYNC, we now concentrate on FSYNC. We first consider a strong model, assuming robots agree on both axis, have a common unit distance, and assuming movements are rigid, before relaxing all hypotheses in Sect. 6. By Theorem 2, it is sufficient to give an algorithm for a one-dimensional space. Figure 1 illustrates the two possible configurations: either the distance between the two robots is greater than one (the common unit distance), or at most one. In the former case both robots move to the middle. In the latter, we can dictate the right robot to move to the left one, and the left robot to move one unit distance to the right of the other robot. Recall that we can distinguish the left and the right robot on the line because we assume the robots agree on both axis in the two-dimensional space.

Algorithm 2: Rendezvous with rigid movements, and a common coordinate system

---

    Data: r : robot executing the algorithm
    d : the distance between the two robots
    if $d > 1$ **then**   move to the middle
    else
       | **if** *r is the left robot* **then**
       | | move to the point at distance one at the right of the other robot
       | **else**
       | | move to the other robot

case $d > 1$ case $d \leq 1$

Fig. 1. The two possible configurations, depending the distance between the two robots

Theorem 4. *Algorithm 2 solves the SUIR problem in FSYNC with rigid movement, and robots agreeing on both axes and unit distance.*

Proof. First we consider the case where a robot crashes. If the left robot crashes, the right robot halves its distance with the other robot each round until its distance is at most one. Then, the right robot move to the other robot and the rendezvous is achieved.

If the right robot crashes, the left robot halves its distance with the other robot each round until its distance is at most one unit. Then, the left robot moves to the right of the other robot, at distance one. It then move to the other robot and the rendezvous is achieved.

Now assume no robot crashes. If the configuration is such that the distance d between the two robots is greater than one, then both robots move to the middle at the same time and the rendezvous is achieved. Otherwise, if the distance d is at most one, after one round the robots are at distance $1 + d > 1$, so that after one more round, the rendezvous is achieved. $\square$

6   SUIR Algorithm for Disoriented Robots in FSYNC

In this section we present Algorithm 3, which works with disoriented robots (robots do not agree on any axis, nor on the unit distance), and non-rigid movements. The algorithm is defined on the line. Each robot sees the line oriented in some way, but robots might not agree on the orientation of the line. However, since the orientation of the line is deduced from the robot own coordinate system, it does not change over time.

Algorithm 3: SUIR Algorithm for disoriented robots

Let d be the distance to the other robot
Let $i \in \mathbb{Z}$ such that $d \in [2^{-i}, 2^{1-i})$
if $i \equiv 0 \mod 2$ **then** move to the middle
if $i \equiv 1 \mod 4$ **then** *left* $\rightarrow$ move to middle ; *right* $\rightarrow$ move to other
if $i \equiv 3 \mod 4$ **then** *left* $\rightarrow$ move to other ; *right* $\rightarrow$ move to middle

---

The different moves of a robot r depend on whether a r sees itself on the left or the right of the other robot, and on its *level*. The level of robot at distance d from the other robot (according its own coordinate system, hence its own unit distance), is the integer $i \in \mathbb{Z}$ such that $d \in [2^{-i}, 2^{1-i})$. Figure 2 summarizes the eight possible views of a robot r, and the corresponding movements. Each line case $i \equiv 0 \mod 4$

case $i \equiv 1 \mod 4$

case $i \equiv 2 \mod 4$

case $i \equiv 3 \mod 4$

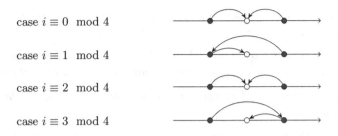

Fig. 2. The four possible configurations, depending the distance between the two robots. We have split the case $i \equiv 0 \mod 2$ into two lines to help the reader.

represents the congruence of the level of the robot modulo four, and on each line, we see the movement of the robot whether it sees itself on the right or on the left of the other robot. A given figure does not necessarily imply that both robots will actually perform the corresponding movement at the same time (since they may *not* have the same view).

For instance, if a robot r_1 has a level i_1 congruent to 1 modulo 4 and sees itself on the right, while the other robot r_2 has a level i_2 congruent to 2 modulo 4, and also sees itself on the right, then r_1 moves to the other robot position, and r_2 moves to the middle. Assuming both robots reach their destinations, then the distance between them is divided by two (regardless of the coordinate system) so their levels increase by one, and they both see the other robot on the other side, so each robot now sees the other robot on its left.

Let C be any configuration and d is the distance (in the global coordinate system Z) between the two robots. Let x, resp. y, be the distance, in Z, traveled by the left robot, resp. the right robot. Since the robots move toward each other, after executing one round, the distance between the robot becomes $f(d, x, y) = |d - x - y|$.

Lemma 2. *If at least one robot is dictated to move to the middle, then we have $f(d, x, y) \leq d - \min(\delta, d/2)$.*

Proof. For any fixed d, using the symmetry of f (with respect to the second and third argument), we have $f(d, x, y) = g_d(x + y)$ with $g_d : w \mapsto |d - w|$. We know that the distance traveled by the robots is either 0 (if one robot crashes), or at least $\min(d/2, \delta)$, but we cannot have $x = y = 0$. Also, since at least one robot moves to the middle, we have either *(i)* $x \leq d/2$ and $y \leq d$, or *(ii)* $x \leq d$ and $y \leq d/2$. Hence, the sum $x + y$ is in the interval $[\min(d/2, \delta), 3d/2]$.

As a convex function, the maximum of g_d is reached at the boundary of its domain

$$f(d, x, y) = g_d(x + y) \leq \max(g_d(3d/2), g_d(\min(d/2, \delta)))$$
$$\leq \max(d/2, d - \min(d/2, \delta)) = d - \min(d/2, \delta)$$

$\square$

The next lemma is a direct consequence of Lemma 2, but due to space constraints, its proof has been ommited.

Lemma 3. *From a configuration where robots are at distance d (in Z), then, after two rounds, the distance between the robots decreases by at least $\min(\delta, d/2)$.*

Lemma 4. *Assuming rigid movement and one robot crash, Algorithm 3 solves the SUIR problem.*

Proof. Let i be the level of the correct robot r. Assume the other robot crashes. Robot r either sees itself on the right or on the left of the other robot.

If r sees itself on the right, then depending on its level, either r moves to the middle, or move to the other robot. In the former case, the level of r increases by one and r continues to see itself on the right. In the latter case, the rendezvous is achieved in one round. After at most three rounds, the level of r is congruent to 1 modulo 4 so that after at most four rounds the rendezvous is achieved.

Similarly, if r sees itself on the left, then after at most four rounds, r level is congruent to 3 modulo 4 and the rendezvous is achieved.   □

If, at round t, one robot sees itself on the right, and the other sees itself on the left, then they agree on the orientation of the line at time t. Since, for each robot, the orientation of the line does not change, then they agree on it during the whole execution (except when they are gathered, as the line is not defined in that case).

Similarly, if at some round, both robots see themselves at the right (resp. at the left), then their orientations of the line are opposite, and remain opposite during the whole execution (again, until they gather). Hence we have the following remark.

Remark 1. Consider two disoriented robots moving on the line L joining them and executing Algorithm 3. Then, either they have a common orientation of L during the whole execution (while they are not gathered), or they have opposite orientations of L during the whole execution (while they are not gathered).

Lemma 5. *Assuming rigid movements, no crash, and robots having **a common orientation** of the line joining them, then, Algorithm 3 solves the SUIR problem.*

Proof. Since the robots have a common orientation, we know there is one robot that sees itself on the right and and one robot that sees itself on the left. Of course the robots are not aware of this, but we saw in the previous remark that a common orientation is preserved during the whole execution (while robots are not gathered).

Let $(i, j) \in \mathbb{Z}^2$ denote a configuration where the robot on the left is at level i, and the robot on the right is at level j, and we write $(i, j) \equiv (k, l) \mod 4$ if and only if $i \equiv k \mod 4$ and $j \equiv l \mod 4$.

To prove the lemma we want to show that for any configuration $(i, j) \in \mathbb{Z}^2$, the robots achieve rendezvous. Take an arbitrary configuration $(i, j) \in \mathbb{Z}^2$. We consider all 16 cases:

1. **if** $(i, j) \equiv (0, 0) \mod 4$: rendezvous is achieved in one round.
2. **if** $(i, j) \equiv (0, 1) \mod 4$: we reach configuration $(j + 1, i + 1) \equiv (2, 1) \mod 4$
3. **if** $(i, j) \equiv (0, 2) \mod 4$: rendezvous is achieved in one round.
4. **if** $(i, j) \equiv (0, 3) \mod 4$: rendezvous is achieved in one round.
5. **if** $(i, j) \equiv (1, 0) \mod 4$: rendezvous is achieved in one round.
6. **if** $(i, j) \equiv (1, 1) \mod 4$: we reach configuration $(j + 1, i + 1) \equiv (2, 2) \mod 4$
7. **if** $(i, j) \equiv (1, 2) \mod 4$: rendezvous is achieved in one round.
8. **if** $(i, j) \equiv (1, 3) \mod 4$: rendezvous is achieved in one round.
9. **if** $(i, j) \equiv (2, 0) \mod 4$: rendezvous is achieved in one round.
10. **if** $(i, j) \equiv (2, 1) \mod 4$: we reach configuration $(j + 1, i + 1) \equiv (2, 3) \mod 4$
11. **if** $(i, j) \equiv (2, 2) \mod 4$: rendezvous is achieved in one round.
12. **if** $(i, j) \equiv (2, 3) \mod 4$: rendezvous is achieved in one round.
13. **if** $(i, j) \equiv (3, 0) \mod 4$: we reach configuration $(j + 1, i + 1) \equiv (1, 0) \mod 4$
14. **if** $(i, j) \equiv (3, 1) \mod 4$: we reach configuration $(j, i) \equiv (1, 3) \mod 4$
15. **if** $(i, j) \equiv (3, 2) \mod 4$: we reach configuration $(j + 1, i + 1) \equiv (3, 0) \mod 4$
16. **if** $(i, j) \equiv (3, 3) \mod 4$: we reach configuration $(j + 1, i + 1) \equiv (0, 0) \mod 4$

In any case, rendezvous is achieved after at most three rounds.   □

Lemma 6. *Assuming rigid movement, no crash, and robots having **opposite orientations** of the line joining them, then, Algorithm 3 solves the SUIR problem.*

Proof. Since the robots have opposite orientations, we know they either both see themselves on the right or they both both see themselves on the left. Of course the robots are not aware of this, but we saw in the previous remark that the opposite orientations are preserved during the whole execution (while robots are not gathered).

In this proof, $R\{i, j\}$ denotes a configuration where both robots see themselves on the right and one of them has level i and the other as level j. Here, the order between i and j does not matter (hence the set notation). Similarly $L\{i, j\}$ denotes a configuration where both robots see themselves on the left, and one of them has level i and the other as level j.

Here, assuming without loss of generality that $i \leq j \mod 4$, we write $R\{i, j\} \equiv (k, l) \mod 4$, resp. $L\{i, j\} \equiv (k, l) \mod 4$, if and only if, $i \equiv k \mod 4$ and $j \equiv l \mod 4$.

To prove the lemma, we want to show that for any configuration $R\{i, j\}$ or $L\{i, j\}$, the robots achieve rendezvous. Take an arbitrary configuration $(i, j) \in \mathbb{Z}^2$. We consider all 20 cases:

1. **if** $L\{i, j\} \equiv (0, 0) \mod 4$: rendezvous is achieved in one round.
2. **if** $L\{i, j\} \equiv (0, 1) \mod 4$: rendezvous is achieved in one round.
3. **if** $L\{i, j\} \equiv (0, 2) \mod 4$: rendezvous is achieved in one round.
4. **if** $L\{i, j\} \equiv (0, 3) \mod 4$: we reach configuration $R\{i+1, j+1\} \equiv (0, 1) \mod 4$.
5. **if** $L\{i, j\} \equiv (1, 1) \mod 4$: rendezvous is achieved in one round.
6. **if** $L\{i, j\} \equiv (1, 2) \mod 4$: rendezvous is achieved in one round.

7. **if** $L\{i,j\} \equiv (1,3) \mod 4$: we reach configuration $R\{i+1,j+1\}\equiv(0,2)$ mod 4.
8. **if** $L\{i,j\} \equiv (2,2) \mod 4$: rendezvous is achieved in one round.
9. **if** $L\{i,j\} \equiv (2,3) \mod 4$: we reach configuration $R\{i+1,j+1\}\equiv(0,3)$ mod 4.
10. **if** $L\{i,j\} \equiv (3,3) \mod 4$: we reach configuration $R\{i,j\}\equiv(3,3) \mod 4$.
11. **if** $R\{i,j\} \equiv (0,0) \mod 4$: rendezvous is achieved in one round.
12. **if** $R\{i,j\} \equiv (0,1) \mod 4$: we reach configuration $L\{i+1,j+1\}\equiv(1,2)$ mod 4.
13. **if** $R\{i,j\} \equiv (0,2) \mod 4$: rendezvous is achieved in one round.
14. **if** $R\{i,j\} \equiv (0,3) \mod 4$: rendezvous is achieved in one round.
15. **if** $R\{i,j\} \equiv (1,1) \mod 4$: we reach configuration $L\{i,j\}\equiv(1,1) \mod 4$.
16. **if** $R\{i,j\} \equiv (1,2) \mod 4$: we reach configuration $L\{i+1,j+1\}\equiv(2,3)$ mod 4.
17. **if** $R\{i,j\} \equiv (1,3) \mod 4$: we reach configuration $L\{i+1,j+1\}\equiv(0,2)$ mod 4.
18. **if** $R\{i,j\} \equiv (2,2) \mod 4$: rendezvous is achieved in one round.
19. **if** $R\{i,j\} \equiv (2,3) \mod 4$: rendezvous is achieved in one round.
20. **if** $R\{i,j\} \equiv (3,3) \mod 4$: rendezvous is achieved in one round.

In any case, rendezvous is achieved after at most three rounds.   □

Theorem 5. *Algorithm 3 solves the SUIR problem with disoriented robots in FSYNC.*

Proof. By Lemma 3, the distance between the two robots decreases by at least $\min(\delta, d/2)$ every two rounds. Hence, eventually, robots are at distance smaller than δ from one another and, from this point in time, movements are rigid. Assume now that movements are rigid. If a robot crashes, then the rendezvous is achieved by using Lemma 4. Otherwise, depending on whether the robots have a common orientation or opposite orientation of the line joining them (see Remark 1), the Theorem follows either by using Lemma 5 or by using Lemma 6.   □

7   Concluding Remarks

We considered the problem of Stand Up Indulgent Rendezvous (SUIR). Unlike classical rendezvous, the SUIR problem is unsolvable in SSYNC even with the strongest assumptions: robots share a common $x - y$ coordinate system, and have access to infinite persistent memory. We demonstrate that it is nevertheless solvable in FSYNC without *any* additional assumptions. A natural open question is related to the optimality (in time) of our algorithm.

Also, we would like to investigate further the possibility of deterministic strong gathering for $n \geq 3$ robots. It is known that executing a single robot at a time in SSYNC is insufficient [8,9], but additional hypotheses may make the problem solvable.

References

1. Agmon, N., Peleg, D.: Fault-tolerant gathering algorithms for autonomous mobile robots. SIAM J. Comput. **36**(1), 56–82 (2006). https://doi.org/10.1137/050645221
2. Balabonski, T., Delga, A., Rieg, L., Tixeuil, S., Urbain, X.: Synchronous gathering without multiplicity detection: a certified algorithm. Theo. Comput. Syst. **63**(2), 200–218 (2017). https://doi.org/10.1007/s00224-017-9828-z
3. Bhagat, S., Gan Chaudhuri, S., Mukhopadhyaya, K.: Fault-tolerant gathering of asynchronous oblivious mobile robots under one-axis agreement. In: Rahman, M.S., Tomita, E. (eds.) WALCOM 2015. LNCS, vol. 8973, pp. 149–160. Springer, Cham (2015). https://doi.org/10.1007/978-3-319-15612-5_14
4. Bouzid, Z., Das, S., Tixeuil, S.: Gathering of mobile robots tolerating multiple crash faults. In: Proceedings of 33rd IEEE International Conference on Distributed Computing Systems (ICDCS), pp. 337–346, July 2013. https://doi.org/10.1109/ICDCS.2013.27
5. Bramas, Q., Tixeuil, S.: Wait-free gathering without chirality. In: Scheideler, C. (ed.) SIROCCO 2014. LNCS, vol. 9439, pp. 313–327. Springer, Cham (2015). https://doi.org/10.1007/978-3-319-25258-2_22
6. Cohen, R., Peleg, D.: Convergence properties of the gravitational algorithm in asynchronous robot systems. SIAM J. Comput. **34**(6), 1516–1528 (2005). https://doi.org/10.1137/S0097539704446475
7. Das, S., Flocchini, P., Prencipe, G., Santoro, N., Yamashita, M.: Autonomous mobile robots with lights. Theor. Comput. Sci. **609**, 171–184 (2016). https://doi.org/10.1016/j.tcs.2015.09.018
8. Défago, X., Gradinariu, M., Messika, S., Raipin-Parvédy, P.: Fault-tolerant and self-stabilizing mobile robots gathering. In: Dolev, S. (ed.) DISC 2006. LNCS, vol. 4167, pp. 46–60. Springer, Heidelberg (2006). https://doi.org/10.1007/11864219_4
9. Défago, X., Potop-Butucaru, M.G., Clément, J., Messika, S., Parvédy, P.R.: Fault and byzantine tolerant self-stabilizing mobile robots gathering - feasibility study -. CoRR abs/1602.05546 (2016). http://arxiv.org/abs/1602.05546
10. Flocchini, P., Prencipe, G., Santoro, N., Widmayer, P.: Gathering of asynchronous robots with limited visibility. Theor. Comput. Sci. **337**(1–3), 147–168 (2005). https://doi.org/10.1016/j.tcs.2005.01.001
11. Heriban, A., Défago, X., Tixeuil, S.: Optimally gathering two robots. In: Proceedings of19th International Conference on Distributed Computing and Networking, ICDCN, pp. 3:1–3:10, January 2018. https://doi.org/10.1145/3154273.3154323
12. Izumi, T., et al.: The gathering problem for two oblivious robots with unreliable compasses. SIAM J. Comput. **41**(1), 26–46 (2012). https://doi.org/10.1137/100797916
13. Okumura, T., Wada, K., Katayama, Y.: Brief announcement: optimal asynchronous rendezvous for mobile robots with lights. In: Proceedings of 19th International Symposium on Stabilization, Safety, and Security of Distributed Systems (SSS), pp. 484–488, November 2017. https://doi.org/10.1007/978-3-319-69084-1_36
14. Suzuki, I., Yamashita, M.: Distributed anonymous mobile robots: formation of geometric patterns. SIAM J. Comput. **28**(4), 1347–1363 (1999). https://doi.org/10.1137/S009753979628292X
15. Viglietta, G.: Rendezvous of two robots with visible bits. In: Proceedings of 9th International Symposium on Algorithms and Experiments for Sensor Systems, Wireless Networks and Distributed Robotics, (ALGOSENSORS), pp. 291–306, September 2013. https://doi.org/10.1007/978-3-642-45346-5_21

Brief Announcement: Gathering in Linear Time: A Closed Chain of Disoriented and Luminous Robots with Limited Visibility

Jannik Castenow$^{(\boxtimes)}$, Jonas Harbig, Daniel Jung, Till Knollmann, and Friedhelm Meyer auf der Heide

Heinz Nixdorf Institute and Computer Science Department, Paderborn University, Paderborn, Germany
{janniksu,jharbig,jungd,tillk,fmadh}@mail.upb.de

Abstract. This work focuses on the following question related to the GATHERING problem of n autonomous, mobile robots in the Euclidean plane: Is it possible to solve GATHERING of disoriented robots with limited visibility in $o(n^2)$ fully synchronous rounds ($\mathcal{F}$SYNC)? The best known algorithm considering the $\mathcal{OBLOT}$ model (oblivious robots) needs $\Theta\left(n^2\right)$ rounds [6]. The lower bound for this algorithm even holds in a simplified closed chain model, where each robot has exactly two neighbors and the chain connections form a cycle. The only existing algorithms achieving a linear number of rounds for disoriented robots assume robots that are located on a two dimensional grid [1] and [5]. Both algorithms consider the $\mathcal{LUMINOUS}$ model.

We show that, considering a closed chain, n disoriented robots with limited visibility in the Euclidean plane can be gathered in $\Theta\left(n\right)$ rounds assuming the $\mathcal{LUMINOUS}$ model. The lights are used to initiate and perform so-called runs along the chain. For the start of such runs, locally unique robots need to be determined. In contrast to the grid [1], this is not possible in every configuration in the Euclidean plane. Based on the theory of isogonal polygons by Grünbaum, we identify the class of isogonal configurations in which – due to a high symmetry – no such locally unique robots can be identified. Our solution combines two algorithms: The first one gathers isogonal configurations; it works without any lights. The second one works for non-isogonal configurations; it identifies locally unique robots to start runs, using a constant number of lights. Interleaving these algorithms solves the GATHERING problem in $\mathcal{O}\left(n\right)$ rounds.

Keywords: Mobile robots · Gathering · Closed chain · Local · Runtime

1 Introduction

The GATHERING problem requires a set of initially scattered point-shaped robots to meet at the same (not predefined) position. This problem has been studied

A full version of this brief announcement is available online [4].

S. Devismes and N. Mittal (Eds.): SSS 2020, LNCS 12514, pp. 60–64, 2020.
https://doi.org/10.1007/978-3-030-64348-5_5

under several different robot and time models all having in common that the capabilities of the individual robots are very restricted. The central questions among all these models are: Which capabilities of robots are needed to solve the GATHERING problem and how do these capabilities influence the runtime? While the question about solvability is quite well understood nowadays, much less is known concerning the question how the capabilities influence the runtime. The best known algorithm – GO-TO-THE-CENTER (see [2] and [6] for the runtime analysis) – for n disoriented robots (no agreement on the coordinate systems) in the Euclidean plane with limited visibility in the $\mathcal{OBLOT}$ model, assuming the $\mathcal{F}$SYNC scheduler (robots operate in fully synchronized Look-Compute-Move cycles), requires $\Theta(n^2)$ rounds. The best known lower in this model is the trivial $\Omega(n)$ bound. Thus, the above mentioned question can be formulated as follows: Is it possible to gather n disoriented robots with limited visibility in the Euclidean plane in $o(n^2)$ rounds, and which capabilities do the robots need?

The $\Omega(n^2)$ lower bound for GO-TO-THE-CENTER examines an initial configuration where the robots form a cycle with neighboring robots having a constant distance. It is shown that GO-TO-THE-CENTER takes $\Omega(n^2)$ rounds until the robots start to see more robots than their initial neighbors. Thus, the lower bound holds even in a simpler closed chain model, where the robots form an arbitrarily winding closed chain, and each robot sees exactly its two direct neighbors. Our main result is that this quadratic lower bound can be beaten for the closed chain model, if we allow each robot a constant number of visible lights, which can be seen by the neighboring robots, i.e.; if we allow $\mathcal{LUMINOUS}$ robots, compare [7]. For this model we present an algorithm with linear runtime.

Related Work. For a comprehensive overview over the area of mobile robots, we refer the reader to [8]. We focus on results about the synchronous setting ($\mathcal{F}$SYNC), where algorithms as well as runtime bounds are known. In the $\mathcal{OBLOT}$ model, there is the GO-TO-THE-CENTER algorithm [2] that solves GATHERING of disoriented robots with local visibility in $\Theta\left(n^2\right)$ rounds assuming the $\mathcal{F}$SYNC scheduler [6]. The same runtime can be achieved for robots located on a two dimensional grid [3]. Interestingly, a lower bound of $\Omega(D_G^2)$ for $D_G \in \Theta\left(\sqrt{\log n}\right)$ has been shown for any *conservative* algorithm, an algorithm that only increments the edge set of the visibility graph. D_G denotes the diameter of the initial visibility graph [10]. Faster runtimes could so far only be achieved by assuming agreement on one or two axes of the local coordinate systems or considering the $\mathcal{LUMINOUS}$ model. In [11], a universally optimal algorithm with runtime $\Theta\left(D_E\right)$ for robots in the Euclidean plane assuming one-axis agreement in the $\mathcal{OBLOT}$ model is introduced. D_E denotes the Euclidean diameter of the initial configuration. Beyond the one-axis agreement, the algorithm crucially depends on the distinction between the viewing range of a robot and its *connectivity range*: Robots only consider other robots within their connectivity range as their neighbors but can see farther beyond. Notably, this algorithm also works under the $\mathcal{A}$SYNC scheduler. Assuming disoriented robots, the algorithms that achieve a runtime of $o(n^2)$ are developed under the $\mathcal{LUMINOUS}$ model and assume robots that are located on a two dimensional grid: There exist two algorithms having an asymptotically optimal runtime of $\mathcal{O}(n)$; one algorithm for *closed chains* [1] and another one for connected swarms [5]. Following the notion of [1], we consider a closed chain of robots in this work. In a closed chain, the robots form a winding, possibly self-intersecting, chain where the distance between two neighbors is upper bounded by the connectivity range and the robots can see a constant distance along the chain in each direction, denoted as the viewing range. The main difference between a closed chain and arbitrary swarms is that in the closed chain, a robot only observes a constant number of its direct neighbors while in arbitrary swarms all robots in the viewing range of a robot are considered. Interestingly, the lower bound of the GO-TO-THE-CENTER algorithm [6], holds also for the closed chain model.

Our Contribution. We give the first asymptotically optimal algorithm that solves GATHERING of disoriented robots in the Euclidean plane. More precisely, we show that a closed chain of disoriented robots with limited visibility located in the Euclidean plane can be gathered in $\mathcal{O}(n)$ rounds assuming the $\mathcal{LUMINOUS}$ model with a constant number of lights and the $\mathcal{F}$SYNC scheduler. This is asymptotically optimal.

Theorem 1. *For any initially connected closed chain of disoriented robots in the Euclidean plane with a viewing range of 4 and a connectivity range of 1, GATH-ERING can be solved in $\mathcal{O}(n)$ rounds under the $\mathcal{F}$SYNC scheduler and a constant number of visible lights. The number of rounds is asymptotically optimal.*

The visible lights help to exploit asymmetries in the chain to identify locally unique robots that generate *runs*. One of the major challenges is the handling of highly symmetric configurations. While it is possible to identify locally unique robots in every connected configuration on the grid [1], this is *impossible* in the Euclidean plane. We identify the class of *isogonal configurations* based on the theory of isogonal polygons by Grünbaum [9] and show that no locally unique robots can be determined in these configurations. We believe that this characterization is of independent interest because highly symmetric configurations often cause a large runtime. For instance, the lower bound of the GO-TO-THE-CENTER algorithm holds for an isogonal configuration [6].

2   Model

We consider n robots $r_0, \ldots, r_{n-1}$ connected in a closed chain. Every robot r_i has two direct neighbors: r_{i-1} and r_{i+1} (modulo n). The connectivity range is assumed to be 1, i.e., two direct neighbors are allowed to have a distance of at most 1. The robots are disoriented, i.e.; they do not agree on any axis of their local coordinate systems and the latter can be arbitrarily rotated. This also means that there is no common understanding of left and right. Except of their direct neighbors, robots have a constant viewing radius along the chain. Each robot can see its 4 predecessors and successors along the chain. We assume the $\mathcal{LUMINOUS}$ model: the robots have a constant number of locally visible states (lights) that can be perceived by all robots in their neighborhood.

3 Algorithm

Our approach consists of two algorithms – one for asymmetric configurations and one for highly symmetric (isogonal) configurations. The main concept of the asymmetric algorithm is the notion of a *run*. A run is a visible state (realized with lights) that is passed along the chain in a fixed direction. Robots with a run perform a movement operation while robots without do not. The movement is sequentialized in a way that in round t the robot r_i executes a move operation (and neither r_{i-1} nor r_{i+1}), the robot r_{i+1} in round $t+1$ and so on, see Fig. 1.

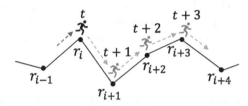

Fig. 1. A run located at r_i in round t is passed in its direction along the chain, i.e.; it is located at r_{i+x} in round $t+x$.

This way, any moving robot does not have to consider movements of its neighbors to ensure connectivity. To preserve the connectivity of the chain only two run patterns are allowed: a robot r_i has a run and either none of its direct neighbors (an *isolated run*) has a run or exactly one of its direct neighbors has a run such that the runs are heading in each other's direction (a *joint run-pair*). This essentially means that there are no sequences of runs of length at least 3.

For robots with a run, there are three kinds of movement operations, the *merge*, the *shorten* and the *hop*. The purpose of the **merge** is to reduce the number of robots in the chain. It is executed by a robot r_i if its neighbors have a distance of at most 1. In this case, r_i is not necessary for the connectivity of the chain and can safely be removed. Removing r_i means, that it jumps onto the position of its next neighbor in the direction of the run, the robots merge their neighborhoods and both continue to behave as a single robot in future rounds. The execution of a merge stops a run. Moreover, some additional care has to be taken here: removing robots from the chain decreases the distance of nearby runs. To avoid that runs come too close, a merge stops all runs that might be present in the neighborhoods of the two merging robots. The goal of a **shorten** is to reduce the length of the chain. Intuitively, if the angle between vectors of r_i pointing to its neighbors is not too large, it can reduce the length of the chain by jumping to the midpoint between its neighbors. The execution of a shorten also stops a run. In case neither a merge nor a shorten can be applied, a **hop** is executed. The purpose of a hop is to exchange two neighboring vectors in the chain. By this, each run is associated with a run vector that is swapped along the chain until it finds a position at which progress can be made. For each operation, there is also a *joint* one which is a similar operation executed by a joint run-pair.

The main question now is *where* runs should be started. For this, we identify robots that are – regarding their local neighborhood – geometrically unique. These robots are assigned an Init-State allowing them to regularly generate new runs. During the generation of new runs it is ensured that a certain distance to other runs is kept. In isogonal configurations, however, it is not possible to identify locally unique robots. To overcome this, we introduce an additional algorithm for these configurations. Isogonal configurations have in common that all robots lie on the boundary of a common circle. We exploit this fact by letting the robots move towards the center of the surrounding circle in every round until they finally gather in its center. Additional care has to be taken in case both algorithms interfere with each other. This can happen if some parts of the chain are isogonal while other parts are asymmetric. We show how to handle such a case and ensure that the two algorithms do not hinder each other later.

References

1. Abshoff, S., Cord-Landwehr, A., Fischer, M., Jung, D., der Heide, F.M.A.: Gathering a closed chain of robots on a grid. In: 2016 IEEE International Parallel and Distributed Processing Symposium, IPDPS 2016, 23–27 May 2016, Chicago, IL, USA, pp. 689–699 (2016)
2. Ando, H., Oasa, Y., Suzuki, I., Yamashita, M.: Distributed memoryless point convergence algorithm for mobile robots with limited visibility. IEEE Trans. Rob. Autom. **15**(5), 818–828 (1999)
3. Castenow, J., Fischer, M., Harbig, J., Jung, D., Meyer auf der Heide, F.: Gathering anonymous, oblivious robots on a grid. Theor. Comput. Sci. **815**, 289–309 (2020)
4. Castenow, J., Harbig, J., Jung, D., Knollmann, T., Meyer auf der Heide, F.: Gathering in linear time: a closed chain of disoriented and luminous robots with limited visibility. CoRR abs/2010.04424 (2020). https://arxiv.org/abs/2010.04424
5. Cord-Landwehr, A., Fischer, M., Jung, D., Meyer auf der Heide, F.: Asymptotically optimal gathering on a grid. In: Proceedings of the 28th ACM Symposium on Parallelism in Algorithms and Architectures, SPAA 2016, Asilomar State Beach/Pacific Grove, 11–13 July 2016, CA, USA, pp. 301–312 (2016)
6. Degener, B., Kempkes, B., Langner, T., Meyer auf der Heide, F., Pietrzyk, P., Wattenhofer, R.: A tight runtime bound for synchronous gathering of autonomous robots with limited visibility. In: SPAA 2011: Proceedings of the 23rd Annual ACM Symposium on Parallelism in Algorithms and Architectures, 4–6 June 2011, San Jose, CA, USA, pp. 139–148. ACM (2011)
7. Di Luna, G., Viglietta, G.: Robots with lights. In: Distributed Computing by Mobile Entities, Current Research in Moving and Computing, pp. 252–277 (2019)
8. Flocchini, P., Prencipe, G., Santoro, N. (eds.): Distributed Computing by Mobile Entities, Current Research in Moving and Computing. Lecture Notes in Computer Science, vol. 11340. Springer (2019). https://doi.org/10.1007/978-3-030-11072-7_1
9. Grünbaum, B.: Metamorphoses of polygons. Lighter Side Math. 35–48 (1994)
10. Izumi, T., Kaino, D., Potop-Butucaru, M.G., Tixeuil, S.: On time complexity for connectivity-preserving scattering of mobile robots. T. C. S. **738**, 42–52 (2018)
11. Poudel, P., Sharma, G.: Universally optimal gathering under limited visibility. In: Proceedings of Stabilization, Safety, and Security of Distributed Systems - 19th International Symposium, SSS 2017, 5–8 November 2017, Boston, MA, USA, pp. 323–340 (2017)

A Discrete and Continuous Study
of the MAX-CHAIN-FORMATION Problem
Slow down to Speed Up

Jannik Castenow[1(✉)], Peter Kling[2], Till Knollmann[1],
and Friedhelm Meyer auf der Heide[1]

[1] Heinz Nixdorf Institute and Computer Science Department, Paderborn University,
Paderborn, Germany
{jannik.castenow,till.knollmann,fmadh}@upb.de
[2] University of Hamburg, Hamburg, Germany
peter.kling@uni-hamburg.de

Abstract. Most existing robot formation problems seek a target formation of a certain *minimal* and, thus, efficient structure. Examples include the GATHERING and the CHAIN-FORMATION problem. In this work, we study formation problems that try to reach a *maximal* structure, supporting for example an efficient coverage in exploration scenarios. A recent example is the NASA Shapeshifter project [22], which describes how the robots form a relay chain along which gathered data from extraterrestrial cave explorations may be sent to a home base.

As a first step towards understanding such maximization tasks, we introduce and study the MAX-CHAIN-FORMATION problem, where n robots are ordered along a winding, potentially self-intersecting chain and must form a connected, straight line of maximal length connecting its two endpoints. We propose and analyze strategies in a discrete and in a continuous time model. In the discrete case, we give a complete analysis if all robots are initially collinear, showing that the worst-case time to reach an ε-approximation is upper bounded by $\mathcal{O}(n^2 \cdot \log(n/\varepsilon))$ and lower bounded by $\Omega(n^2 \cdot \log(1/\varepsilon))$. If one endpoint of the chain remains stationary, this result can be extended to the non-collinear case. If both endpoints move, we identify a family of instances whose runtime is unbounded. For the continuous model, we give a strategy with an optimal runtime bound of $\Theta(n)$. Avoiding an unbounded runtime similar to the discrete case relies crucially on a counter-intuitive aspect of the strategy: slowing down the endpoints while all other robots move at full speed. Surprisingly, we can show that a similar trick does not work in the discrete model.

Keywords: Mobile robots · Max-chain formation · Continuous time

---

A Brief Announcement of this paper appeared at SPAA'20 [4]; a full version is available online [3].

S. Devismes and N. Mittal (Eds.): SSS 2020, LNCS 12514, pp. 65–80, 2020.
https://doi.org/10.1007/978-3-030-64348-5_6

1 Introduction

Robot coordination problems deal with systems consisting of many autonomous but simple, mobile robots that try to achieve a common task. The robots' capabilities are typically quite restricted (e.g., they have no common coordinate system or sense of direction). Among the most well-studied tasks are GATHERING problems, in which robots are initially scattered and must gather at one point. Another class of important tasks are CHAIN-FORMATION problems, where robots take the role of communication relays that, initially, form a winding chain connecting two distinguished robots. The relays are to move such that the chain becomes straight, allowing for a more energy-efficient communication. Applications of such chain formations can be found in the exploration of difficult terrain that restricts normal communication (e.g., cave systems) [19,22].

Both GATHERING and CHAIN-FORMATION problems can be described as *contracting*: starting from an initially scattered formation, they seek to reach a smaller, more efficient (communication) structure. A natural complement to such contracting formation primitives are *extension problems*. The general idea is to spread a set of distinguished robots such that their convex hull is maximized, while maintaining a suitable connection network of simple relay robots. We initiate the theoretical study of such problems for the case of two distinguished robots connected by a chain of relay robots. Already this comparatively simple scenario turns out to be non-trivial to analyze.

Movement Model and Time Notions. We consider n identical, oblivious, mobile robots with a limited viewing range (normalized to 1) scattered in the Euclidean plane. The robots form a communication chain, such that each robot has a specific predecessor and successor in distance at most 1. We assume no common coordinate systems. Instead, a robot may only measure its relative position (distances and angles) to its two neighbors. We seek a simple, deterministic[1] movement strategy that, when executed simultaneously by all robots, causes them to converge towards a straight chain of (maximal) length $n - 1$. This movement strategy takes the relative positions of the (at most) two neighboring robots and specifies where the robot moves next. It is crucial that the distance between two neighboring robots never exceeds 1, since otherwise we cannot guarantee that the (oblivious) robots will be able to reconnect the chain.

We refer to this as the MAX-CHAIN-FORMATION problem and study it in two different time models, the classical (synchronous) LOOK-COMPUTE-MOVE *(LCM)* model and the *continuous time* model. In the LCM model, time is divided into discrete *rounds* in which all robots simultaneously perform a cycle of a LOOK, a COMPUTE, and a MOVE operation. During the LOOK operation, each robot takes a *snapshot* of its neighbors' current relative positions. Afterward,

---

[1] Determinism implies that from certain, very symmetrical system states, robots will not be able to form a maximum length chain (e.g., when all robots start in the same position). This can be resolved with a very limited and small amount of randomness. (e.g., having the outer robots move in a random direction in such a situation).

all robots start the COMPUTE operation, during which they use their snapshot to compute a *target point*. Finally, all robots perform the MOVE operation by moving to the target point. Together with our simple type of (oblivious and communication-less) robots, this is also known as the $\mathcal{OBLOT}$ model [10].

The above described model is inherently discrete, which severely limits the accuracy of information on which movements are based. The situation observed by a robot at the beginning of a round might be very different from the end of the round, when all other robots performed their movement. This effect can be compensated, e.g., by limiting how far a robot may move towards its target point during a round. [13] considered such a model and studied how it evolves in the limit, such that robots move an infinitesimal distance per round. This gives rise to the *continuous time* model. Here, each robot perpetually measures its neighbors' positions and, at the same time, adjusts the target point towards which it moves. This model exhibits fundamentally different properties, as was already experimentally observed in [13] and later analytically proven in [6] (see our detailed discussion of related work).

While the continuous model is certainly idealized, it also abstracts away the "loss of discretization" and allows one to focus on the complexity of the formation problem. In a sense, it showcases the best possible improvement one can hope for when approaching LCM cycles of length zero in practical implementations.

Related Work. The following overview focuses on robot formation strategies with known runtime bounds. In particular, we do not cover semi- or asynchronous variants of the LCM model, in which the robots' LCM cycles are not necessarily synchronized. In such systems, already achieving a task like GATHERING may be impossible [8] or requires additional robot properties [11,20]. The synchronous setting allows us to concentrate on the runtime analysis and to better compare the discrete and continuous models. See [10] for a quite complete and very recent survey on robot coordination problems.

The GATHERING problem has been considered in both the discrete and continuous setting. Here, there is no predecessor/successor relation between the robots, and the snapshot from the LOOK operation contains all robot positions within viewing range. A natural strategy is to move towards the center of the smallest enclosing circle spanning all robots in viewing range. In the discrete setting, [1] proved that this strategy gathers all robots in finite time; a runtime bound of $\Theta(n^2)$ was proven later in [7]. Up to now, this strategy achieves the asymptotically fastest (and conjectured optimal) runtime in this model. Taking a look at the continuous setting yields a very different situation: [13] proposed a simple, continuous strategy, in which robots try and decide locally whether they are at a vertex of the global convex hull formed by all robots. If a robot concludes that it is at such a vertex, it moves along the angle bisector towards the inside of the (supposed) convex hull. This strategy was shown to gather all robots in finite time. Later, [15] proved that the strategy's worst-case runtime is $\Theta(n)$; a considerable improvement about the $\Theta(n^2)$ bound for discrete GATHERING. For an overview over continuous strategies for GATHERING, see [17].

The CHAIN-FORMATION problem was introduced and analyzed by [9] in the discrete setting. The authors proposed the natural GO-TO-THE-MIDDLE (GTM) strategy, in which each robot moves towards the midpoint between its two neighbors. It is proven that GTM requires $\mathcal{O}(n^2 \cdot \log(n/\epsilon))$ rounds to reach an ϵ-approximation (w.r.t. the length) of the straight chain between the base stations. [16] gave an almost matching lower bound of $\Omega(n^2 \cdot \log(1/\epsilon))$ and generalized these bounds to a class of (linear) strategies related to GTM. Note that while there are some discrete CHAIN-FORMATION strategies, specifically [18], that achieve a better (linear) asymptotic runtime, such strategies are known only for relaxed models and goals (e.g., reaching only a $\Theta(1)$-approximation). The continuous setting was analyzed by [6], who suggested the MOVE-ON-BISECTOR (MoB) strategy (robots move along the angle bisector formed by their two neighbors) and proved a runtime of $\Theta(n)$. Similar to the GATHERING problem, we see a linear improvement when going from the discrete to the continuous setting.

Scenarios related to the idea of extension problems have been considered in other settings (like on discrete graphs) under the name uniform scattering or deployment [2,21]. The general problem of forming a line in a distributed system has been studied in many different contexts, see e.g. [12,14,19]. While the presented theoretical models are certainly idealized (ignoring, e.g., collisions of physical robots), such algorithms can be adapted for practical systems [23].

Our Contribution. We adapt the known (contracting) CHAIN-FORMATION strategies GTM (discrete setting) and MoB (continuous setting) such that they still straighten the chain but, at the same time, keep extending its length. The basic idea is to let inner robots perform the contracting strategy while the two outer robots extend the chain by moving away from their respective neighbor. While this seems to be a small modification of the contracting strategies on a conceptual level, we identify a much more complex behavior of the robots caused by the extension part. This also affects the analysis – we use several different techniques: among others, we make use of discrete Fourier transforms, the mixing time of Markov chains and the stability theory of dynamical systems.

Section 3 considers the discrete setting, for which we distinguish the one-dimensional case (all robots are initially collinear) and the general two-dimensional case. In the one-dimensional case, we already see that very symmetric configurations are problematic for any (deterministic) strategy. This is obvious for the trivial configuration (all robots start in the same spot). But also from less contrived starting positions (e.g., when the initial chain is symmetrical around the origin), any deterministic strategy results in a non-maximal chain (that potentially keeps moving) (see Theorem 4). Still, in the case of our proposed MAX-GTM strategy, we can show:

Theorem 1. *Under the MAX-GTM strategy on the line, the robot movement reaches in time $\Omega(n^2 \cdot \log(1/\epsilon))$ and $\mathcal{O}(n^2 \cdot \log(n/\epsilon))$ an ϵ-approximation of: a stationary, max-chain of length $n - 1$, if the outer robots move in different directions after $n - 2$ rounds or a chain of non-maximal length moving at speed*

$1/n$ (marching chain), *if the outer robots move in the same direction after $n-2$ rounds.*

While this gives a pretty complete picture of the one-dimensional case, the two-dimensional case exhibits a much more complex behavior. We can still prove convergence in finite time and derive a lower bound (which now depends also on the outer robots' initial distance) but an upper bound remains elusive.

Theorem 2. *Under the* MAX-GTM *strategy, the robot movement reaches an ε-approximation either of the max-chain or of a one-dimensional marching chain. There are configurations for which this takes $\Omega(n^2 \cdot \log(1/\delta))$ rounds, where δ denotes the initial distance between the outer robots.*

Interestingly, however, fixing the position of one of the two outer robots enables us to employ tools from Markov Chain theory (as used in previous results [16]), yielding again the same almost tight runtime bound as in the one-dimensional case (see Theorem 9). Given this and some simple experimental evaluations, we conjecture that the lower worst-case bound stated in Theorem 2 is tight.

Section 4 considers the continuous setting. As in the discrete setting, very symmetric configurations again lead to unavoidable problems for deterministic strategies. Moreover, a naïve translation of the MoB strategy results in the same dependency on the outer robots' initial distance δ. However, the continuous model allows for an interesting tweak which, as we show in Sect. 5, cannot be done in the discrete model. Namely, it turns out that decreasing the speed of outer robots by a small constant τ gets rid of the dependency on δ and yields an optimal, linear runtime bound. As a byproduct, this also causes symmetrical initial positions to collapse to a single point instead of becoming a marching chain. Summarized, we get the following result for the continuous setting:

Theorem 3. MAX-MoB *reaches in worst-case optimal time $\Theta(n)$ a stationary, maximum chain of length $n-1$ or the chain collapses to a single point.*

Our results show that the idealized continuous model yields again a linear speed-up for the MAX-CHAIN-FORMATION problem, similar as for contracting robot formation problems. The major open problem is to find an upper runtime bound for MAX-GTM in the discrete setting where both endpoints move. Moreover, while very symmetrical initial configurations pose a problem for deterministic algorithms, both, experiments with a simple, custom simulator and looking at our processes from the perspective of dynamical systems suggest that such configurations are few and unstable. Thus, minor, random perturbations usually yield a configuration in which the robots reach the desired maximal chain. We analyze this observation formally by proving that the marching chain is an unstable fixed point of the related dynamical system. We discuss this in more detail towards the end of Sect. 3. Due to space constraints, all proofs can be found in the full version of this paper [3].

2   Model and Problem Description

We follow the robot model of the CHAIN-FORMATION problem [5,6,9,15,16]: We consider n robots, $r_1, \ldots, r_n$ that are connected in a chain topology positioned in the Euclidean plane. The robots r_1 and r_n are denoted as *outer robots* and all other robots are *inner robots*. Each inner robot r_i can distinguish its two neighbors r_{i-1} and r_{i+1} while the robots do not have a common understanding of left and right. The outer robots have only a single neighbor: the neighbor of r_1 is r_2 and r_n's neighbor is r_{n-1}. Based on their neighborhoods, robots can detect whether they are an inner or an outer robot. Each robot has a uniform viewing range of one. Apart from their direct neighbors, robots cannot see any other robot that might be present in their viewing range. Initially, at time t_0, the chain topology is *connected*, i.e., the distance between a robot and its neighbors is less than or equal to one. The position of r_i at time t is denoted by $p_i(t) \in \mathbb{R}^2$ and for all $2 \leq i \leq n$, the vector $w_i(t) := p_i(t) - p_{i-1}(t)$ is the vector pointing from robot r_{i-1} to robot r_i at time t. Starting at robot r_1, a *configuration* of robots at time t can be written as $w(t) := (w_2(t), w_3(t), \ldots w_n(t))^T$. The *length* of a configuration at time t is denoted by $L(t) := \sum_{i=2}^{n} \|w_i(t)\|$, where $\|w_i(t)\|$ denotes the Euclidean norm of vector $w_i(t)$. For a vector $w_i(t)$, we denote the normalized vector $\frac{1}{\|w_i(t)\|} w_i(t)$ by $\widehat{w}_i(t)$. The Euclidean distance between two robots r_i and r_j at time t is denoted by $\Delta_{i,j}(t) = \|p_i(t) - p_j(t)\|$.

Next, we introduce a characterization of configurations that is relevant for our analyses. In *one-dimensional* configurations, the positions of all robots are collinear. In *two-dimensional* configurations, there exists a set of at least 3 robots whose positions are not collinear. Our analyses distinguish two special kinds of one-dimensional configurations: *Opposed configurations* and *marching configurations*. In opposed configurations, the outer robots are on different sides of their neighbors, i.e., $\widehat{w}_2(t) = \widehat{w}_n(t)$. In marching configurations, the outer robots are on the same side of their neighbors, i.e., $\widehat{w}_2(t) = -\widehat{w}_n(t)$. For $2 \leq i \leq n-1$, we denote by $\alpha_i(t)$ the smaller of the two angles created by anchoring $w_{i+1}(t)$ at the terminal point of $w_i(t)$. Our goal is to reach a configuration with $\Delta_{1,n}(t) = n-1$. More precisely, each vector w_i should have a length of 1 and $w_i(t) = w_{i+1}(t)$ for $2 \leq i \leq n-1$. We call this configuration a *max-chain*. We say that we have reached an ε-approximation of the max-chain if $\Delta_{1,n}(t) \geq (1 - \varepsilon)(n-1)$ and $\|w_i(t)\| > 1 - \varepsilon$ for all $2 \leq i \leq n$.

We assume a very restricted robot model, namely robots having the capabilities of the $\mathcal{OBLOT}$ model with disoriented coordinate systems and limited visibility. Thus, the robots neither have a global coordinate system nor a common compass. A robot can only observe the position of its neighbors relative to its own. We assume that the robots can measure distances precisely and have a common notion of unit distance. Additionally, the robots are *oblivious* and cannot rely on any information from the past. Furthermore, the robots cannot communicate. Throughout this work, we consider two different notions of time, the $\mathcal{F}$SYNC time model and the continuous time model. In $\mathcal{F}$SYNC all robots operate in fully synchronous LOOK-COMPUTE-MOVE (LCM) cycles (rounds), i.e.; robots observe their environment, compute a target point and finally move there. The *continuous time model* can be seen as a continuous variant of the $\mathcal{F}$SYNC model for an infinitesimal small movement distance for each robot per round [13]. In this model, robots continuously observe their environment and adjust their own movement. There is no delay between observing the environment and adjusting the movement. At every point in time, the movement of each robot r_i can be expressed by a *velocity vector* $v_i(t)$ with $0 \leq \|v_i(t)\| \leq 1$, i.e., the maximal speed of a robot is bounded by 1. The function $p_i \colon \mathbb{R}_{>0} \to \mathbb{R}^2$ is the *trajectory* of r_i. The trajectories p_i are continuous but not necessarily differentiable because robots are able to change their speed and direction non-continuously. However, natural movement strategies, such as the strategy presented in this paper, have (right) differentiable trajectories. Thus, the velocity vector of a robot $v_i \colon \mathbb{R}_{>0} \to \mathbb{R}^2$ can be seen as the (right) derivative of p_i and we can write $v_i(t) = p_i{}'(t)$.

3   The Discrete Case

In this section, we describe MAX-GTM for the $\mathcal{F}$SYNC time model. Intuitively, the strategy solves two tasks concurrently. The first task is to arrange all robots on a straight line while the second task is to lengthen the chain by moving the outer robots away from each other. For the first task, we adapt the GTM-strategy for CHAIN-FORMATION in which all inner robots move to the midpoint between their neighbors in every round. For the second task, the outer robots move as far as possible away from their neighbors while keeping the chain connected.

3.1   MAX-GO-TO-THE-MIDDLE (MAX-GTM)

MAX-GTM works as follows: Every inner robot moves to the midpoint between its neighbors. The new position of an inner robot r_i can thus be computed as $p_i(t+1) = \frac{1}{2}p_{i-1}(t) + \frac{1}{2}p_{i+1}(t)$. An outer robot moves as far possible away from its neighbor by imagining a virtual robot. At time t, the outer robot r_1 normalizes the vector $w_2(t)$, imagines a virtual robot r_0 positioned at $p_1(t) - \widehat{w}_2(t)$ and moves to the midpoint between r_0 and r_2. Thus, $p_1(t+1) = \frac{1}{2}p_1(t) + \frac{1}{2}p_2(t) - \frac{1}{2}\widehat{w}_2(t)$. The procedure works analogously for $r_n \colon p_n(t+1) = \frac{1}{2}p_{n-1}(t) + \frac{1}{2}p_n(t) + \frac{1}{2}\widehat{w}_n(t)$. In the special case $p_1(t) = p_2(t)$ ($p_{n-1}(t) = p_n(t)$ respectively) r_1 (r_n) does not move. Similarly, we can derive formulas for $w(t + 1)$: $w_2(t + 1) = \frac{1}{2}\widehat{w}_2(t) +$

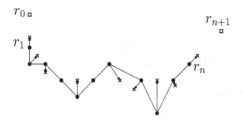

$$S(t) =$$

$$\begin{bmatrix} \frac{1}{2 \cdot \|w_2(t)\|} & \frac{1}{2} & 0 & 0 & \dots & 0 & 0 \\ \frac{1}{2} & 0 & \frac{1}{2} & 0 & \dots & 0 & 0 \\ 0 & \frac{1}{2} & 0 & \frac{1}{2} & \dots & 0 & 0 \\ \vdots & & \vdots & \vdots & \vdots & & \vdots \\ 0 & & 0 & 0 & \dots & \frac{1}{2} & \frac{1}{2 \cdot \|w_n(t)\|} \end{bmatrix}$$

(a) A visualization of MAX-GTM. The target point of each robot is marked by a cross.

(b) The strategy matrix $S(t)$ that depends on $w_2(t)$ and $w_n(t)$.

Fig. 1. A visualization of MAX-GTM and the strategy matrix $S(t)$.

$\frac{1}{2}w_3(t), w_i(t+1) = \frac{1}{2}w_{i-1}(t) + \frac{1}{2}w_{i+1}(t)$ and $w_n(t+1) = \frac{1}{2}w_{n-1}(t) + \frac{1}{2}\widehat{w}_n(t)$. Simplified, we can compute $w(t+1)$ as a matrix-vector product: $w(t+1) = S(t) \cdot w(t) = \prod_{i=0}^{t} S(i) \cdot w(0)$ with the strategy matrix $S(t)$. See Figs. 1a and 1b for a visualization of MAX-GTM and the strategy matrix $S(t)$.

3.2   One-Dimensional Analysis

Next, we investigate the performance of MAX-GTM in a one-dimensional configuration. These configurations already reveal an interesting behavior of MAX-GTM: some configurations do not converge to a max-chain but to a different structure, we will denote as the marching chain. The two classes of configurations that play a role in this analysis are marching and opposed configurations.

Lemma 1. MAX-GTM *switches at most once between opposed and marching configurations. The switch is executed after at most $n - 2$ rounds.*

The switch between opposed and marching configurations can only happen if $w_3(t_0) = -\widehat{w}_2(t_0)$ or $w_{n-1}(t_0) = -\widehat{w}_n(t_0)$. In this case, r_1 and r_2 or r_{n-1} and r_n move to the same position and it can take up to $n - 2$ rounds until this is resolved. See [3] for a more comprehensive discussion of this special case. For the ease of notation, we say in the following that a configuration is an opposed configuration if it is an opposed configuration after applying MAX-GTM for $n-2$ rounds (similar for marching configurations).

As a consequence of Lemma 1, starting from a marching configuration, MAX-GTM does not converge to a max-chain. For some highly symmetric configurations, for instance the configuration depicted in Fig. 2, this even cannot be obtained by any deterministic strategy due to a high symmetry.

Theorem 4. *There are marching configurations that cannot be transformed into a max-chain by any deterministic strategy.*

For opposed configurations, we can show that MAX-GTM converges towards an ε-approximation of the max-chain. Define $m_i(t) = 1 - w_i(t)$. For the analysis, we use the potential function $\phi_1(t) = \sum_{i=2}^{n} m_i(t)^2$ as a progress measure. Intuitively, $\phi_1(t)$ measures how close the configuration is to the max-chain, since in the max-chain, $w_i(t) = 1$ for all i. For $m_1(t) = 0$ and $m_{n+1}(t) = 0$, it holds $\phi_1(t+1) = \sum_{i=2}^{n} m_i(t+1)^2 = \sum_{i=2}^{n} \left(\frac{m_{i-1}(t)+m_{i+1}(t)}{2}\right)^2$. Inspired by [5], we can analyze $\phi_1(t)$ with the help of discrete Fourier Transforms. Discrete Fourier Transforms are useful here, because they allow us to decouple the computations of the $m_i(t+1)$'s. By now, we can express $m_i(t+1)$ based upon $m_{i-1}(t)$ and $m_{i+1}(t)$. Discrete Fourier Transforms remove this dependency such that we get a single (non-recursive) formula for each $m_i(t+1)$ and can bound $\phi_1(t+1)$ by the slowest decreasing $m_i(t)$, resulting in an upper runtime bound of $\mathcal{O}\left(n^2 \cdot \log\left(n/\varepsilon\right)\right)$.

Theorem 5. *Started in an opposed configuration, MAX-GTM needs at most $\mathcal{O}(n^2 \cdot \log\left(n/\varepsilon\right))$ rounds to achieve an ε-approximation of the max-chain.*

The analysis of the *mixing time* of a Markov Chain allows us to prove a close lower bound of $\Omega\!\left(n^2 \cdot \log\left(1/\varepsilon\right)\right)$. Here, we can rewrite $w(t)$ and $S(t)$ slightly such that the resulting strategy matrix is a stochastic matrix that could also be the transition matrix of a Markov Chain with a single absorbing state. Markov Chains with a single absorbing state have a unique stationary distribution, such that some bounds for the mixing times of Markov Chains can be applied.

Theorem 6. *There exist opposed configurations such that* MAX-GTM *needs at least* $\Omega\!\left(n^2 \cdot \log(1/\varepsilon)\right)$ *rounds to achieve an ε-approximation of the max-chain.*

Marching configurations do not converge to a max-chain but have a different convergence behavior, they converge to the *marching chain*. It is called marching chain because the robots all together move into the same direction and never stop. The configuration $w_M(t)$ defines the marching chain. $w_M(t) = (1 - \frac{2}{n}, 1 - \frac{4}{n}, \ldots, \frac{2}{n}, 0, -\frac{2}{n}, -\frac{4}{n}, \ldots, -(1 - \frac{2}{n}))^T$. Figure 2 visualizes this marching chain. Observe that $S(t) \cdot w_M(t) = w_M(t)$ ($w_M(t)$ is an eigenvector of $S(t)$ to the eigenvalue 1). In the marching chain, each robots moves distance $\frac{1}{n}$ per round.

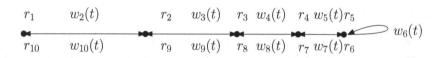

Fig. 2. A marching chain for $n = 10$. Every position is occupied by two robots.

Starting in a marching configuration, the convergence time until all vectors only differ up to ε from their corresponding vector in the marching chain is equal to the runtime bound for opposed configurations. Here, we can again use the analysis of the mixing time of a Markov Chain for a slightly different transition matrix. We consider the vectors $z_i(t) = p_i(t+1) - p_i(t)$ pointing from the current position of a robot to its next position. The changes of the vectors $z_i(t)$ can be calculated via a matrix-vector product of a doubly stochastic transition matrix and the current vectors. A doubly stochastic matrix can also be the transition matrix of an irreducible, aperiodic and reversible Markov Chain. These Markov Chains converge to a unique stationary distribution and the runtime bounds depend on the second largest eigenvalue of the transition matrix. By analyzing this second largest eigenvalue we can prove the following runtime bounds:

Theorem 7. *Given a marching configuration,* MAX-GTM *needs at most* $\mathcal{O}(n^2 \cdot \log(n/\varepsilon))$ *and at least* $\Omega(n^2 \cdot \log(1/\varepsilon))$ *rounds to achieve an ε-approximation of the marching chain.*

3.3 Two-Dimensional Analysis

Next, we prove a convergence result for two-dimensional configurations, stating that an initial configuration either converges to the max-chain or to the marching chain. In the analysis, we again consider the vectors $z_i(t) = p_i(t+1) - p_i(t)$ and the function $\phi_2(t) = \sum_{i=1}^{n} \|z_i(t)\|^2$. $\phi_2(t)$ is a monotonically decreasing function and the analysis of $\phi_2(t+1) - \phi_2(t)$ allows us to conclude that two-dimensional configurations either converge to a marching chain or to the max-chain.

Theorem 8. *Given an arbitrary connected chain in the Euclidean plane, Max-GtM converges either to the marching chain or to the max-chain.*

Interestingly, when assuming that only one of the outer robot moves while the other one remains stationary, we can prove the same upper runtime bound as for one-dimensional configurations. The proof relies on the analysis of $\phi_2(t)$ for this case. Again, the analysis of a transition matrix plays a role here – since only one outer robot moves we obtain a *substochastic* transition matrix where every row except of one sums up to 1. The last row only sums up to 1/2 such that high powers of this matrix converge to the 0-matrix. A diagonalization of the transition matrix yields the following runtime bound:

Theorem 9. *In case one of the outer robots is stationary and all other robots move according to Max-GtM, an ε-approximation of the max-chain is achieved after $\mathcal{O}(n^2 \cdot \log(n/\varepsilon))$ rounds.*

In case both outer robots move, we identify a certain class of configurations that lead to an arbitrarily high runtime based on a parameter δ which can be seen as the width of the configuration. Before defining the configurations, we give some intuition about their construction: Applying Max-GtM to the configuration of robots can be interpreted as a discrete time *dynamical system*. In Sect. 3.2, we have seen that this dynamical system has two different (classes of) *fixed points*, i.e., a configuration that remains unchanged when applying Max-GtM. These fixed points are the max-chain and the marching chain. We can prove that the marching chain is an *unstable* fixed point. Unstable means that a small perturbation in the configuration results in a different behavior – the dynamical system moves away from this fixed point. In our case this means that a small perturbation in the marching chain leads to a configuration that converges to the max-chain. For a formal description of the relation to dynamical systems and a proof that the marching chain is an unstable fixed point, we refer the reader to [3]. We use the property that marching chains are unstable fixed points to define configurations which are very close to the marching chain, *discrete δ-V-configurations*. For $\delta = 0$, discrete δ-V-configurations and marching chains coincide. Choosing *any* $\delta > 0$ changes the behavior such that the configuration converges to the max-chain. The runtime, however, can be arbitrarily high depending on δ.

Definition 1. *For n even, a discrete δ-V-configuration is defined by the vectors $w_i(t) := (\frac{\delta}{n-1}, 1 - \frac{2(i-1)}{n})^T$ for $i = 2, \ldots, n$.*

Theorem 10. *Starting in a discrete δ-V-configuration, Max-GtM needs at least $\Omega(n^2 \cdot \log(1/\delta))$ rounds to achieve an ε-approximation of the max-chain.*

Hence, the runtime of MAX-GTM can be arbitrarily high depending on δ. The dependence on δ can be removed in the continuous time model by an (at the first sight) counter-intuitive approach: The outer robots move slower than the inner robots. The same approach does not work in $\mathcal{F}$SYNC (see Sect. 5).

4   The Continuous Case

This section is dedicated to the MAX-MOB strategy that transforms a connected chain into a max-chain in the continuous time model. After introducing the strategy, we continue with some preliminaries in Sect. 4.1 and provide an intuitive explanation of the strategy combined with a proof outline in Sect. 4.2.

MAX-MOVE-ON-BISECTOR (MAX-MOB)   Outer robots move with a maximal speed of $(1 - \tau)$ for a constant $0 < \tau \le 1/2$ as follows: In case $\|w_2(t)\| < 1$: $v_1(t) = -(1 - \tau) \cdot \widehat{w}_2(t)$. Similarly, in case $\|w_n(t)\| < 1$: $v_n(t) = (1 - \tau) \cdot \widehat{w}_n(t)$. In other words, outer robots move with a speed of $(1 - \tau)$ away from their direct neighbors. Otherwise, provided $\|w_2(t)\| = 1$ ($\|w_n(t)\| = 1$ respectively), an outer robot adjusts its own speed and tries to stay in distance 1 to its neighbor while moving with a maximal speed of $1 - \tau$. An inner robot r_i with $0 < \alpha_i(t) < \pi$ moves only if at least one of the following three conditions holds: $\|w_i(t)\| = 1$, $\|w_{i+1}(t)\| = 1$ or $\alpha_i(t) < \psi$ for $\psi := 2 \cdot \cos^{-1}(1 - \tau)$. Otherwise an inner robot does not move at all. In case one of the conditions holds, an inner robot moves with speed 1 along the angle bisector formed by the vectors pointing to its neighbors. As soon as the position of the robot and the positions of its neighbors are collinear it continues to move with speed 1 towards the midpoint between its neighbors while ensuring to stay collinear. Once it has reached the midpoint it adjust its own speed to stay on the midpoint. See Fig. 3 for a visualization.

4.1   Preliminaries

For both outer robots we determine the index of the first robot that is not collinear with its neighbors and the respective outer robot.

Definition 2. $\ell(t)$ *is the index, s.t. for all* $2 < j \le \ell(t)$ *either* $w_j(t) = (0,0)$ *or* $\widehat{w}_j(t) = \widehat{w}_2(t), w_{\ell(t)+1}(t) \ne (0,0)$ *and* $\widehat{w}_{\ell(t)+1}(t) \ne \widehat{w}_2(t)$. *Similarly, define* $r(t)$ *to be the index such that for all* $r(t) < j < n$ *either* $w_j(t) = (0,0)$ *or* $\widehat{w}_j(t) = \widehat{w}_n(t)$ *and* $w_{r(t)}(t) \ne (0,0)$ *and* $\widehat{w}_{r(t)}(t) \ne \widehat{w}_n(t)$. *In case there is no such an index define* $\ell(t) = r(t) = 0$. $\alpha_{\ell(t)}(t)$ *and* $\alpha_{r(t)}(t)$ *are denoted as outer angles.*

We omit the time parameter t when it is clear from the context, e.g., we write $\alpha_\ell(t)$ instead of $\alpha_{\ell(t)}(t)$. In addition to the indices $\ell(t)$ and $r(t)$, we define the last indices of robots (starting to count at $\ell(t)$ and $r(t)$) that are collinear with their neighbors and the corresponding robot with index $\ell(t)$ or $r(t)$.

Definition 3. *Let* $\ell^+(t)$ *be the smallest index larger than* $\ell(t)$ *such that* $\alpha_{\ell+}(t) < \pi$. *Similarly let* $r^+(t)$ *be the largest index less than* $r(t)$ *such that* $\alpha_{r+}(t) < \pi$.

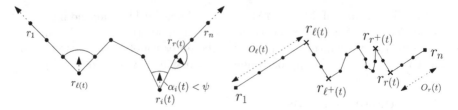

Fig. 3. A chain visualizing the movements of MAX-MoB. The velocity vectors are depicted by dashed arrows.

Fig. 4. A visualization of $\ell(t), \ell^+(t), r(t),$ $r^+(t)$, $O_\ell(t)$ and $O_r(t)$.

Definition 4. *The* left outer length $O_\ell(t) := \sum_{i=2}^{\ell(t)} \|w_i(t)\|$ *and the* right outer length $O_r(t) := \sum_{i=r(t)+1}^{n} \|w_i(t)\|$. *The maximal values of the left and right outer length are denoted by* $\gamma_\ell(t) := \ell(t) - 1$ *and* $\gamma_r(t) := n - r(t)$. *Additionally, the* inner length *is defined as* $I(t) := \sum_{i=\ell(t)+1}^{r(t)} \|w_i(t)\|$. *See Fig. 4.*

4.2   Intuition and Proof Outline

The main idea of MAX-MoB is to flatten and stretch the chain starting at the outer robots towards the inside of the chain. At first, $\|w_2(t)\| = 1$ and $\|w_n(t)\| = 1$ is ensured, afterwards the angles $\alpha_2(t)$ and $\alpha_{n-1}(t)$ should reach a size of π and so on until finally all vectors have length 1 and all angles have a size of π. Figure 5 visualizes this core idea. To achieve this behavior, one of the two cases in which an inner robot r_i moves demands either $\|w_i(t)\| = 1$ or $\|w_{i+1}(t)\| = 1$, because locally it can assume that it is already located on the straight line to one outer robot and all vectors into the direction of the outer robot have a length of 1. In addition, an inner robot r_i moves if $\alpha_i(t) < \psi$. In Sect. 5 we see that this property is crucial for the linear runtime of the strategy by introducing configurations that have a high runtime that not only depends on the number of robots in case the property is ignored. To express the behavior of flattening and stretching the chain starting at the outer robots towards the inside of the chain, we have introduced the indices $\ell(t)$ and $r(t)$. For each of the two sets of robots $r_1, \ldots, r_{\ell(t)}$ and $r_{r(t)}, \ldots, r_n$ it always holds that these robots continue to stay collinear for the rest of the execution.

The outer angles $\alpha_\ell(t)$ and $\alpha_r(t)$ play an important role in the analysis. We divide the possible sizes into three intervals, $[0, \psi), [\psi, \frac{3}{4}\pi]$ and $(\frac{3}{4}\pi, \pi]$. For an outer angle $\alpha_i(t) \in [0, \psi)(i \in \{\ell(t), r(t)\})$ two properties hold: $I(t)$ decreases with speed at least $1 - \tau$ and the corresponding outer length decreases since the outer robots move with speed at most $1 - \tau$. As $I(t)$ decreases with a constant speed of at least $1 - \tau$, the total time in such a case is upper bounded by $\mathcal{O}(n)$. Given $\alpha_i(t) \in [\psi, \frac{3}{4}\pi]$, the strategy is designed such that r_i only moves if $O_i(t) = \gamma_i(t)$. Thus, as long as $O_i(t) < \gamma_i(t)$, $O_i(t)$ increases with speed $1 - \tau$. As soon as $O_i(t) = \gamma_i(t)$ holds, the robot r_i starts moving along its bisector.

This movement causes a decrease of $I(t)$ with speed at least $\cos\left(\frac{3}{8}\pi\right)$ while the length of $O_i(t)$ does not change. Since, $I(t)$ decreases with constant speed, this case can hold for time at most $\mathcal{O}(n)$. For the last interval, $\alpha_i(t) \in \left(\frac{3}{4}, \pi\right]$ we use a different progress measure since large angles cause a very slow decrease of $I(t)$ which cannot be bounded by a constant anymore. Therefore, we consider the height $H_i(t)$. Assume $i = \ell(t)$, then $H_\ell(t)$ denotes the distance between $r_{\ell(t)}$ and the line segment connecting r_1 and $r_{\ell+(t)}$. Intuitively, if we consider the line segment connecting r_1 and $r_{\ell+(t)}$ as a line parallel to the x-axis, the robot r_ℓ moves with a velocity vector that has a small angle to the y-axis towards this line segment. Thus, $H_i(t)$ decreases with constant speed. All in all, we can prove that the total time of outer angles in any of these intervals is bounded by $\mathcal{O}(n)$ such that finally $\ell(t) = r(t)$ holds. Lastly, we analyze the case $\ell(t) = r(t)$ and prove a linear runtime for the strategy (Theorem 11).

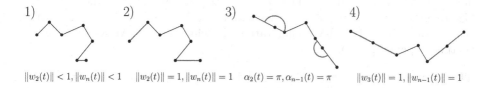

1) $\|w_2(t)\| < 1, \|w_n(t)\| < 1$ 2) $\|w_2(t)\| = 1, \|w_n(t)\| = 1$ 3) $\alpha_2(t) = \pi, \alpha_{n-1}(t) = \pi$ 4) $\|w_3(t)\| = 1, \|w_{n-1}(t)\| = 1$

Fig. 5. A visualization of the core idea of MAX-MoB. 1) depicts an initial configuration. 2) visualizes the configuration after stretching $w_2(t)$ and $w_n(t)$. In 3), $\alpha_2(t) = \pi$ and $\alpha_{n-1}(t) = \pi$. In 4) r_1 and r_2 as well as r_{n-1} and r_n have moved such that $\|w_2(t)\| = \|w_3(t)\| = \|w_{n-1}(t)\| = \|w_n(t)\| = 1$.

Theorem 11. *Starting* MAX-MoB *in a two-dimensional configuration, the initial chain is either transformed into a straight line of length $n - 1$ or all robots are located at the same position after time $\mathcal{O}(n)$.*

Our simulations support the following conjecture.

Conjecture 1. The set of initial two-dimensional configurations that result in a configuration where all robots are located on the same position when applying MAX-MoB has Lebesgue measure 0.

5   On the Speed of the Outer Robots

We close by a brief discussion on the influence of the speed of the outer robots. An elaboration on it can be found in the full version [3]. It turns out there exists a special class of configurations – called continuous δ-V-configurations – parameterized in the initial distance of the two outer robots δ for which MAX-MoB needs a runtime independent of δ, see Theorem 12.

Theorem 12. *Starting in a continuous δ-V-configuration, MAX-MOB needs time at most $n \cdot \left(\frac{1}{\tau} + \frac{1}{1-\tau}\right)$ to transform the configuration into a max-chain.*

One might suspect that an algorithm in which the outer robots move at full speed stretches the chain faster. Interestingly, this is not true! For such an algorithm – let it be called NAIVE-MAX-MOB – we can show that the runtime for continuous δ-V-configurations is lower bounded dependent on δ.

Theorem 13. *NAIVE-MAX-MOB transforms a continuous δ-V-configuration into a max-chain in time $\Omega(n \cdot \log(1/\delta))$.*

Slowing down the outer robots actually allows us to achieve a runtime independent of the initial configuration in the continuous case. Can we apply the same idea in the discrete model? Unfortunately not! Consider the algorithm $(1 - \tau)$-MAX-GTM that behaves as MAX-GTM except that the outer robots move always by a distance of $(1 - \tau)$ times the distance they would in MAX-GTM. Similar to discrete δ-V-configurations for MAX-GTM, there exists the class of discrete $(\delta, 1 - \tau)$-V-configurations in which $(1 - \tau)$-MAX-GTM has a runtime depending on δ.

Theorem 14. *Starting in a discrete $(\delta, 1 - \tau)$-V-configuration, $(1 - \tau)$-MAX-GTM needs at least $\Omega(n^2 \cdot \log(1/\delta))$ rounds to achieve an ε-approximation of the max-chain.*

References

1. Ando, H., Oasa, Y., Suzuki, I., Yamashita, M.: Distributed memoryless point convergence algorithm for mobile robots with limited visibility. IEEE Trans. Rob. Autom. **15**(5), 818–828 (1999)
2. Barrière, L., Flocchini, P., Barrameda, E.M., Santoro, N.: Uniform scattering of autonomous mobile robots in a grid. In: 23rd IEEE International Symposium on Parallel and Distributed Processing, IPDPS 2009, Rome, Italy, 23–29 May 2009, pp. 1–8. IEEE (2009)
3. Castenow, J., Kling, P., Knollmann, T., Meyer auf der Heide, F.: A Discrete and Continuous Study of the Max-Chain-Formation Problem. CoRR abs/2010.02043 (2020). https://arxiv.org/abs/2010.02043
4. Castenow, J., Kling, P., Knollmann, T., Meyer auf der Heide, F.: A discrete and continuous study of the max-chain-formation problem: slow down to speed up. In: 32nd ACM Symposium on Parallelism in Algorithms and Architectures, SPAA 2020, USA, 15–17 July 2020, pp. 515–517. ACM (2020)
5. Cohen, R., Peleg, D.: Local algorithms for autonomous robot systems. In: Flocchini, P., Gasieniec, L. (eds.) SIROCCO, vol. 4056, pp. 29–43. Springer, Heidelberg (2006). https://doi.org/10.1007/11780823_4
6. Degener, B., Kempkes, B., Kling, P., Meyer auf der Heide, F.: Linear and competitive strategies for continuous robot formation problems. TOPC **2**(1), 2:1–2:18 (2015)

7. Degener, B., Kempkes, B., Langner, T., Meyer auf der Heide, F., Pietrzyk, P., Wattenhofer, R.: A tight runtime bound for synchronous gathering of autonomous robots with limited visibility. In: Proceedings of the 23rd Annual ACM Symposium on Parallelism in Algorithms and Architectures, SPAA 2011, San Jose, CA, USA, 4–6 June 2011, pp. 139–148. ACM (2011)

8. Dieudonné, Y., Petit, F.: Self-stabilizing Deterministic Gathering. In: Dolev, S. (ed.) ALGOSENSORS 2009. LNCS, vol. 5804, pp. 230–241. Springer, Heidelberg (2009). https://doi.org/10.1007/978-3-642-05434-1_23

9. Dynia, M., Kutyłowski, J., Lorek, P., auf der Heide, F.M.: Maintaining communication between an explorer and a base station. In: Pan, Y., Rammig, F.J., Schmeck, H., Solar, M. (eds.) BICC 2006. IIFIP, vol. 216, pp. 137–146. Springer, Boston, MA (2006). https://doi.org/10.1007/978-0-387-34733-2_14

10. Flocchini, P., Prencipe, G., Santoro, N. (eds.): Distributed Computing by Mobile Entities, Current Research in Moving and Computing, Lecture Notes in Computer Science, vol. 11340. Springer, Heidelberg (2019). https://doi.org/10.1007/978-3-030-11072-7

11. Flocchini, P., Prencipe, G., Santoro, N., Widmayer, P.: Gathering of asynchronous robots with limited visibility. Theor. Comput. Sci. **337**(1–3), 147–168 (2005)

12. Gmyr, R., et al.: Forming tile shapes with simple robots. In: Doty, D., Dietz, H. (eds.) DNA 2018. LNCS, vol. 11145, pp. 122–138. Springer, Cham (2018). https://doi.org/10.1007/978-3-030-00030-1_8

13. Gordon, N., Wagner, I.A., Bruckstein, A.M.: Gathering multiple robotic a(ge)nts with limited sensing capabilities. In: Dorigo, M., et al. (eds.) ANTS 2004. LNCS, vol. 3172, pp. 142–153. Springer, Heidelberg (2004). https://doi.org/10.1007/978-3-540-28646-2_13

14. Jiang, Z., Wang, X., Yang, J.: Distributed line formation control in swarm robots. In: IEEE International Conference on Information and Automation, ICIA 2018, Wuyishan, China, 11–13 August 2018, pp. 636–641. IEEE (2018)

15. Kempkes, B., Kling, P., Meyer auf der Heide, F.: Optimal and competitive runtime bounds for continuous, local gathering of mobile robots. In: 24th ACM Symposium on Parallelism in Algorithms and Architectures, SPAA 2012, Pittsburgh, PA, USA, 25–27 June 2012, pp. 18–26. ACM (2012)

16. Kling, P., Meyer auf der Heide, F.: Convergence of local communication chain strategies via linear transformations: or how to trade locality for speed. In: Proceedings of the 23rd Annual ACM Symposium on Parallelism in Algorithms and Architectures, SPAA 2011, San Jose, CA, USA, 4–6 June 2011, pp. 159–166. ACM (2011)

17. Kling, P., Meyer auf der Heide, F.: Continuous protocols for swarm robotics. In: Distributed Computing by Mobile Entities, Current Research in Moving and Computing, pp. 317–334 (2019)

18. Kutyłowski, J., Meyer auf der Heide, F.: Optimal strategies for maintaining a chain of relays between an explorer and a base camp. Theor. Comput. Sci. **410**(36), 3391–3405 (2009)

19. Nguyen, H.G., Pezeshkian, N., Raymond, S.M., Gupta, A., Spector, J.M.: Autonomous communication relays for tactical robots. In: Proceedings of the 11th International Conference on Advanced Robotics (ICAR), pp. 35–40 (2003)

20. Poudel, P., Sharma, G.: Universally optimal gathering under limited visibility. In: Spirakis, P., Tsigas, P. (eds.) SSS 2017. LNCS, vol. 10616, pp. 323–340. Springer, Cham (2017). https://doi.org/10.1007/978-3-319-69084-1_23

21. Shibata, M., Mega, T., Ooshita, F., Kakugawa, H., Masuzawa, T.: Uniform deployment of mobile agents in asynchronous rings. J. Parallel Distrib. Comput. **119**, 92–106 (2018)
22. Tagliabue, A., Schneider, S., Pavone, M., Agha-mohammadi, A.: Shapeshifter: A multi-agent, multi-modal robotic platform for exploration of titan. CoRR abs/2002.00515 (2020). https://arxiv.org/abs/2002.00515
23. Yun, X., Alptekin, G., Albayrak, O.: Line and circle formation of distributed physical mobile robots. J. Field Rob. **14**(2), 63–76 (1997)

Invited Paper: Reactive PLS
for Distributed Decision

Jiaqi Chen[1]([⊠]), Shlomi Dolev[2], and Shay Kutten[1]

[1] Technion - Israel Institute of Technology, Haifa, Israel
sjackie@campus.technion.ac.il, kutten@technion.ac.il
[2] Ben-Gurion University of the Negev, 84105 Beer-Sheva, Israel
dolev@cs.bgu.ac.il

Abstract. We generalize the definition of Proof Labeling Schemes to reactive systems, that is, systems where the configuration is supposed to keep changing forever. As an example, we address the main classical test case of reactive tasks, namely, the task of token passing. Different RPLSs are given for the cases that the network is assumed to be a tree or an anonymous ring, or a general graph, and the sizes of RPLSs' labels are analyzed. We also address the question whether an RPLS exists. Interestingly, for the anonymous ring, it is known that no token passing algorithm is possible even if the number n is known. Nevertheless, we show that an RPLS *is* possible. We show that if one drops the assumption that n is known, the construction becomes impossible.

Keywords: Proof Labeling Schemes · Distributed proofs · Distributed reactive systems

1 Introduction

Proof Labeling Schemes [37] as well as most later variations of distributed "local" verification of global properties, were suggested for "input/output" distributed algorithms, where the (distributed) input is available at the nodes before an algorithm is executed, and the algorithm uses it to generate the final output and terminate. (Those tasks are also called "static" or "one-shot"; they are "silent" in terms of self-stabilization [21]). As discussed in Subsect. 1.1, they were shown useful in the context of the development of algorithms, as well as in the context of distributed computability and complexity. This motivates the extension formalized in the current paper – that of local checking in *reactive systems*, where

---

S. Dolev—work is supported by the Rita Altura Trust Chair in Computer Science, and is partially supported by a grant from the Ministry of Science and Technology, Israel & the Japan Science and Technology Agency (JST), and the German Research Funding (DFG, Grant#8767581199).
S. Kutten—The research of Shay Kutten was supported in part by a grant from the Hiroshi Fujiwara Cyber Security Research Center at the Technion.

S. Devismes and N. Mittal (Eds.): SSS 2020, LNCS 12514, pp. 81–96, 2020.
https://doi.org/10.1007/978-3-030-64348-5_7

the configuration is supposed to keep changing forever. We note that reactive systems are by far, the bulk of distributed systems.

In the current paper, we define and initiate the development of efficient Reactive Proof Labeling Schemes (RPLS) tailored to specific tasks. As a test case, we address the main classical problem of self-stabilization – token passing. Different RPLSs are given for the cases that the network is assumed to be a tree, or a ring, or a general graph, and the sizes of their labels are analyzed.

We also address the question whether an RPLS exists. Interestingly, it is known that no token passing algorithm exists for the anonymous ring even if the number n of nodes is known [17] (except for special cases [13]). Nevertheless, we show that there exists an RPLS for token passing on anonymous rings, provided n is known, for any value of n. We also show that no RPLS exists for the anonymous token ring if we remove the assumption that n is known.

Recall that by coupling a non-self-stabilizing algorithm with a "one-shot" task, one could develop a self-stabilizing algorithm for that task. Similarly, coupling such an RPLS with a non-stabilizing token passing algorithm would ease the task of designing a self-stabilizing token passing algorithm. Another possible direct application is to allow the use of a randomized algorithm until stabilization when deterministic algorithms do not exist, and to cease the use of potentially expensive randomness when the RPLS does not indicate an illegal state [24].

1.1 Background

The notion of local checking started in the context of self-stabilization, and later motivated further research. Early self-stabilizing algorithms [13,17–19,39] as if by magic, managed to *stabilize* (see definitions in [20]) without explicitly declaring the detection of a faulty global state. In [17], the configuration (the collection of all nodes' state) is "legal" iff the state of exactly one node is "holding a token". A node holding a token takes a certain step to "pass the token" to the next node, whether the configuration is legal or not. This seems a much more elegant design than addressing multiple cases. Looking at this very elegant algorithm, one is amazed how extra tokens (if such exist) somehow disappear. Similarly, in [18,19], a node repeatedly chooses to point at the neighbor who has the lowest value in the "distance-from-the-root" variable whether the current configuration is legal or not. This, again, elegantly avoids cluttering the algorithm with various cases and subcases. Surprisingly, if the configuration is not already a Breadth First Search tree rooted at the assumed special node, then it converges to being such a tree.

Despite the elegance of the above approach, it also has the disadvantage of requiring the ingenuity to especially design an algorithm for different tasks from scratch. On the other hand, the opposite approach, that of detecting an illegal state and then addressing it, also seems difficult at first. As Dijkstra described [17], the difficulty in self-stabilization arises from the fact that the configuration is not known to anybody. A node is not aware of the current state of far away nodes. It can know only about some past states that may have changed. The seminal works of [14,34] suggested a way to overcome this inherent difficulty in distributed systems. A "snapshot" algorithm collects a "consistent" (see [14]) representation of the configuration to some given "leader" node, who can then check whether the configuration follows some given *legality predicate*, e.g., "exactly one node has a token", and detect the case where it does not. (A self-stabilizing algorithm to elect such a leader was suggested at the same time in [2]). This could simplify the task of designing self-stabilizing algorithms by practically automating the part of the task involving the detection of a faulty state. This allows the modular approach of designing a part of the algorithm especially for addressing faults. However, this kind of checking, requiring that a fault even at very remote nodes is detected, was named "global checking" [7] for a good reason. It carried a lot of overhead in time and communication (even if the communication was reduced by tailoring the global checking to a specific task and even assuming that the checking process itself could not suffer faults but only the checked task may be faulty [11,36,40]).

The notion of *local* checking and detecting a violation of the legality predicate on the configuration was thus suggested in [2]. The main new idea (the "local detection paradigm") was that if the configuration was *not* legal, then *at least one node* detects the violation using *only* its own state and the state of its neighbors. (If the configuration was legal, then *no node* detects a violation). Note that the locality was made possible by allowing the detection to be made by a single and arbitrary node, which is enough in the context of self-stabilization, since then correction measures (of various kinds [1,2,5–8,22,31,38]) can be taken. An example of a predicate that can be checked that way is equality – in the election algorithm of [2], every node checked repeatedly whether the unique identity of the node it considered to be the leader was indeed the same as the identity its neighbors considered to belong to the leader. Predicates that can be computed that way are *locally checkable*.

Some other properties, e.g., the acyclicity of a subgraph, required enhancing the output (and the legality predicate) by the addition of information at each node to ensure local checkability of the enhanced predicate. This was generalized by various other papers, at first, still in the context of self-stabilization. In [6,8], general functions were discussed. In [21], this allowed the definition of a stronger form of self-stabilization by reaching a state in which communication can be saved. In [10], the generalization was to consider different faults at different distances in time or in graph distance, and thus invest more only in cases where more exact pinpointing of the faults was needed.

The first impact of local checkability outside of the realm of self-stabilization is due to [43]. Indeed, they motivated the study of locally checkable predicates by the fact that local detection is useful in self-stabilization. Still, the subject of their paper was the not-necessarily-self-stabilizing *computation* of a language (viewing the output configuration as a word in a distributed language) for which a locally checkable predicate holds. They asked which of those could also be computed in $O(1)$ time. This started a long line of research, eventually characterizing exactly what (distributed time) complexity classes exist for languages whose predicate is locally checkable. See e.g., [9].

A further development beyond self-stabilization was the notion of Proof Labeling Schemes (PLS) [37]. Those were suggested as a distributed counterpart of the classical foundation of computing notion of Nondeterministic Polynomial (NP) time. Recall that a language is in NP iff there exists a "verifier" such that for each word in the language, there exists a "witness" that can be checked "easily" by the verifier. The word is in the language if and only if the verifier accepts. In a PLS, the *distributed* "witness" is some label, possibly different for each node. The verifier is distributed in the sense that in each node, it can check only variables and the label of the node, and the labels of the node's neighbors.[1] Here, "easily" checking means $O(1)$ distributed time, rather than the polynomial non-distributed time of the classical NP notion.

This led to an ongoing rich area of research by adapting notions, and hopefully also results of classical non-distributed theory into the distributed realm, see e.g., [27,30,35,42]. In some cases, there are multiple interactions between a prover and the distributed verifier at each node. This is still very different from the subject of the current paper. For example, in [35,42], multiple interactions are useful even in the verification of one, non-changing configuration, while here, we deal with a configuration that is supposed to be changed in an ongoing manner. Moreover, in [27,35,42], an all-powerful and non-distributed prover is assumed, that can communicate with the nodes. This is very different from the current paper where the idea is to make the proof labeling "long-lived" by having it maintained by a *distributed* mechanism such that the configuration changed by a correct algorithm can still be verified as correct after the change. *Static* checking in a network where edges may be inserted or removed was addressed in [29]. However, the labels there are never changed, so this can support only the case that both the graph before the change and the graph after the change obey the predicate. Additional complexity related results [45] were obtained by expanding the research into less local forms of checking.

The local detection paradigm was also applied to obtain results in distributed testing, hardness of approximation, Peer-to-Peer networks, and more [24,26,44, 45]. Many generalizations were suggested, such as verifiers who are allowed to communicate to a distance larger than 1, or to a given number of nodes [30]. Another line of research generalized the verifier to output some general function rather than a binary one (of either accepting or rejecting) [25].

Unfortunately, verification schemes were designed for a single configuration, e.g., the output of "one time" distributed tasks. Algorithms that updated the labels were used, e.g., after a system reset [2,7] or even every step [2] (to verify that a certain global condition (a cycle) did not occur in the new configuration). Still, there, the tasks addressed were not reactive (e.g., leader election [2]).

Checking is indeed often intermixed in the design of reactive systems, to identify situations that may need a repair. More directly, we are inspired by the local stabilizer [1] – a general fault recovery mechanism. Essentially, it reduces

---

[1] Under some other definitions, then the predicate is computed on the whole state of the neighbor [3,21,33] or just variables on a specific port of an edge at a node and the neighbor at the other endpoint of that edge [6–8].

the distributed network into a non-distributed one by having each node v maintain copies of the local states of every other node in the system (but the copy of the state of a remote node is from an earlier time in history since it takes time until v can hear about it). Hence, each node can simulate nodes in the whole system up to some time ago. The simulations in neighboring nodes are compared each step so inconsistencies (caused by a fault) are detected one time unit after the fault. The detection part of [1] can be viewed as a reactive counterpart of the non - reactive "universal" PLS [37] since it too sends all the information available and can detect an illegal configuration for any predicate. However, in our context, the scheme of [1] can be viewed more as a proof of possibility rather than an efficient algorithm, because of the large communication and memory required for replicating every piece of information everywhere continuously. We still follow the idea that the algorithm configuration, as well as the set of node labels, resulted from the recent history of the execution. This is useful, for example, when coming to define the labels of the "initial configuration" (e.g., in the case some outside corrective measure resets the system to a legal configuration plus legal labels). One approach for generating such an initial labeling is to simulate some fault free history invented for that.[2]

Other Related Notions. Another notion that sounds similar to an RPLS – *local observer* was proposed in [11,12]. A local observer algorithm enables the detection of stabilization (versus the detection of illegality addressed here) through one observer at a unique node, as opposed to requiring one observer at every node in [40]. An observer, similar to an RPLS, does not interfere with the observed algorithm. However, the observers of [11,12,40] relied on the extra assumption that the variables and messages of the observing algorithm itself (at least in one node [11]) cannot be corrupted (possibly analogous to relying on extra assumptions when consulting an oracle in complexity theory). Such an assumption is not used here. Note that all these observers were *not* "local" in the sense used here, of detection time and distance (see Sect. 1.1).

2   Preliminaries

We represent a distributed system by a connected graph $G = (V, E)$ that belongs to a graph family F. The topology of G does not change. Let $n = |V|$. For any vertex $v \in V$, $N(v)$ is the set of neighbors of v and $deg(v)$ is the number of neighbors of v. Either the nodes may be identical or each node v may have a unique identifier $id(v)$ (that cannot be changed by the environment, nor by an algorithm). For the sake of simple presentation only, we use the "KT1 model" [4] (knowing neighbor's id) for trees and general graphs. We use the sense-of-direction model [28] for anonymous rings.

---

[2] This may sound not unlike the claim that the world was created only a few thousand years ago, but created with a history built in, e.g., looks as if there were dinosaurs in time much older than a few thousand years ago.

The communication model is synchronous and time is divided into rounds. In each round, a node goes through three stages in the following order: sending messages to neighbors, receiving messages sent by neighbors in the same round, and local computation. In the computation stage, each node v follows some distributed algorithm Alg; let v's *algorithm state* include all v's variables and constants used by Alg, as well as the received messages. (We use just "state" when no confusion arises).

Let s be a mapping from the nodes to the set of algorithm states. The collection of nodes' states is called the configuration. Denote by $V_s(r)$ (or just V when s is clear from the context) the set of nodes together with their local states for round r. Specifically, unless stated otherwise, when any value at round r is mentioned, the reference is to the end of the computation stage of round r. For example, for each variable var in a local state, we may use the notation var_r to denote the value of var at (the end of) round r. A *legality predicate* f is defined on $G_s(r) = (V_s(r), E)$. The value of f may change when the configuration is changed by the environment (see "fault" below). Alg may also change the configuration in the computation stage.

Often, it is convenient to define the legality predicate only for some subset of the variables of Alg (see [5]), called the *interface variables* (for example, any token passing algorithm can be revised slightly to also maintain a variable that says whether the node has the token or not). This way, predicate f (and the RPLS defined below) can be made independent of the specific algorithm used (e.g., the specific token passing algorithm used) as long as it follows the specification for the problem it solves. Henceforward, when speaking of the (algorithm) state of a node v, we may be speaking only of the interface variables at v. Similarly, the configuration may include just the interface variable of every node v.

Definition 1. *A **fault** occurs when the environment (rather than Alg) changes the value of at least one variable at one or more nodes. This can be either a variable in the algorithm state or a variable of the RPLS.*

We assume that the environment may choose one round r (only) such that every fault occurs just before round r starts (and after round $r - 1$ ends, if such exists). This constraint on the faults guarantees that any message a node v receives from neighbor u is indeed the one that u sends to v at the beginning of the same round. The constraint can be relaxed using known techniques (see the discussion of atomicity in [23] and elsewhere, e.g., [3]), but is used here for the ease of the exposition.

The execution of a reactive task is, in general, infinite. However, the definitions are simplified by choosing some arbitrary round r_0 and calling it the initial round, and calling its configuration the initial configuration.

Definition 2. *The execution is **correct** until r (excluding r) if $\forall r_0 \leq r_k < r$, no faults occur before the beginning of r_k and $f_{r_k} = 1$.*

Definition 3. *The configuration of Alg becomes **incorrect** in r if the execution is correct until r (excluding r) but some faults occur just before r and change the configuration such that $f_r = 0$.*

2.1   Definition of Reactive Proof Labeling Schemes

Definition 4. *An RPLS for a predicate f on some given set of interface variables is a pair $\pi = (\mathcal{M}, \mathcal{V})$ of two algorithms. The marker $\mathcal{M}$ algorithm consists of a **node marker** $\mathcal{M}(v)$ in each node v that maintains a variable $L(v)$ called a **label** of v which is stored in v but is not considered to be a part of the (algorithm) state of v and is not manipulated by Alg. The marker is distributed[3] but its specification includes the specification of its initial values at r_0 as a function of the values of the interface variables at round r_0.*

Similarly, the verifier *algorithm $\mathcal{V}$ consists of a **node** verifier $\mathcal{V}(v)$ in each node v. After Alg completes its computation for round r, the marker $\mathcal{M}(v)$ may change the value of $L(v)$. Then, $L(v)$ is sent to v's neighbors at the beginning of round $r + 1$. In the computation stage of $r + 1$, verifier $\mathcal{V}(v)$ is executed before Alg. The verifier uses v's Alg state, as well as $L(v)$ and $L(u)$ received from each neighbor u, to output either 0 or 1.*

If the interface variables are only a subset of the Alg state then, before the execution of $\mathcal{M}(v)$ in each round r, Alg updates the values of v's interface variables which are accessible for $\mathcal{M}(v)$ in r and for $\mathcal{V}(v)$ at the beginning of $r + 1$.

An RPLS $\pi = (\mathcal{M}, \mathcal{V})$ is correct if the following two properties hold.

i. For every $G \in F$, if the execution is correct until $r + 1$ (excluding $r + 1$) and the labels were labeled according to the specification of $\mathcal{M}$ at round r_0, then $\forall r_0 < r_k \leq r + 1$, $\mathcal{V}(v)$ outputs 1 for all v.

ii. For every $G \in F$, if the configuration becomes incorrect in r, then $\exists v \in G$ such that $\mathcal{V}(v)$ outputs 0 at the beginning of $r + 1$.

The interface between the RPLS and the algorithm is shown in Algorithm 1. Intuitively, for each round, the RPLS implies a PLS (the specification of the initial labels plus the verifier's action). The addition here is that $\mathcal{M}$ updates the PLS each round to accommodate the actions of *Alg*. If *f* does not hold even in r_0, then definition 4 implies that this is detected already in the next round (if no further faults occurred) and further updates of the labels do not interest us here. Hence, let us now treat the case that *f* holds for the initial state.

Definition 5. *A **legal initial labeling** for a given RPLS is the assumed set of labels of $\mathcal{M}(v)$ in a given initial configuration of Alg at round r_0, defined if f holds for that configuration.*

Definition 6. *The **size** of an RPLS $\pi = (\mathcal{M}, \mathcal{V})$ is the maximum number of bits in the label that $\mathcal{M}(v)$ assigns over all the nodes v in G and all $G \in F$ and all the rounds. For a family F and a legality predicate f, we say that the **proof size** of F and f is the smallest size of any RPLS for f over F.*

---

[3] A non-distributed marker may also be useful in some settings.

Note that the size of a label does not depend on Alg (as long as f does not depend on Alg, e.g., when f is defined over the interface variables only).

Algorithm 1: Implementation of an RPLS and Alg at v at round r.

1: Send $L_{r-1}(v)$ to neighbors
2: Receive $L_{r-1}(u)$ from each neighbor u
3: Call $\mathcal{V}(v)$
4: Call Alg to update the values of the interface variables
5: Call $\mathcal{M}(v)$

---

3   RPLS for Token Passing Algorithms

Definition 7. *Given a graph $G = (V, E)$ that belongs to a graph family F, a **token predicate** f_{token} is a predicate evaluated at each node on its state and the messages it has received which determines whether it* holds *a token.*

Definition 8. *A **Token Passing Algorithm (TPA)** is any distributed algorithm for which a token predicate is defined.*
The legality predicate f for token passing evaluates to 1 if there is exactly one token-holder on G and evaluates to 0 otherwise.
We assume that TPA may cause f_{token} at some node v to cease to hold at some round r and to start to hold at some neighbor u of v. We then say that the token is passed from v to u in r.

We assume that the token can be passed exactly once from the token-holder v to exactly one of v's neighbors in each round until $r + 1$ (excluding $r + 1$) if the execution is correct until $r + 1$ (excluding $r + 1$). In each round r, at line 4 of Algorithm 1, TPA sets the interface variable $s_r(v) = 1$ if v holds the token in r and $s_r(v) = 0$ otherwise. We assume $\mathcal{M}(v)$ knows whether the token is passed from some $u \in N(v)$ to v or from v to u in r when such a movement happens in r (implemented easily using the notion of interface variables). For the 3 RPLSs given in the next sections of this paper we show (in the longer version of this extended abstract) the following:

Theorem 1. *Given (1) the RPLS is executed with Alg and with a legal initial labeling at time r_0, and (2) the execution is correct until $r + 1$ (excluding $r + 1$), then all the verifiers output 1 in $r + 1$.*

Theorem 2. *If the configuration becomes incorrect in r, then at least one verifier outputs 0 in $r + 1$.*

4   RPLS for Token Passing on Trees

Let us start with a simple example of a token passing over a tree. The new RPLS is a "link reversal" algorithm. Such algorithms maintain and change virtual directions on the edges, thus creating a directed graph rooted at objects such as

a token, see e.g., [15,16,32,46]. In the RPLS, the violation of f can be detected
by some node not having an edge directed out but still not having the token,
or a node having an edge directed out but having a token, or by a node having
two such edges. When the token is passed from a node v to its neighbor u, the
marker algorithm updates the direction of edge (u, v) so that the tree is still
rooted at the token. The nice property is that this update can be performed in
the same round as the one in which a token is passed and only at the two nodes
that are anyhow involved in the token pass.

The RPLS maintains the following variables for each node v:

•$weight(v)_u = 0$, 1 or 2, $\forall u \in N(v)$—Informally, each edge (u, v) is directed
from the endpoint with the "higher"(modulo 3) weight to the endpoint with the
"lower" (modulo 3) weight. See Definition 9 and Definition 10 below.

•$L(v) = (id(v), (id(u), weight(v)_u)|\forall u \in N(v))$.

Definition 9. *For 0, 1 and 2, define:* $0 \prec 1, 1 \prec 2, 2 \prec 0$.

Definition 10. *We say that edge (v, u) is an **incoming** edge for v and an
outgoing edge for u in round r, denoted by $v \xleftarrow{r} u$ if the following holds: for
the pair of neighbors u and v, there is one pair $(id(x), weight_r(v)_x)$ in $L_r(v)$
and one pair $(id(y), weight_r(u)_y)$ in $L_r(u)$ such that: $id(u) = id(x) \wedge id(v) =
id(y) \wedge weight_r(v)_x \prec weight_r(u)_y$.*

Specification of the legal initial labeling: denote the token holder of r_0 by
v. The nodes are labeled such that v has only incoming edges in r_0 while every
node $u \neq v$ has exactly 1 outgoing edge and every other edge of u is incoming
to u. Clearly, the verifiers output 1 in $r_0 + 1$ (See C1 and C2 of Algorithm 2).

Algorithm 2: $\mathcal{V}(v)$ in round r

1: **IF** any of the following conditions is violated, output 0:
2:     C1: if $s_{r-1}(v) = 1$, then $\forall u \in N(v), v \xleftarrow{r-1} u$
3:     C2: if $s_{r-1}(v) = 0$, then $\exists u \in N(v) : u \xleftarrow{r-1} v \wedge \forall w \in N(v) \wedge w \neq u : v \xleftarrow{r-1} w$
4: **ELSE**, output 1

Algorithm 3: $\mathcal{M}(v)$ in round r

1: **IF** token is passed from $u \in N(v)$ to v in r:
2:     $weight(v)_u \Leftarrow (weight_{r-1}(u)_v - 1) \bmod 3$

---

5   RPLS for Anonymous Rings

Consider a ring $R = (V, E)$ of $n > 2$ identical nodes (without unique identifiers).
We assume that n is known to every node. Each node has two edges, each
connecting to one neighbor. Each node can distinguish between its two neighbors
– a *successor* (the clockwise neighbor) and a *predecessor* (the counterclockwise
neighbor). The label of each node v is an integer in $[0, n^2 - 1]$.

Definition 11. *For integers x, y in $[0, n^2 - 1]$, let $x \prec y$ if $y = (x + 1) \bmod n^2$.*

Specification of the Legal Initial Labeling: For any $k = 0,1,...,\ n-1$, let node $v_{k+1 \bmod n}$ be the successor of v_k, such that $s(v_{n-1}) = 1$. (The numbering is not known to the nodes). The legal initial labeling is: For every $0 \le k \le n-1$, let $L(v_k) = k$. This gives us $L(v_{n-1}) = (L(v_0) + n - 1) \bmod n^2$ and $L(v_0) \prec L(v_1) \prec ... \prec L(v_{n-1})$. Clearly, $f_{r_0} = 1$ and all the verifiers output 1 in $r_0 + 1$ if no fault occurs before $r_0 + 1$. See C1 and C2 of Algorithm 4.

Algorithm 4: $\mathcal{V}(v)$ in round r

1: /*Denote the successor of v by w*/
2: **IF** any of the following conditions is violated, output 0:
3:    C1: if $s_{r-1}(v) = 1$, then $L_{r-1}(v) = (L_{r-1}(w) + n - 1) \bmod n^2$
4:    C2: if $s_{r-1}(v) = 0$, then $L_{r-1}(v) \prec L_{r-1}(w)$
5: **ELSE** output 1

Algorithm 5: $\mathcal{M}(v)$ in round r

1: /*Denote the predecessor of v by u*/
2: **IF** token is passed from $u \in N(v)$ to v in r: /*token passed clockwise*/
3:    $L(v) \Leftarrow (L_{r-1}(u) + 1) \bmod n^2$
4: **IF** token is passed from v to $u \in N(v)$ in r: /*token passed counterclockwise*/
5:    $L(v) \Leftarrow (L_{r-1}(u) - (n - 1)) \bmod n^2$

5.1   Necessity of Assumptions for Any RPLS for Anonymous Rings

We prove that without the assumption that each node knows the size of the ring, no RPLS exists for token passing on anonymous rings.

Lemma 1. *If there exists an RPLS for f on anonymous rings, then there exists a PLS for f on anonymous rings.*

Lemma 2. *If there exists a (static) PLS for f on anonymous rings, then there exists a (static) PLS which verifies whether the size of an anonymous ring is n.*

Lemma 3. (Lemma 3.1 in [37]). *There is no (static) on anonymous of an anonymous ring is n.*

The proof of these lemmas, as well as the other omitted proofs, can be found in the longer version of this extended abstract. Combining Lemma 1, Lemma 2, and Lemma 3 completes the proof that there does not exist an RPLS for token passing on anonymous rings without some additional assumption such that the size of the anonymous ring is known to every node.

6   RPLS for General Graphs G

In this section, we assume that each node has a unique identifier. We also assume that each node knows the number n of nodes in the graph.

Definition 12. *A round r is a **checkpoint** if $r \bmod n = 0$.*

Informally, the first idea behind the RPLS for general graphs was to try to use the PLS of [37] for a *static* rooted spanning tree. That is, as long as the token resides in one node v and does not move, let v be the root and verify that v holds a token. Every other node $u \neq v$ then needs to verify it does not hold a token, and (using the PLS) that a root does exist. Unfortunately, consider the case that the token is passed from v to some other node w over an edge that does *not* belong to the tree. While it may be easy to update the tree so that w is the new root, it may take diameter (of the tree) time for a distributed marker to update the PLS of [37]. Hence, the next idea is to have a second part of the RPLS to verify the movements of the token at any round after r_0. Whenever a node u passes the token to some other node w, both u and w record this move, as well as the round number when this happened. Thus, had we allowed them to remember all the history, they could simulate their actions in the execution. As shown later by induction (in the long version of this paper), if the records of all the nodes match those of their neighbors, and match the assumption that at time r_0, the token holder was v, then there exists indeed a single token.

Unfortunately, the above method would have used unbounded history, as well as an unbounded round number. The main new idea is how to truncate the history from time to time. Specifically, at each round r that is not a checkpoint, the static tree is one that verifies the place of the token at the *one before last* checkpoint r_1. Any part of the history before r_1 is forgotten and the count of the time starts by setting $r_1 = 0$. The static tree is replaced every checkpoint, to prove the location of the token in the *previous* checkpoint (n rounds earlier) since it takes a diameter time to construct such a PLS.

$L(v)$ is a collection of the local variables at v maintained by the RPLS:

- *static_root(v)*—id of the root node of the static tree.
- *static_parent(v)*—$NULL$ or the id of v's parent on the static tree.
- *static_dist(v)*—distance from the root of the static tree.
- *cand_root(v)*—id of the root of the candidate tree.
- *cand_dist(v)*—distance from the root of the candidate tree.
- *cand_parent(v)*—$NULL$ or the id of v's parent on the candidate tree.
- *dynamic_parent(v)*—$NULL$ or the id of v's parent on the dynamic tree.
- *token_in(v)*—set of pairs of $(r, id(u))$ which logs the round r at which the token was sent from v's neighbor u to v.
- *token_out(v)*—set of pairs of $(r, id(u))$ which logs the round r at which the token was sent from v to v's neighbor u.

Definition 13. *The **static tree** of round r is the collection of edges pointed by the static_parent$_r$(v) pointers at each node v. The **dynamic tree** of round r is the collection of edges pointed by the dynamic_parent$_r$(v) pointers at each node v. The **candidate tree** of checkpoint r is the dynamic tree of checkpoint r.*

Specification of the Legal Initial Labeling: Let r_0 be a checkpoint. Select any spanning tree rooted at the token holder v and set $static\_dist_{r_0}(v) \Leftarrow 0$, $static\_parent_{r_0}(v) \Leftarrow NULL$. Then $\forall w \neq v$ that is k hops away from v, denote w's parent by some neighbor p that is $k-1$ hops away from v. Also, set $static\_dist_{r_0}(w) \Leftarrow k$, $static\_parent_{r_0}(w) \Leftarrow id(p)$. In addition, for every $u \in G$, $static\_root_{r_0}(u) \Leftarrow id(v)$, $cand\_root_{r_0}(u) \Leftarrow id(v)$, $cand\_parent_{r_0}(u) \Leftarrow static\_parent_{r_0}(u)$, $cand\_dist_{r_0}(u) \Leftarrow static\_dist_{r_0}(u)$, and $dynamic\_parent_{r_0}(u) \Leftarrow static\_parent_{r_0}(u)$. At last, set $token\_in_{r_0}(u) \Leftarrow \emptyset$, $token\_out_{r_0}(u) \Leftarrow \emptyset$.

See the RPLS in Algorithm 6 and Algorithm 7. The entries in $token\_in(v)$ and $token\_out(v)$ are sorted in a linearly increasing order over the rounds of the entries. Denote the kth entries of $token\_in(v)$ and $token\_out(v)$ by en_k^{in} and en_k^{out} respectively. Let $|token\_in(v)| = a$ and $|token\_out(v)| = b$. Given any $(r_1, id(u))$ and $(r_2, id(w))$, define $(r_1, id(u)) \prec (r_2, id(w))$ if $r_1 < r_2$. Note that for simplicity of exposition, the round r looks unbounded in the algorithm. However, a "bounded timestamp" can be easily implemented by encoding the rounds modulo $2n$, since the history before the previous checkpoint is forgotten.

Algorithm 6: $\mathcal{V}(v)$ in round r

1: **IF** $checkST(v) = 0$, output 0 /* see Algorithm 8*/
2: **ELSE IF** any of the following conditions is violated, output 0:
3:     /*H0-H3 check consistency of the history logs*/
4:     /*Denote checkpoint $r - 1 - ((r - 1) \bmod n) - n$ by r_{c_1} */
5:     H0: $\forall (r_k, id(u)) \in token\_out_{r-1}(v)$ or $token\_in_{r-1}(v)$, $r_{c_1} < r_k < r \wedge u \in N(v)$
6:     H1: $\forall (r_1, id(u)), (r_2, id(w)) \in token\_in_{r-1}(v) \vee token\_out_{r-1}(v), r_1 \neq r_2$
7:     H2.1: if $static\_root_{r-1}(v) = id(v) \wedge s_{r-1}(v) = 0$, then $b = a + 1 > 0$ and $en_1^{out} \prec en_1^{in} \prec en_2^{out} \prec en_2^{in} \prec ... \prec en_a^{out} \prec en_a^{in} \prec en_{a+1}^{out}$
8:     H2.2: if $static\_root_{r-1}(v) \neq id(v) \wedge s_{r-1}(v) = 1$, then $a = b + 1 > 0$ and $en_1^{in} \prec en_1^{out} \prec en_2^{in} \prec en_2^{out} \prec ... \prec en_b^{in} \prec en_b^{out} \prec en_{b+1}^{in}$;
9:     H2.3: if $static\_root_{r-1}(v) = id(v) \wedge s_{r-1}(v) = 1$, then $a = b$ and $en_1^{out} \prec en_1^{in} \prec en_2^{out} \prec en_2^{in} \prec ... \prec en_b^{out} \prec en_b^{in}$
10:    H2.4: if $static\_root_{r-1}(v) \neq id(v) \wedge s_{r-1}(v) = 0$, then $a = b$ and $en_1^{in} \prec en_1^{out} \prec en_2^{in} \prec en_2^{out} \prec ... \prec en_a^{in} \prec en_a^{out}$
11:    H3.1: $\forall u \in N(v)$, if $\exists (r, id(u)) \in token\_out_{r-1}(v)$, then $\exists (r, id(v)) \in token\_in_{r-1}(u)$
12:    H3.2: $\forall u \in N(v)$, if $\exists (r, id(u)) \in token\_in_{r-1}(v)$, then $\exists (r, id(v)) \in token\_out_{r-1}(u)$
13: **ELSE** output 1

Algorithm 7: $\mathcal{M}(v)$ in round r

1: **IF** $cand\_parent(v) \neq NULL \wedge cand\_dist_{r-1}(cand\_parent(v)) \neq NULL$:
2:   $cand\_root(v) \Leftarrow cand\_root_{r-1}(cand\_parent(v))$
3:   $cand\_dist(v) \Leftarrow cand\_dist_{r-1}(cand\_parent(v)) + 1$
4: **IF** token is passed from $u \in N(v)$ to v in r:
5:   $dynamic\_parent(v) \Leftarrow NULL$, add $(r, id(u))$ in $token\_in(v)$
6: **ELSE IF** token is passed from v to $u \in N(v)$ in r:
7:   $dynamic\_parent(v) \Leftarrow id(u)$, add $(r, id(u))$ in $token\_out(v)$
8: **IF** $r \mod n = 0$:
9:   remove all the entries with $r_k \leq r - n$ in $token\_out(v)$ and $token\_in(v)$.
10:   $static\_root(v) \Leftarrow cand\_root(v)$
11:   $static\_parent(v) \Leftarrow cand\_parent(v)$
12:   $static\_dist(v) \Leftarrow cand\_dist(v)$
13:   $cand\_parent(v) \Leftarrow dynamic\_parent(v)$
14:   **IF** $s(v) = 1$:
15:     $cand\_root(v) \Leftarrow id(v)$, $cand\_dist(v) \Leftarrow 0$
16:   **ELSE** $cand\_root(v) \Leftarrow NULL$, $cand\_dist(v) \Leftarrow NULL$

Algorithm 8: Procedure $checkST(v)$ in round r

1: **IF** any of the following conditions is violated, return 0:
2:   /*S1-S3 check the correctness of the static spanning tree*/
3:   S1: $\forall u \in N(v)$: $static\_root_{r-1}(v) = static\_root_{r-1}(u)$
4:   S2: if $static\_root_{r-1}(v) = id(v)$, then
        $static\_parent_{r-1}(v) = NULL \wedge static\_dist_{r-1}(v) = 0$
5:   S3: if $static\_root_{r-1}(v) \neq id(v)$, then $\exists u \in N(v) : static\_parent_{r-1}(v) = id(u) \wedge$
        $static\_dist_{r-1}(v) = static\_dist_{r-1}(u) + 1$
6: **ELSE** return 1

---

7 Concluding Remarks

In the introduction, we mentioned quite a few implications of PLSs, the study of which for the notion of RPLS may yield interesting results. We now consider an additional implication of the new notion. Under the PLS generalization studied here, the predicate was changed together with (legitimate) changes in the configuration. Still, the examples addressed only the case that the legality predicate is defined for a single configuration at a time. One could define predicates involving more than a single configuration (intuitively this makes it easier to check also *liveness* properties while checking only one configuration at a time is better suited for checking *safety*). For example, given a universal scheme that saves the whole reachable history and user inputs at the nodes (constructing a universal RPLS based on the approach of [1]), one can verify that a token moved only clockwise in the last t rounds for some t (unless the history in all the nodes is fake, note that the history in each node is eventually updated to be correct).

Another generalization addresses the "locality" of the marker, and especially that of the initial configuration. Informally, Linial [41] asked "from which distance must the information arrive to compute a given function". One could ask similar questions for the checking, rather than for computing. In particular, using the approach of [1], the labels of some node v are influenced even by an event that happened at some large distance t from a node v. It may take at least t time after the event for the label at v to be impacted. Let a t-semi-universal RPLS be one where a node maintains all the history of all the nodes (as in [1]) but only the history of the last t (rather than n) rounds. It would be interesting to characterize the hierarchy of (reactive/interactive) distributed tasks. The *distributed task class-t* consists of all distributed tasks with legality predicates that can be checked by a t-semi-universal RPLS but not by a $(t-1)$-semi-universal RPLS.

References

1. Afek, Y., Dolev, S.: Local stabilizer. J. Parallel Distrib. Comput. **62**(5), 745–765 (2002)
2. Afek, Y., Kutten, S., Yung, M.: Memory-efficient self stabilizing protocols for general networks. In: van Leeuwen, J., Santoro, N. (eds.) WDAG 1990. LNCS, vol. 486, pp. 15–28. Springer, Heidelberg (1991). https://doi.org/10.1007/3-540-54099-7_2
3. Afek, Y., Kutten, S., Yung, M.: The local detection paradigm and its applications to self-stabilization. Theoret. Comput. Sci. **186**(1–2), 199–229 (1997)
4. Awerbuch, B., Goldreich, O., Vainish, R., Peleg, D.: A trade-off between information and communication in broadcast protocols. J. ACM (JACM) **37**(2), 238–256 (1990)
5. Awerbuch, B., Kutten, S., Mansour, Y., Patt-Shamir, B., Varghese, G.: Time optimal self-stabilizing synchronization. In: Proceedings of the Twenty-Fifth Annual ACM Symposium on Theory of Computing, pp. 652–661 (1993)
6. Awerbuch, B., Patt-Shamir, B., Varghese, G.: Self-stabilization by local checking and correction. FOCS. **91**, 268–277 (1991)
7. Awerbuch, B., Patt-Shamir, B., Varghese, G., Dolev, S.: Self-stabilization by local checking and global reset. In: Tel, G., Vitányi, P. (eds.) WDAG 1994. LNCS, vol. 857, pp. 326–339. Springer, Heidelberg (1994). https://doi.org/10.1007/BFb0020443
8. Awerbuch, B., Varghese, G.: Distributed program checking: a paradigm for building self-stabilizing distributed protocols. FOCS **91**, 258–267 (1991)
9. Balliu, A., Brandt, S., Olivetti, D., Suomela, J.: How much does randomness help with locally checkable problems? In: Proceedings of the 39th Symposium on Principles of Distributed Computing, pp. 299–308 (2020)
10. Beauquier, J., Delaët, S., Dolev, S., Tixeuil, S.: Transient fault detectors. In: Kutten, S. (ed.) DISC 1998. LNCS, vol. 1499, pp. 62–74. Springer, Heidelberg (1998). https://doi.org/10.1007/BFb0056474
11. Beauquier, J., Pilard, L., Rozoy, B.: Observing locally self-stabilization. J. High Speed Netw. **14**(1), 3–19 (2005)
12. Beauquier, J., Pilard, L., Rozoy, B.: Observing locally self-stabilization in a probabilistic way. In: Fraigniaud, P. (ed.) DISC 2005. LNCS, vol. 3724, pp. 399–413. Springer, Heidelberg (2005). https://doi.org/10.1007/11561927_29
13. Burns, J.E., Pachl, J.K.: Uniform self-stabilizing rings. ACM Trans. Programm. Lang. Syst. (TOPLAS) **11**(2), 330–344 (1989)
14. Chandy, K.M., Lamport, L.: Distributed snapshots: determining global states of distributed systems. ACM Trans. Comput. Syst. **3**(1), 63–75 (1985)

15. Défago, X., Emek, Y., Kutten, S., Masuzawa, T., Tamura, Y.: Communication efficient self-stabilizing leader election. arXiv preprint arXiv:2008.04252 (2020)
16. Demmer, M.J., Herlihy, M.P.: The arrow distributed directory protocol. In: Kutten, S. (ed.) DISC 1998. LNCS, vol. 1499, pp. 119–133. Springer, Heidelberg (1998). https://doi.org/10.1007/BFb0056478
17. Dijkstra, E.W.: Self-stabilization in spite of distributed control. In: Selected Writings on Computing: A Personal Perspective, pp. 41–46. Springer (1982). https://doi.org/10.1007/978-1-4612-5695-3_7
18. Dolev, S., Israeli, A., Moran, S.: Self stabilization of dynamic systems. In: Proceedings of the MCC Workshop on Self-Stabilizing Systems, Microelectronics and Computer Technology Corporation. Technical report Number STP-379-89, Austin (1989)
19. Dolev, S., Israeli, A., Moran, S.: Self stabilization of dynamic systems assuming only read write atomicity. Distrib. Comput. **7**, 3–16 (1993)
20. Dolev, S.: Self-Stabilization. MIT press, Cambridge (2000)
21. Dolev, S., Gouda, M.G., Schneider, M.: Memory requirements for silent stabilization. Acta Informatica **36**(6), 447–462 (1999)
22. Dolev, S., Herman, T.: Superstabilizing protocols for dynamic distributed systems. In: Proceedings of the Fourteenth ACM PODC, p. 255 (1995)
23. Dolev, S., Israeli, A., Moran, S.: Self-stabilization of dynamic systems assuming only read/write atomicity. Distrib. Comput. **7**(1), 3–16 (1993)
24. Dolev, S., Tzachar, N.: Randomization adaptive self-stabilization. Acta informatica **47**(5–6), 313–323 (2010)
25. Emek, Y., Fraigniaud, P., Korman, A., Rosén, A.: Online computation with advice. Theoret. Comput. Sci. **412**(24), 2642–2656 (2011)
26. Even, G., et al.: Three notes on distributed property testing. In: 31st International Symposium on Distributed Computing (DISC 2017). Schloss Dagstuhl-Leibniz-Zentrum fuer Informatik (2017)
27. Feuilloley, L., Fraigniaud, P., Hirvonen, J.: A hierarchy of local decision. arXiv preprint arXiv:1602.08925 (2016)
28. Flocchini, P., Mans, B., Santoro, N.: Sense of direction in distributed computing. Theoret. Comput. Sci. **291**(1), 29–53 (2003)
29. Foerster, K.T., Richter, O., Seidel, J., Wattenhofer, R.: Local checkability in dynamic networks. In: Proceedings of the 18th International Conference on Distributed Computing and Networking, pp. 1–10 (2017)
30. Fraigniaud, P., Korman, A., Peleg, D.: Towards a complexity theory for local distributed computing. J. ACM (JACM) **60**(5), 1–26 (2013)
31. Ghosh, S., Gupta, A., Herman, T., Pemmaraju, S.V.: Fault-containing self-stabilizing algorithms. Proc. ACM PODC **1996**, 45–54 (1996)
32. Ginat, D., Sleator, D.D., Tarjan, R.E.: A tight amortized bound for path reversal. Inf. Process. Lett. **31**(1), 3–5 (1989)
33. Göös, M., Suomela, J.: Locally checkable proofs in distributed computing. Theory Comput. **12**(1), 1–33 (2016)
34. Katz, S., Perry, K.J.: Self-stabilizing extensions for meassage-passing systems. Distrib. Comput. **7**(1), 17–26 (1993)
35. Kol, G., Oshman, R., Saxena, R.R.: Interactive distributed proofs. In: Proceedings of the 2018 ACM PODC, pp. 255–264 (2018)
36. Kor, L., Korman, A., Peleg, D.: Tight bounds for distributed MST verification (2011)
37. Korman, A., Kutten, S., Peleg, D.: Proof labeling schemes. Distrib. Comput. **22**(4), 215–233 (2010)

38. Kutten, S., Patt-Shamir, B.: Time-adaptive self stabilization. In: Proceedings of the Sixteenth ACM PODC, pp. 149–158 (1997)

39. Lamport, L.: The mutual exclusion problem: Part II-statement and solutions. J. ACM **33**(2), 327–348 (1986). https://doi-org.ezlibrary.technion.ac.il/10.1145/5383.5385

40. Lin, C., Simon, J.: Observing self-stabilization. In: Proceedings of the Eleventh ACM PODC, pp. 113–123 (1992)

41. Linial, N.: Locality in distributed graph algorithms. SIAM J. Comput. **21**(1), 193–201 (1992)

42. Naor, M., Parter, M., Yogev, E.: The power of distributed verifiers in interactive proofs. In: Proceedings of the Fourteenth Annual ACM-SIAM Symposium on Discrete Algorithms, pp. 1096–1115. SIAM (2020)

43. Naor, M., Stockmeyer, L.: What can be computed locally? SIAM J. Comput. **24**(6), 1259–1277 (1995)

44. Onus, M., Richa, A., Scheideler, C.: Linearization: locally self-stabilizing sorting in graphs. In: 2007 Proceedings of the Ninth Workshop on Algorithm Engineering and Experiments (ALENEX), pp. 99–108. SIAM (2007)

45. Sarma, A.D., et al.: Distributed verification and hardness of distributed approximation. SIAM J. Comput. **41**(5), 1235–1265 (2012)

46. Welch, J.L., Walter, J.E.: Link reversal algorithms. Synth. Lect. Distrib. Comput. Theory **2**(3), 1–103 (2011)

k-Immediate Snapshot and x-Set Agreement: How Are They Related?

Carole Delporte[1], Hugues Fauconnier[1], Sergio Rajsbaum[2],
and Michel Raynal[3,4(✉)]

[1] IRIF, Université Paris Diderot, Paris, France
[2] Instituto de Matemáticas, UNAM, México D.F 04510, Mexico
[3] Univ Rennes IRISA, CNRS, INRIA, Rennes, France
raynal@irisa.fr
[4] Polytechnic University Hong Kong, Kowloon, Hong Kong

Abstract. An immediate snapshot object is a high level communication object, built on top of a read/write distributed system in which all except one processes may crash. This object provides the processes with a single operation, denoted write_snapshot(), which allows the invoking process to write a value and obtain a set of pairs ⟨process id, value⟩ satisfying some set containment properties, that represent a snapshot of the values written to the object, occurring immediately after the write step.

Considering an n-process model in which up to t processes may crash, this paper introduces first the k-resilient immediate snapshot object, which is a natural generalization of the basic immediate snapshot (which corresponds to the case $k = t = n - 1$). In addition to the set containment properties of the basic immediate snapshot, a k-resilient immediate snapshot object requires that each set returned to a process contains at least $(n - k)$ pairs.

The paper first shows that, for $k, t < n-1$, k-resilient immediate snapshot is impossible in asynchronous read/write systems. Then it investigates a model of computation where the processes communicate with each other by accessing k-immediate snapshot objects, and shows that this model is stronger than the t-crash model. Considering the space of x-set agreement problems (which are impossible to solve in systems such that $x \leq t$), the paper shows then that x-set agreement can be solved in read/write systems enriched with k-immediate snapshot objects for $x = \mathsf{max}(1, t+k-(n-2))$. It also shows that, in these systems, k-resilient immediate snapshot and consensus are equivalent when $1 \leq t < n/2$ and $t \leq k \leq (n-1) - t$. Hence, the paper establishes strong relations linking fundamental distributed computing objects (one related to communication, the other to agreement), which are impossible to solve in pure read/write systems.

Keywords: Asynchronous system · Atomic read/write register · Computability · Distributed algorithm · Immediate snapshot · Impossibility · k-set agreement · Linearizability · Lower/upper bounds · Process crash · Snapshot object · t-resilience

© Springer Nature Switzerland AG 2020
S. Devismes and N. Mittal (Eds.): SSS 2020, LNCS 12514, pp. 97–112, 2020.
https://doi.org/10.1007/978-3-030-64348-5_8

1   Introduction

Context. This article considers the t-crash model consisting of n asynchronous processes, among which any subset of at most t processes may crash, and communicate through a shared memory composed of single writer/multi reader (SWMR) atomic registers. The $(n-1)$-crash model is also called *wait-free* model [12]. We keep the term t-resilience for algorithms. This article focuses on algorithms for distributed tasks in which every non-failed process has to produce an output value (*wait-freedom* progress condition[1]).

A task is defined in terms of (a) possible inputs to the processes, and (b) valid outputs for each assignment of input values (tasks are precisely defined in [6,15,17]). Of special importance is the family of *x-set agreement* tasks [8], one for each integer value of x, $1 \leq x \leq n$. Set agreement was introduced to show a hierarchy of tasks whose solvability depends on t, the number of processes that may crash. In the x-set agreement task, processes decide at most x different values, out of their input assignments. When $x = 1$, x-set agreement is the celebrated *consensus* (CONS) task. Consensus is impossible even in the presence of a single process crash [19], and $(n-1)$-set agreement is wait-free impossible, namely, in the presence of $n-1$ process crashes [3,17,23], a result proved using algebraic topology. More generally, x-set agreement is solvable if and only if $t < x$, as implied by the simulation in [6]. There are characterizations of the solvability of any given task, in the t-crash model, and in others (for an overview of results see [13]).

Immediate Snapshot Object. The *immediate snapshot* (IS) object was first used in [4,23], and then further investigated as an "object" in [3]. This object is at the heart of the *iterated immediate snapshot* (IIS) model [5,16], which consists of n asynchronous wait-free processes, communicating through IS objects. In an *iterated* model [21], the processes execute a sequence of asynchronous rounds, and each round is provided with exactly one object, which allows the processes to communicate only during this round. In the IIS model, for any $r > 0$, a process accesses the rth immediate snapshot object only when it executes the r-th round, and accesses it only once.

From an abstract point of view, an IS object *IS*, can be seen as an initially empty set, which can then contain up to n pairs (one per process), each made up of a process index and a value. This object provides each process with a single operation denoted write_snapshot(), that it can invoke once. The invocation *IS*.write_snapshot(v) by a process p_i adds the pair $\langle i, v \rangle$ to *IS* and returns a set of pairs belonging to *IS* such that the sets returned to the processes that invoke write_snapshot() satisfy specific inclusion properties. It is important to notice that, in the IIS model, the processes access the sequence of IS objects one after the other, in the same order, and asynchronously. The power of the IIS model with respect to task solvability is the same as the one of the classical read/write

---

[1] Weaker progress conditions, such as obstruction-freedom [14] and non-blocking [18] have been proposed for $(n-1)$-resilient algorithms.

model, its interest lies in the fact that it provides a higher abstraction layer than the read/write model; a survey including simulations between iterated and classical models can be found in [15].

Contribution of the Paper. This work continues and generalizes the work started in [9] where a preliminary result was presented. Roughly speaking, while [9] considered the case $k = t$, the present article addresses the more general case $k \leq t$.

As previously said, the IS object was designed for the wait-free model (i.e., $t = n-1$). This paper considers it in the context of the t-crash n-process system models where $t < n - 1$. To this end it generalizes the IS object by introducing the notion of a *k-immediate snapshot* (*k-IS*) object. Such an object provides the processes with a single operation denoted write_snapshot$_k$() which, in addition to the properties of an IS object, returns a set including at least $(n - k)$ pairs. Hence, for $k < n-1$, due to the implicit synchronization implied by the constraint on the minimal size of the sets it returns, a k-IS object allows processes to obtain more information from the whole set of processes than a simple IS object (which may return sets containing less than $(n - k)$ pairs).

The obvious question is then the implementability of a k-IS object in the t-crash n-process asynchronous read/write model. The paper shows first that, differently from the basic IS object which can be implemented in the wait-free model, no k-IS object where $k < n-1$, can be implemented in a 1-crash n-process read/write system.

This impossibility result is far from being the first impossibility result in the presence of asynchrony and process crashes, e.g. see the monograph [2]. We already mentioned the impossibility of Consensus (CONS) in the presence of even a single process crash and the impossibility of x-set agreement (x-SA) when $x \leq t$. These agreement objects are at the heart of the theory of fault-tolerant distributed computing. Hence, a second natural question: Are there relations linking the previous "impossible" objects, namely k-IS and x-SA, and if the answer is "yes", under which conditions? The paper provides the following answers to this question[2].

- Let $1 \leq k \leq t < n$. It is possible to implement a k-IS object in a t-crash n-process read/write system enriched with consensus objects.
- Let $1 \leq t < n/2$ and $t \leq k \leq (n - 1) - t$. k-IS and Consensus are equivalent in a t-crash n-process read/write system. (A and B are equivalent if A can be implemented in the t-crash n-process read/write system enriched with B, and reciprocally.)

---

[2] As already indicated, this work was initiated in [9]. Considering k-IS in a system in which up to k processes may crash, this preliminary result showed that, somehow surprisingly, while there is a deterministic $(n - 1)$-resilient algorithm implementing an $(n - 1)$-IS object in an $(n - 1)$-crash read/write system, there is no t-resilient algorithm that implements a t-IS object when $1 \leq t < n - 1$.

Table 1. From k-IS to x-SA for $n = 11$ and $x = \mathsf{max}(1, t + k - (n - 2))$

$k \rightarrow$	1	2	3	..	..	..	$n-4$	$n-3$	$n-2$	$n-1$
$t \downarrow$	1	2	3	4	5	6	7	8	9	10
1	1-SA	1-SA	1-SA	1-SA	1-SA	1-SA	1-SA	1-SA	1-SA	2-SA
2		1-SA	1-SA	1-SA	1-SA	1-SA	1-SA	1-SA	2-SA	3-SA
3			1-SA	1-SA	1-SA	1-SA	1-SA	2-SA	3-SA	4-SA
4				1-SA	1-SA	1-SA	2-SA	3-SA	4-SA	5-SA
$5 < n/2$					1-SA	2-SA	3-SA	4-SA	5-SA	6-SA
$6 \geq n/2$						3-SA	4-SA	5-SA	6-SA	7-SA
$7 = n-4$							5-SA	6-SA	7-SA	8-SA
$8 = n-3$								7-SA	8-SA	9-SA
$9 = n-2$									9-SA	10-SA
$10 = n-1$										11-SA

– Let $(n-1)/2 \leq k \leq n-1$ and $(n-1) - k \leq t \leq k$. It is possible to implement an x-SA object, where $x = t + k - (n - 2)$, in a t-crash n-process read/write system enriched with k-IS objects.

An illustration of the results is presented in Table 1, which considers a system of $n = 11$ processes. As an example, the entry $\langle 4, n-4 \rangle$ states that, in the presence of up to $t = 4$ crashes, $(n - 4)$-IS allows to solve 2-SA.

Roadmap. The paper is made up of 7 sections. Section 2 presents the basic t-crash n-process asynchronous read/write model, and the definitions of the IS, x-SA, and k-IS objects. Section 3 proves the impossibility for the k-IS object in the previous basic model. The other sections are on the power of k-IS with respect to x-SA. Section 4 shows that x-SA can be built in the t-crash n-process asynchronous read/write model enriched with k-IS objects, for $x = \mathsf{max}(1, t + k - (n - 2))$. Section 5 shows that t-IS and CONS are equivalent in the t-crash n-process asynchronous read/write model when $1 \leq t < n/2$. Section 6 shows that CONS is stronger than k-IS when $n/2 \leq t \leq k < n - 1$. Finally, Sect. 7 concludes the paper.

2   The Model and the Problems

2.1   Basic Read/Write System Model

Processes. The computing model is composed of a set of $n \geq 3$ sequential processes denoted $p_1, ..., p_n$. Each process is asynchronous which means that it proceeds at its own speed, which can be arbitrary and remains always unknown to the other processes.

A process may halt prematurely (crash failure), but executes correctly its local algorithm until it possibly crashes. The model parameter t denotes the maximal number of processes that may crash in a run. A process that crashes in a run is said to be *faulty*. Otherwise, it is *correct* or *non-faulty*. Let us notice that, as a faulty process behaves correctly until it crashes, no process knows if it is correct or faulty. Moreover, due to process asynchrony, no process can know if another process crashed or is very slow.

It is assumed that (a) $t < n$ (at least one process does not crash), and (b) any process, until it possibly crashes, executes correctly the algorithm assigned to it. Moreover, each process is assumed to participate in the algorithm.

Communication Layer. The processes cooperate by reading and writing Single-Writer Multi-Reader (SWMR) atomic read/write registers. This means that the shared memory can be seen as a set of variables $A[1..n]$ where, while $A[i]$ can be read by all processes, it can be written only by p_i.

Notation. The previous model is denoted $\mathcal{CARW}_{n,t}[\emptyset]$ (which means "Crash Asynchronous Read/Write with n processes, among which up to t may crash"). A model constrained by a predicate on t (e.g. $t < a$) is denoted $\mathcal{CARW}_{n,t}[t < a]$. $\mathcal{CARW}_{n,t}[t = n-1]$ is a synonym of $\mathcal{CARW}_{n,t}[\emptyset]$, which (as already indicated) is called *wait-free* model. When considering t-crash models, $\mathcal{CARW}_{n,t}[t < a]$ is less constrained than $\mathcal{CARW}_{n,t}[t < a - 1]$. More generally, $\mathcal{CARW}_{n,t}[P, T]$ denotes the system model $\mathcal{CARW}_{n,t}[\emptyset]$ restricted by the predicate P, and enriched with any number of shared objects of the type T (e.g., consensus objects).

Shared objects are denoted with capital letters. The local variables of a process p_i are denoted with lower case letters, sometimes suffixed by the process index i.

2.2 Immediate Snapshot (IS)

The immediate snapshot (IS) object [3] was informally presented in the introduction. Defined in the context of the wait-free model (i.e., $t = n - 1$), it can be seen as a variant of the snapshot object introduced in [1]. While a snapshot object provides the processes with two operations (write() and snapshot()) which can be invoked separately by a process (usually a process invokes write() before snapshot()), a one-shot immediate snapshot object provides the processes with a single operation write_snapshot() (one-shot means that a process may invoke write_snapshot() at most once).

Definition. Let *IS* be an IS object. It is a set, initially empty, that will contain pairs made up of a process index and a value. Let us consider a process p_i that invokes *IS*.write_snapshot(v). This invocation adds the pair $\langle i, v \rangle$ to *IS* (contribution of p_i to *IS*), and returns to p_i a set, called view and denoted $view_i$, such that the sets returned to processes (that return from their invocation of write_snapshot()) collectively satisfy the following properties.

- Termination. The invocation of write_snapshot() by a correct process terminates.

- Self-inclusion. $\forall i : \langle i, v \rangle \in view_i$.
- Validity. $\forall i : (\langle j, v \rangle \in view_i) \Rightarrow p_j$ invoked write_snapshot(v).
- Containment. $\forall i, j : (view_i \subseteq view_j) \vee (view_j \subseteq view_i)$.
- Immediacy. $\forall i, j : (\langle i, v \rangle \in view_j) \Rightarrow (view_i \subseteq view_j)$.[3]

Implementations of an IS object in the wait-free model $\mathcal{CARW}_{n,t}[t = n - 1]$ are described in [3,22]. While both a one-shot snapshot object and an IS object satisfy the Self-inclusion, Validity and Containment properties, only an IS object satisfies the Immediacy property. This additional property creates an important difference, from which follows that, while a snapshot object is atomic (operations on a snapshot object can be linearized [18]), an IS object is not atomic (its operations cannot always be linearized). However, an IS object is set-linearizable (set-linearizability allows several operations to be linearized at the same point of the time line [7,20]).

The Iterated Immediate Snapshot (IIS) Model. This model (introduced in [5]) considers $t = n - 1$. Its shared memory is composed of a (possibly infinite) sequence of IS objects: $IS[1], IS[2], ...,$ which are accessed sequentially and asynchronously by the processes according to the following round-based pattern executed by each process p_i. The variable r_i is local to p_i; it denotes its current round number.

> $r_i \leftarrow 0$; $\ell s_i \leftarrow$ initial local state of p_i (including its input, if any);
> **repeat forever** % asynchronous IS-based rounds
> $r_i \leftarrow r_i + 1$;
> $view_i \leftarrow IS[r_i].$write_snapshot$(\ell s_i)$;
>   computation of a new local state ℓs_i (which contains $view_i$)
> **end repeat**.

As indicated in the Introduction, when considering distributed tasks (as formally defined in [6,15,17]), the IIS model and $\mathcal{CARW}_{n,t}[t = n - 1]$ have the same computability power [5,11,15].

2.3 x-Set Agreement (x-SA)

x-Set agreement was introduced by S. Chaudhuri [8] to investigate the relation linking the number x of different values that can be decided in an agreement problem, and the maximal number of faulty processes t. It generalizes consensus which corresponds to the instance $x = 1$.

An x-set agreement (x-SA) object is a one-shot object that provides the processes with a single operation denoted propose$_x()$. This operation allows the invoking process p_i to propose a value, which is called *proposed* value, and is passed as an input parameter. It returns a value, called *decided* value. The object is defined by the following set of properties.

---

[3] An equivalent formulation of the Immediacy property is: $\forall i, j : ((\langle i, - \rangle \in view_j) \wedge (\langle j, - \rangle \in view_i)) \Rightarrow (view_i = view_j)$.

- Termination. The invocation of $\mathsf{propose}_x()$ by a correct process terminates.
- Validity. A decided value is a proposed value.
- Agreement. No more than x different values are decided.

It is shown in [4,17,23] that $(n-1)$-SA is impossible to implement in $\mathcal{CARW}_{n,t}[t = n-1]$, and in [6] that x-SA is impossible to implement in $\mathcal{CARW}_{n,t}[x \leq t]$.

2.4 k-Immediate Snapshot

Definition of k-Immediate Snapshot. A k-immediate snapshot (k-IS) object is an immediate snapshot object with the following additional property.

- Output size. The set *view* obtained by a process is such that $|view| \geq n - k$.

This means that in addition to the Self-inclusion, Validity, Containment, and Immediacy properties, the set returned by a process contains at least $(n-k)$ pairs. The associated operation is denoted $\mathsf{write\_snapshot}_k()$.

k-Immediate Snapshot vs x-Set Agreement. When considering a k-IS object and a x-SA object, we have the following differences.

- On concurrency. An x-SA object is atomic (linearizable), while a k-IS object is not (it is only set-linearizable [7,20]). In other words, k-IS objects "accept" concurrent accesses (this is captured by the Immediacy property), while x-SA objects do not.
- On the values returned. When considering an x-SA object, each process p_i knows that each other process p_j (which returns from its invocation of $\mathsf{propose}_x()$) obtains a single value, but it does not know which one (uncertainty); p_i knows only that at most k values are decided by all processes (certainty).
  When considering a k-IS object, each process p_i knows that each other process p_j (which returns from its invocation of $\mathsf{write\_snapshot}_k()$) obtains a set of pairs $view_j$ that is included in, is equal to, or includes its own set of pairs (certainty due to the containment property), but it does not know the size of $view_j$ (uncertainty).

A Property Associated with k-IS Objects. The next theorem (stated and proved in [9]) characterizes the power of a k-IS object in term of its Output size and Containment properties.

Theorem 1. *Let us consider a k-IS object, and assume that all correct processes invoke $\mathsf{write\_snapshot}_k()$. If the size of the smallest view obtained by a process is ℓ ($\ell \geq n-k$), there is a set S of processes such that $|S| = \ell$ and each process of S obtains the smallest view or crashes during its invocation of $\mathsf{write\_snapshot}_k()$.*

Proof. It follows from the Output size property of the k-IS object that no view contains less than $\ell \geq n - k$ pairs. Let min_view be the smallest view returned by a process; hence $\ell = |min\_view|$.

Let us consider a process p_i such that $(\langle i, - \rangle \in min\_view)$, which returns a view. Due to (a) the Immediacy property (namely $(\langle i, - \rangle \in min\_view) \Rightarrow (view_i \subseteq min\_view))$ and (b) the minimality of min_view, it follows that $view_i = min\_view$. As this is true for each process whose pair participates in min_view, it follows that there is a set S of processes such that $|S| = \ell \geq n - k$, and each of these processes obtains min_view, or crashes during its invocation of write$\_$snapshot$_k()$. Due to the Containment property, the others processes crash or obtain views which are a superset of min_view. $\square_{Theorem\,1}$

An Impossibility Result. The following theorem first stated and proved in [9] establishes an important property of a k-IS object.

Theorem 2. *A k-IS object cannot be implemented in $CARW_{n,t}[k < t]$.*

Proof. To satisfy the output size property, the view obtained by a process p_i must contain pairs from $(n - k)$ different processes. If t processes crash (e.g., initial crashes), a process can obtain at most $(n-t)$ pairs. If $t > k$, we have $n-t < n - k$. It follows that, after it has obtained pairs from $(n-t)$ processes, a process can remain blocked forever waiting for the $(t - k)$ missing pairs. $\square_{Theorem\,2}$

3 t-Resilience Impossibility of k-Immediate Snapshot

Theorem 3. *It is impossible to implement a k-IS object in the model $CARW_{n,t}[1 \leq t \leq k < n - 1]$.*

Proof. The proof considers the case $1 = t \leq k < n - 1$ (this constraint explains the model assumption $n \geq 3$, Sect. 2.1). If, for $k \leq n - 1$, there is no implementation of a k-IS object in $CARW_{n,t}[t = 1]$, there is no implementation either for $t \geq 1$. The proof is by contradiction, namely, assuming an implementation of a k-IS object, where $k < n - 1$, in $CARW_{n,t}[t = 1]$, we show that it is possible to solve consensus in $CARW_{n,t}[t = 1, k\text{-IS}]$. As consensus cannot be solved in $CARW_{n,t}[t = 1]$, it follows that k-IS cannot be implemented in $CARW_{n,t}[1 \leq t \leq k]$.

Let us recall the main property of k-IS (captured by Theorem 1). Let ℓ be the size of the smallest view (min_view) returned by a process. There is a set S of ℓ processes such that any process of S returns min_view or crashes, and $\ell \geq n - k$. As $k < n - 1$ (theorem assumption), we have $\ell \geq 2$, which means that at least two processes obtain min_view. It follows that, if a process obtains the views returned by the k-IS object to $(n - 1)$ processes, one of these views is necessarily min_view. This constitutes Observation O.

Let us now consider Algorithm 1. In addition to a k-IS object denoted IS, the processes access an array $VIEW[1..n]$ of SWMR atomic registers, initialized to $[\bot, \cdots, \bot]$. The aim of $VIEW[i]$ is to store the view obtained by p_i from the k-IS object IS. When it calls propose$_1(v)$, a process p_i invokes first the k-IS object, in which it deposits the pair $\langle i, v \rangle$ and obtains a view from it (line 1), that it writes in $VIEW[i]$ to make it publicly known (line 2). Then, it waits

---

operation propose$_1(v)$ **is**
(1) $view_i \leftarrow IS.\text{write\_snapshot}_k(v)$;
(2) $VIEW[i] \leftarrow view_i$;
(3)   wait($|\{ j$ such that $VIEW[j] \neq \perp\}| = n - t$);
(4)   **let** $view$ **be** the smallest of the previous $(n - t)$ views;
(5)   return(smallest proposed value in $view$)
end operation.

Algorithm 1: Solving consensus in $\mathcal{CARW}_{n,t}[t = 1, k\text{-IS}]$ (code for p_i)

until it sees the views of at least $(n - 1)$ processes (line 3). Finally, p_i extracts from these views the one with the smallest cardinality (line 4), and returns the smallest value contained in this smallest view (line 5).

We show that this reduction algorithm solves consensus in $\mathcal{CARW}_{n,t}[t = 1, k\text{-IS}]$. As at least $(n - 1)$ processes do not crash, and write in their entry of the array $VIEW[1..n]$, no correct process can block forever at line 2, proving the Termination property of consensus.

As $\ell \geq n - k \geq 2$, it follows from Observation O that at least one of the views obtained by a process at line 3 is necessarily min_view. It follows that each process that executes line 3 obtains min_view and returns its smallest value at line 4), proving the Agreement property of consensus.

The consensus Validity property follows directly from k-IS Validity property, and the observation that any set $view$ contains only proposed values line 4).
$\square_{Theorem\ 3}$

Remark. When considering Algorithm 1, let us observe that, as $n - k \leq n - t$, the array $VIEW[1..n]$ can be replaced by a k-immediate snapshot object $IS2$. We obtain then the following algorithm.

> **operation** propose$_1(v)$ **is**
> $\quad view1_i \leftarrow IS.\text{write\_snapshot}_k(v)$;
> $\quad view2_i \leftarrow IS2.\text{write\_snapshot}_k(view1_i)$;
> $\quad$**let** $view$ **be** the smallest view in $view2_i$;
> $\quad$return(smallest proposed value in $view$)
> **end operation.**

4   From k-Immediate Snapshot to x-Set Agreement

This section proves the content of Table 1, namely x-SA can be implemented in the system model $\mathcal{CARW}_{n,t}[t \leq k < n - 1]$, for $x = \max(1, t + k - (n - 2))$. Interestingly, the algorithm providing such an implementation is Algorithm 1, whose operation name is now propose$_x()$ (instead of propose$_1(v)$).

Theorem 4. *Let* $x = \max(1, k + t - (n - 2))$. *Algorithm 1 implements an x-SA object in the system model* $\mathcal{CARW}_{n,t}[1 \leq t \leq k < n - 1, k\text{-IS}]$.

Proof. The x-SA Termination follows directly from the Termination property of the underlying k-IS object IS, the fact that there are at least $(n - t)$ correct processes, and the assumption that all correct processes invoke propose$_x()$. The x-SA Validity property follows directly from the Validity property of the IS.

As far as the x-SA Agreement property is concerned, we have the following. Due to Theorem 1, a set of $\ell \geq n - k$ processes obtain the smallest possible view min_view, which is such that $|min\_view| = \ell \geq n - k$. It follows that, at most k processes obtain a view different from min_view. In the worst case, these k views are different. Consequently, there are at most $k + 1$ different views, namely min_view, $V(1)$, ..., $V(k)$, and due to their Containment property, we have $min\_view \subset V(1) \subset \cdots \subset V(k)$. The rest of the proof is a case analysis according to the value of $(n - t)$ with respect to k.

- $n - t > k$. In this case, a process obtains views from $(n - t)$ processes (line 3), and in the first case it obtains the views $V(1)$, ..., $V(k)$. But as $n - t > k$ it also obtains min_view from at least one process. It follows that, all processes see min_view, and consequently decide the same value at line 5. Hence, $(n - t > k) \Rightarrow (x = 1)$.
- $n - t = k$. In this case, it is possible that some processes do not obtain min_view at line 3. But, if this occurs, they necessarily obtain the views from the $n - t = k$ processes that deposited $V(1)$, ..., $V(k)$ in $VIEW[1..n]$. Hence, all these processes obtains $V(1)$ at line 3, and decide consequently the same value from $V(1)$. As the decided values are decided from the views min_view and $V(1)$, we have $(n - t = k) \Rightarrow (x = 2)$.
- $n - t = k - 1$. In this case, it is possible that, at line 3, some processes obtain not only min_view, but also $V(1)$ and decide the smallest value of $V(2)$. As the decided values are then decided from the views min_view, $V(1)$, and $V(2)$, we have $(n - t = k - 1) \Rightarrow (x = 3)$.
- Applying the same reasoning to the general case $n - t = k - c$, we obtain $(n - t = k - c) \Rightarrow (x = 2 + c)$.

Abstracting the previous case analysis, we obtain $x = 1$ (consensus) for $n - t > k$, and $x = k + t - (n - 2)$, i.e., when $n - t = k - x + 2$, from which follows that $x = \mathsf{max}(1, k + t - (n - 2))$, which completes the proof of the theorem. $\square_{Theorem\ 4}$

The next corollary is a re-statement of Theorem 4 for $x = 1$.

Corollary 1. *Algorithm 1 implements a* CONS *object in the system model* $\mathcal{CARW}_{n,t}[1 \leq t < n/2, t \leq k \leq (n - 1) - t, k\text{-IS}]$.

5 An Equivalence Between k-Immediate Snapshot and Consensus

This section shows first that consensus is strong enough to implement a k-IS object when $t \leq k$. Combining this result with the fact consensus can be implemented from a k-IS object in $\mathcal{CARW}_{n,t}[1 \leq t < n/2, t \leq k \leq (n - 1) - t]$ (Corollary 1), we obtain that consensus and k-IS are equivalent in $\mathcal{CARW}_{n,t}[1 \leq t < n/2, t \leq k \leq (n - 1) - t]$.

5.1 From CONS to k-IS in $\mathcal{CARW}_{n,t}[t \leq k \leq n-1]$

Algorithm 2 describes a reduction of k-IS to consensus in $\mathcal{CARW}_{n,t}[0 < t \leq k \leq n-1]$. This algorithm uses three shared data structures. The first is an array $REG[1..n]$ of SWMR atomic registers (where $REG[i]$ is associated with p_i), the second is a consensus object denoted CS, and the third is an immediate snapshot object denoted IS (let us recall that such an object can be implemented in $\mathcal{CARW}_{n,t}[t \leq n-1]$).

```
operation write_snapshot_k(v_i) is
(1)    REG[i] ← v_i;
(2)    wait (|j such that REG[j] ≠ ⊥}| ≥ n − k);
(3)    aux_i ← {⟨j, REG[j]⟩ such that REG[j] ≠ ⊥};
(4)    view_i ← CS.propose_1(aux_i);
(5)    if (⟨i, v_i⟩ ∈ view_i)
(6)       then return(view_i)
(7)       else aux_i ← IS.write_snapshot(v_i);
(8)            view_i ← view_i ∪ aux_i;
(9)            return(view_i)
(10) end if
end operation.
```

Algorithm 2: Building k-IS in $\mathcal{CARW}_{n,t}[0 < t \leq k \leq n-1, \text{CONS}]$ (code for p_i)

The behavior of a process p_i can be decomposed in three parts.

- When it invokes write_snapshot$_k(v_i)$, p_i first deposits its value v_i in $REG[i]$, in order all processes to know it, and waits until at least $(n - k)$ processes have deposited their input value in $REG[1..n]$ (lines 1–2).
- Then p_i proposes to the underlying consensus object CS, the set of all the pairs $\langle j, REG[j] \rangle$ such that $REG[j] \neq \bot$ (lines 3–4). Let us notice that this set contains at least $(n - k)$ pairs. Hence, the consensus object returns to p_i a view $view_i$, which contains at least $(n - k)$ pairs.
- Finally, p_i returns a view (of at least $(n - k)$ pairs).
  - If $view_i$ contains its own pair $\langle i, v_i \rangle$, p_i returns $view_i$ (line 6).
  - If $view_i$ does not contain $\langle i, v_i \rangle$, p_i proposes v_i to the underlying immediate snapshot object from which it obtains a set of pairs aux_i (line 7). Let us notice that, due to the properties of the immediate snapshot object IS, aux_i contains the pair $\langle i, v_i \rangle$. Process p_i then adds aux_i to $view_i$ (line 8) and returns it (line 9).

Theorem 5. *Algorithm 2 implements a k-IS object in the system model* $\mathcal{CARW}_{n,t}[0 < t \leq k \leq n-1, \text{CONS}]$.

Proof. Proof of k-IS Self-inclusion. If p_i returns at line 6, self-inclusion follows directly from the predicate of line 5. If this predicate is not satisfied, p_i invokes the underlying immediate snapshot object IS with the value v_i it initially proposed (line 7). It then follows from the self-inclusion property of IS that aux_i contains $\langle i, v_i \rangle$, and due to line 8, the set $view_i$ that is returned at line 9 contains $\langle i, v_i \rangle$.

Proof of k-IS Validity. This property follows from (a) the fact that a process p_i assigns to $REG[i]$ the value it wants to deposit in the k-IS object, (b) this atomic variable is written at most once (line 1), and (c) the predicate $REG[j] \neq \perp$ is used at line 3 to extract values from $REG[1..n]$.

The Output size property follows from (a) the predicate of line 2, which ensures that the set $view_i$ obtained at line 4 from the underlying consensus object contains at least $n - t \geq n - k$ pairs, and the fact that a set $view_i$ cannot decrease (line 3).

Proof of k-IS Containment. Let P6 (resp., P9) be the set of processes that terminate at line 6 (resp., 9). Let $view$ be the set of pairs decided by the underlying consensus object CS (line 4). Hence, all the processes in P6 return $view$. Due to line 8, the set $view_i$ returned by a process that terminates at line 9 includes $view$. It follows that $\forall\, p_j \in$ P6, $p_i \in$ P9, we have $view_j = view \subset view_i$.

Let us now consider two processes p_i and p_j belonging to P9. It then follows from the IS Containment property of the underlying IS object, that we have $aux_i \subseteq aux_j$ or $aux_j \subseteq aux_i$ (where the value of aux_i and aux_j are the ones at line 7). Consequently, at line 8 we have $view_i \subseteq view_j$ or $view_j \subseteq view_i$, which completes the proof of the k-IS Containment property.

Proof of k-IS Immediacy. Let p_i and p_j be two processes that return $view_i$ and $view_j$, respectively, such that $\langle i, v \rangle \in view_j$. We have to show that $view_i \subseteq view_j$. Let us considering the sets P6 and P9 defined above. There are three cases.

- Both p_i and p_j belong to P6. In this case, due to line 4, we have $view_i = view_j$.
- p_i belongs to P6, while p_j belong to P9. In this case, due to line 8, we have $view_i \subset view_j$.
- Both p_i and p_j belong to P9. In this case, due to the IS Immediacy property of IS we have (at line 8) $\langle i, - \rangle \in aux_j \Rightarrow aux_i \subseteq aux_j$ (and $\langle j, - \rangle \in aux_i \Rightarrow aux_j \subseteq aux_i$). Let $view$ the set of pairs returned by the consensus object line 4. As, due to line 9, we have $view_i \leftarrow view \cup aux_i$ and $view_j \leftarrow view \cup aux_j$, the k-IS Immediacy property follows.

Proof of k-IS Termination. Let p be the number of processes that deposit a value in REG. As $t \leq k$, we have $n - k \leq n - t \leq p \leq n$. It follows that no correct process can wait forever at line 2. The fact that no correct process blocks forever at line 4 and line 7 follows from the termination property of the underlying consensus and immediate snapshot objects. $\square_{Theorem\,5}$

5.2   When Consensus and k-IS Are Equivalent

Let us consider the right triangular matrix defined by the entries are marked "x-SA" in Table 1. Theorem 5 states that it is possible to implement k-IS from

CONS for any entry (t, k) belonging to this triangular matrix. Combined with Corollary 1, we obtain the following theorem.

Theorem 6. CONS *objects and and k-IS objects are equivalent in the system model* $\mathcal{CARW}_{n,t}[0 < t < n/2, t \leq k \leq (n-1) - t]$.

6 When Consensus Is Stronger Than k-Immediate Snapshot

Section 4 investigated the power of k-IS to implement x-SA objects, namely x-SA can be implemented in $\mathcal{CARW}_{n,t}[1 \leq t \leq k < n - 1, k\text{-IS}]$ where $x = \max(1, t + k - (n - 2))$, see Theorem 4. As we have seen, considering the other direction, Sect. 5 has shown that k-IS can be implemented in $\mathcal{CARW}_{n,t}[1 \leq t \leq k < n - 1, \text{CONS}]$ (Theorem 5). The combination of these results showed that Consensus and k-IS are equivalent in $\mathcal{CARW}_{n,t}[0 < t = k < n/2]$ (Theorem 6).

This section shows an upper bound on the power of k-IS to implement x-SA objects, namely, k-IS objects are not powerful enough to implement consensus in the system model $\mathcal{CARW}_{n,t}[n/2 \leq t \leq k < n - 1]$.

Preliminary: A Simple Lemma. Let us remark that, as immediate snapshot objects, k-immediate snapshot objects are not linearizable. As a k-IS object IS contains values from at least $(n - k)$ processes, at least $(n - k)$ processes must have invoked the operation $IS.\mathsf{write\_snapshot}_k()$ for any invocation of $\mathsf{write\_snapshot}_k()$ to be able to terminate. It follows that there is a time τ at which $n - k$ processes have invoked $IS.\mathsf{write\_snapshot}_k()$ and have not yet returned. We then say that these $(n - k)$ processes are *"inside IS"*. Hence the following lemma.

Lemma 1. *If an invocation of* $\mathsf{write\_snapshot}_k()$ *on a k-immediate snapshot object IS terminates, there is a time τ at which at least $(n - k)$ processes are inside IS.*

Theorem 7. *There is no algorithm implementing a CONS object in the system model* $\mathcal{CARW}_{n,t}[n/2 \leq t \leq k < n - 1, k\text{-IS}]$.

Due to page limitation, the reader will find the proof of this theorem in [10].

7 Conclusion

The aim and content of the paper. The paper has first introduced the notion of a k-immediate snapshot (k-IS) object, which generalizes the notion of immediate snapshot (IS) objects to t-crash n-process systems (the IS object corresponds to the case $k = t = n - 1$). It has then shown that k-IS objects cannot be implemented in asynchronous read/write systems for $k < n - 1$.

The paper considered then the respective power of k-IS objects and x-set agreement objects (x-SA) in t-crash-prone systems. As both these families of objects are impossible to implement in read/write systems for $t, k < n - 1$ or $x \leq t$, respectively, the paper strove to establish which of k-IS and x-SA objects are the most "impossible to solve". The main results are the following where the zones A, B, C, D, refer to Fig. 1.

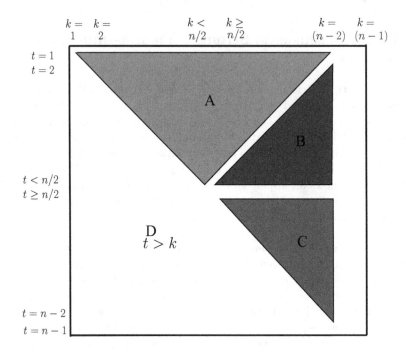

Fig. 1. Summarizing the results

- Even if we have CONS objects, it is not possible to implement k-IS objects in a t-crash system where $t > k$ (Zone D).
- It is possible to implement x-SA objects, where $x = \mathsf{max}(1, t + k - (n - 2))$, from k-IS objects in systems where $1 \leq t \leq k < n - 1$ (Zone A + B + C).
- It is possible to implement k-IS objects from 1-SA objects (CONS) in read/write systems where $1 \leq t \leq k \leq n - 1$ (Zone A + B + C).
- 1-SA objects (CONS) and k-IS objects are equivalent in read/write systems where $1 \leq t < n/2$ and $t \leq k \leq (n - 1) - t$ (Zone A).
- It is not possible to implement 1-SA (consensus) from k-IS objects in read/write systems when $n/2 \leq t \leq k < n - 1$ (Zone C).

Stated in a more operational way, these results exhibit the price of the synchronization hidden in k-IS object (which requires that the view returned to a process contains at least $(n - k)$ pairs, (where a pair is made up of a value plus the id of the process that deposited it in the k-IS object).

More generally, the previous results establish a computability map relating important problems, which are impossible to solve in pure read/write systems.

Open Problems. The following problems remain to be solved to obtain a finer relation linking k-IS and x-SA, when $t > 1$.

- Direction "from k-IS to x-SA". Is it possible to implement x-SA objects, with $1 \leq x < t+k-(n-2)$ in t-crash n-process systems enriched with k-IS objects (Zone B)? We conjecture that the answer to this question is *no*.
- Direction "from x-SA to k-IS". Given an x-SA object, which k-IS objects can be implemented from it? More generally, is there a "k-IS-like" communication object such that x-SA and this "k-IS-like" object are computationally equivalent (by "k-IS-like" we mean an object possibly weaker than a k-IS object)?

Acknowledgments. The authors want to thank the referees for their constructive comments. This work has been partially supported by the French ANR project DESCARTES (16-CE40-0023-03) devoted to layered and modular structures in distributed computing, and the UNAM-PAPIIT projects IN107714, IN106520.

References

1. Afek, Y., Attiya, H., Dolev, D., Gafni, E., Merritt, M., Shavit, N.: Atomic snapshots of shared memory. J. ACM **40**(4), 873–890 (1993)
2. Attiya, H., Ellen, F.: Impossibility Results for Distributed Computing. Synthesis Lectures on Distributed Computing Theory. Morgan & Claypool, San Rafael (2014). 162 p
3. Borowsky, E., Gafni, E.: Immediate atomic snapshots and fast renaming. In: Proceedings of the 12th ACM Symposium on Principles of Distributed Computing (PODC 1993), pp. 41–50. ACM Press (1993)
4. Borowsky, E., Gafni, E.: Generalized FLP impossibility results for t-resilient asynchronous computations. In: 25th ACM Symposium on Theory of Computing, pp. 91–100. ACM Press (1993)
5. Borowsky, E., Gafni, E.: A simple algorithmically reasoned characterization of wait-free computations. In: Proceedings of the 16th ACM Symposium on Principles of Distributed Computing (PODC 1997), pp. 189–198. ACM Press (D1997)
6. Borowsky, E., Gafni, E., Lynch, N., Rajsbaum, S.: The BG distributed simulation algorithm. Distrib. Comput. **14**, 127–146 (2001). https://doi.org/10.1007/PL00008933
7. Castañeda, A., Rajsbaum, S., Raynal, M.: Unifying concurrent objects and distributed tasks: interval-linearizability. J. ACM **65**(6), 42 (2018). Article 45
8. Chaudhuri, S.: More choices allow more faults: set consensus problems in totally asynchronous systems. Inf. Comput. **105**(1), 132–158 (1993)
9. Delporte, C., Fauconnier, H., Rajsbaum, S., Raynal, M.: t-resilient immediate snapshot is impossible. In: Suomela, J. (ed.) SIROCCO 2016. LNCS, vol. 9988, pp. 177–191. Springer, Cham (2016). https://doi.org/10.1007/978-3-319-48314-6_12
10. Delporte, C., Fauconnier, H., Rajsbaum, S., Raynal, M.: t-resilient k-immediate snapshot and its relation with agreement problems. Technical report, ArXiv:2010.00096, 15 p. (2020)
11. Gafni, E., Rajsbaum, S.: Distributed programming with tasks. In: Lu, C., Masuzawa, T., Mosbah, M. (eds.) OPODIS 2010. LNCS, vol. 6490, pp. 205–218. Springer, Heidelberg (2010). https://doi.org/10.1007/978-3-642-17653-1_17

12. Herlihy, M.P.: Wait-free synchronization. ACM Trans. Program. Lang. Syst. **13**(1), 124–149 (1991)
13. Herlihy, M.P., Kozlov, D., Rajsbaum, S.: Distributed Computing Through Combinatorial Topology. Morgan Kaufmann/Elsevier, Amsterdam (2014). 336 p. ISBN 9780124045781
14. Herlihy, M.P., Luchangco, V., Moir, M.: Obstruction-free synchronization: double-ended queues as an example. In: Proceedings of the 23th International IEEE Conference on Distributed Computing Systems (ICDCS 2003), pp. 522–529. IEEE Press (2003)
15. Herlihy, M., Rajsbaum, S., Raynal, M.: Power and limits of distributed computing shared memory models. Theoret. Comput. Sci. **509**, 3–24 (2013)
16. Herlihy, M.P., Shavit, N.: A simple constructive computability theorem for wait-free computation. In: 26th ACM Symposium on Theory of Computing, pp. 243–252. ACM Press (1994)
17. Herlihy, M.P., Shavit, N.: The topological structure of asynchronous computability. J. ACM **46**(6), 858–923 (1999)
18. Herlihy, M.P., Wing, J.M.: Linearizability: a correctness condition for concurrent objects. ACM Trans. Program. Lang. Syst. **12**(3), 463–492 (1990)
19. Loui, M., Abu-Amara, H.: Memory requirements for agreement among unreliable asynchronous processes. Adv. Comput. Res. **4**, 163–183 (1987)
20. Neiger, G.: Set-linearizability. In: Brief Announcement in Proceedings of the 13th ACM Symposium on Principles of Distributed Computing (PODC 1994), p. 396. ACM Press (1994)
21. Rajsbaum, S.: Iterated shared memory models. In: López-Ortiz, A. (ed.) LATIN 2010. LNCS, vol. 6034, pp. 407–416. Springer, Heidelberg (2010). https://doi.org/10.1007/978-3-642-12200-2_36
22. Raynal, M.: Concurrent Programming: Algorithms, Principles and Foundations. Springer, Heidelberg (2013). https://doi.org/10.1007/978-3-642-32027-9. 515 p. ISBN 978-3-642-32026-2
23. Saks, M., Zaharoglou, F.: Wait-free k-set agreement is impossible: the topology of public knowledge. SIAM J. Comput. **29**(5), 1449–1483 (2000)

Brief Announcement: Local Deal-Agreement Based Monotonic Distributed Algorithms for Load Balancing in General Graphs

Yefim Dinitz, Shlomi Dolev, and Manish Kumar[✉]

Ben-Gurion University of the Negev, Be'er Sheva, Israel
{dinitz,dolev}@cs.bgu.ac.il, manishk@post.bgu.ac.il

Abstract. In computer networks, participants may cooperate in processing tasks, balancing working loads among them. The distributed load balancing problem is well-known. We present local algorithms solving it based on a short *deal-agreement communication*. Unlike the previous algorithms, they converge *monotonically*, always providing a better feasible state as the execution progresses. Our synchronous algorithms achieve ϵ-Balanced state for the continuous setting in time $O(nD\log(nK/\epsilon))$ and 1-Balanced state for the discrete setting in time $O(nD\log(nK/D) + nD^2)$, for *general graphs* in the worst case, where n is the number of nodes, K is the initial discrepancy, and D is the graph diameter. We also suggest an *asynchronous* load balancing algorithm solving the problem in time $O(nK^2)$ for general graphs, and its *self-stabilizing* version.

Keywords: Distributed algorithms · Deterministic · Load balancing · Self-stabilization · Monotonic

1   Introduction

The distributed load balancing problem is defined when there is an undirected network (graph) of computers (nodes), each one assigned a non-negative working load, and they like to balance their loads. If nodes u and v are connected by an edge, then any part of the load of u may be transferred over that edge from u to v, and similarly from v to u. The information at the nodes is local, and the only way to get more knowledge on the graph is by communicating with its neighbors. The application and scope include grid computing, clusters, and clouds.

The accepted global measure for the deviation of a current state from being balanced is its *discrepancy*, defined as $K = L_{max} - L_{min}$, where L_{max} (L_{min})

This research was (partially) funded by the Office of the Israel Innovation Authority of the Israel Ministry of Economy under Genesis generic research project, the Rita Altura trust chair in computer science, and by the Lynne and William Frankel Center for Computer Science.

© Springer Nature Switzerland AG 2020
S. Devismes and N. Mittal (Eds.): SSS 2020, LNCS 12514, pp. 113–117, 2020.
https://doi.org/10.1007/978-3-030-64348-5_9

is the currently maximum (minimum) node load in the graph. An alternative, local way to measure the deviation is the maximal difference of loads between neighboring nodes: a state is ϵ-*Balanced* if that difference is at most ϵ. In the discrete problem setting, all loads and thus also all transfer amounts should be integers; in the continuous one, transfer amounts are arbitrary. In this paper, we concentrate on *deterministic* algorithms solving the problem in *a worst case time polynomial in the global input size*, that is in the number n of graph nodes and in the logarithm of the maximal load (though we deviate from polynomiality for the asynchronous model).

The research on the load balancing problem began from the papers of Cybenko [4] and Boillat [3]. Both are based on the concept of *diffusion*: at any synchronized round, every node divides its load *equally* among its neighbors and itself. As a rule, the case of *d-regular* graphs is considered; only laconic remarks on a possibility to generalize the results to the case of general graphs appear in the literature. Markov chains and ergodic theory are used for deriving the rate of convergence. In the discrete setting, diffusion methods require rounding of every transferred amount, which makes the analysis harder; Rabani et al. [9] made a substantial advancement in that direction; their the time bound for reaching the discrepancy of ϵ in the worst case is $O\left(\frac{\ln(Kn^2/\epsilon)}{(1-\lambda)}\right)$, where λ is the second largest eigen-value of the diffusion matrix. The diffusion approach is popular in the literature. The alternative methods are *mathching* (see, e.g., [10]) and *balancing circuits* (see, e.g., [2,9]). For the discrete setting and the considered computational model, all those approaches do not achieve neither a constant final discrepancy, nor a constant-balanced state. Many suggested algorithms cannot be stopped at any time, since intermediate solutions either might include negative node loads, or might be worse than previous ones. Almost all papers on load balancing use the synchronous distributed model. The only theoretically based approach suggested for *asynchronous* distributed setting is turning it to synchronous by appropriately enlarging the time unit, see e.g., [1].

We suggest using the distributed computing approach based on *short agreement between neighboring nodes* in load balancing. We develop *local* distributed algorithms, with no global information collected at the nodes; the advantage is that the actual time of an algorithm run can be quite small, if the problem instance is lucky. We say that a load balancing algorithm is *monotonic* if the maximal load value never increases and the minimal load value never decreases. Such algorithms produce a not worse feasible state at each step of the execution, and thus are *anytime* in the sense of [5,8]. Our main results on load balancing are as follows, where D is the graph diameter, and ϵ is an arbitrary constant.

- In the continuous setting, the first synchronized deterministic algorithm for *general graphs*, which is monotonic and works in time $O(nD\log(nK/\epsilon))$.
- In the discrete setting, the first deterministic algorithms for *general graphs* achieving a 1-Balanced state in time depending on the initial discrepancy logarithmically. It is monotonic and works in time $O(nD\log(nK/D) + nD^2)$.
- The first asynchronous anytime algorithm, and its self-stabilizing version.

The full version of this paper can be found in arXiv [6].

2   Monotonic Distributed Load Balancing Algorithms

Algorithm 1: Synchronous Single-Proposal Algorithm: Continuous

---

Input: An undirected graph $G = (V, E, load)$

1 **Execute forever do**

2 | **for** *every node u* **do**

3 | | **if** *u has at least one neighbor with a strictly smaller load* **then**

4 | | | find the neighbor, v, with the minimal load

5 | | | u sends to v a transfer proposal of $(load(u) - load(v))/2$

6 | **for** *every node u* **do**

7 | | **if** *there is at least one transfer proposal to u* **then**

8 | | | find a neighbor, w, proposing to u the transfer of maximum value

9 | | | node u makes a deal: increases its load by the value proposed by w and informs node w on accepting its proposal

10 | **for** *every node u* **do**

11 | | node u updates its load w.r.t. the deal made on its proposal, if accepted, and sends the current value of $load(u)$ to every its neighbor

---

Let us begin with the synchronous model. Algorithm 1 solves the continuous load balancing problem. It is composed of three-phase rounds, one phase upon the global clock tick, cyclically. At each round, each node sends a transfer proposal to *at most one* of its neighbors. In reply, each node accepts a single proposal among those sent to it, if any. (Each node may finally both send and get load at same round.)

The analysis of Algorithm 1 is based on node potentials. Let L_{avg} be the average value of *load* over V. We define potentials $p(u) = (load(u) - L_{avg})^2$ for any node u, and $p(G) = \sum_{u \in V} p(u)$ for entire G. Any transfer of load l from u to v in our algorithms satisfies $load(u) - load(v) \geq 2l > 0$. For any such transfer, we prove that it decreases $p(G)$ by at least $2l^2$. The central point of our analysis is the following statement.

Lemma 1. *If the discrepancy of G at the beginning of some round is K, the potential of G decreases after that round by at least $K^2/2D$.*

Proof. Consider an arbitrary round. Let x and y be nodes with load L_{max} and L_{min}, respectively, and let P be a *shortest* path from y to x, $P = (y = v_0, v_1, v_2, \ldots, v_k = x)$. Note that $k \leq D$. Consider the sequence of edges (v_{i-1}, v_i) along P, and choose its sub-sequence S consisting of all edges with $\delta_i = load(v_i) - load(v_{i-1}) > 0$. Let $S = (e_1 = (v_{i_1-1}, v_{i_1}), e_2 = (v_{i_2-1}, v_{i_2}), \ldots, e_{k'} = (v_{i_{k'}-1}, v_{i_{k'}}))$, $k' \leq k \leq D$. Observe that by the definition of S, interval $[L_{min}, L_{max}]$ on the load axis is covered by intervals $[load(v_{i_j-1}), load(v_{i_{j}-1})]$, since $load(v_{i_1-1}) = L_{min}$, $load(v_{i_{k'}}) = L_{max}$, and for any $2 \leq j \leq k'$, $load(v_{i_{j-1}}) \geq load(v_{i_j-1})$. As a consequence, the sum of load differences $\sum_{j=1}^{k'} \delta_{i_j}$ over S is at least $L_{max} - L_{min} = K$.

Since for every node v_{i_j}, its neighbor v_{i_j-1} has a strictly lesser load, the condition of the first **if** in Algorithm 1 is satisfied for each v_{i_j}. Thus, each v_{i_j} proposes a transfer to its minimally loaded neighbor; denote that neighbor by w_j. Note that the transfer amount in that proposal is at least $\delta_{i_j}/2$. Hence, *the sum of load proposals issued by the heads of edges in S is at least $K/2$.* By the algorithm, each node w_i accepts the biggest proposal sent to it, which value is at least $\delta_{i_j}/2$. Consider the simple case when all nodes w_j are different. Then, the total decrease of the potential at the round, Δ, is at least $\sum_j 2(\delta_{i_j}/2)^2$. By simple algebra, for a set of at most D numbers with a sum bounded by K, the sum of numbers' squares is minimal if there are exactly D equal numbers summing to K. We obtain $\Delta \geq D \cdot 2(K/2D)^2 = K^2/2D$, as required.

The rest of the proof reduces the general case to the simple case as above.

We prove that Algorithm 1 is monotonic, and that it arrives at the discrepancy of at most ϵ in time $O(nD \log(nK/\epsilon))$.

The algorithm for the discrete setting differs by the rounding of proposal values only. Its analysis up to the arrival at a discrepancy of at most $2D$ is similar; the rest of its execution is analyzed separately. Also that algorithm is monotonic, and it arrives at a 1-Balanced state in time $O(nD \log(nK/D)+nD^2)$.

We believe that the running time bounds of deal-agreement distributed algorithms for load balancing could be improved by future research. This is since the current bounds are based on analyzing only a single path at each iteration.

Multiple-Proposal Load Balancing Algorithm. We suggest also the monotonic synchronous deal-agreement algorithm based on *multiple proposals*. There, each node may propose load transfers to several of its neighbors with smaller load, aiming to *equalize the loads* in its neighborhood as much as possible. We formalize this as follows. Consider node p and the part $\mathcal{V}_{less}(p)$ of its neighbors with loads smaller than $load(p)$. Node p proposes load transfers to some of the nodes in $\mathcal{V}_{less}(p)$ in such a way that if all its proposals would be accepted, then the resulting minimal load in the node set $\mathcal{V}_{less}(p) \cup \{p\}$ will be maximal. (Compare with the scenario, where we pour water into a basin with unequal heights at its bottom: the flat water surface will cover the deepest pits.) Performing deals in parallel with several neighbors has a potential to yield faster convergence in practice, as compared with the single-proposal algorithm.

Asynchronous Load Balancing Algorithm. The asynchronous version of the load balancing algorithm is based on repeated enquiries of the load of the neighbors and whenever proposing a deal to a neighbor with a lower load, wait for the acknowledgment of the proposal acceptance or rejection prior to reexamination. In more detail, our asynchronous load balancing algorithm is based on distributed proposals. There, each node may propose load transfers to several of its neighbors by computing $\mathcal{PV}_{less}(p)$, which is part of $\mathcal{V}_{less}(p)$. $\mathcal{PV}_{less}(p)$ is the resulting minimal loaded node set whose load is less than *TentativeLoad* after all proposal gets accepted. While sending the proposal, each node sends the value of *LoadToTransfer* (load which can be transferred to neighboring node) and *TentativeLoad* (load of the node after giving loads to its neighbors) with all set of nodes in $\mathcal{PV}_{less}(p)$. After receiving a proposal, the node sends an acknowledgment to the sender node; the sender node waits for an acknowledgment from all nodes of $\mathcal{PV}_{less}(p)$. The asynchronous algorithm ensures that the local computation between two nodes is assumed to be before the second communication starts. Consider an example where a node q of $\mathcal{PV}_{less}(p)$ receives a proposal and the deal happens between node p and node q. In this case, *TentativeLoad(p)* is always greater than the load of node q (when q responds to the deal) because node p is waiting for acknowledgments from all nodes of $\mathcal{PV}_{less}(p)$.

Self-stabilizing Load Balancing Algorithm. The self-stabilizing load balancing algorithm is based on the asynchronous version, where a self-stabilizing data link algorithm is used to verify that eventually (after the stabilization of the data-link) whenever a neighbor sends and acknowledge accepting a deal, the invariant of load transfer, from a node with load higher than the load of the acknowledging node, holds. This solution can be extended to act as a superstabilizing algorithm [7], gracefully, dealing with dynamic settings, where nodes can join/leave the graph anytime, as well as handle received/dropped loads.

References

1. Aiello, W., Awerbuch, B., Maggs, B.M., Rao, S.: Approximate load balancing on dynamic and asynchronous networks. In: 25th Annual ACM Symposium on Theory of Computing (STOC), pp. 632–641 (1993)
2. Aspnes, J., Herlihy, M., Shavit, N.: Counting networks and multi-processor coordination. In: 23rd Annual ACM Symposium on Theory of Computing (STOC), pp. 348–358 (1991)
3. Boillat, J.E.: Load balancing and Poisson equation in a graph. Concurrency Pract. Experience **2**(4), 289–314 (1990)
4. Cybenko, G.: Dynamic load balancing for distributed memory multiprocessors. J. Parallel Distrib. Comput. **7**(2), 279–301 (1989)
5. Dean, T.L., Boddy, M.S.: An analysis of time-dependent planning. In: 7th National Conference on Artificial Intelligence, pp. 49–54 (1988)
6. Dinitz, Y., Dolev, S., Kumar, M.: Local deal-agreement based monotonic distributed algorithms for load balancing in general graphs. CoRR abs/2010.02486 (2020)
7. Dolev, S., Herman, T.: Superstabilizing protocols for dynamic distributed systems. Chic. J. Theor. Comput. Sci. **1997**, 3.1–3.15 (1997)
8. Horvitz, E.: Reasoning about beliefs and actions under computational resource constraints. Int. J. Approx. Reason. **2**(3), 337–338 (1988)
9. Rabani, Y., Sinclair, A., Wanka, R.: Local divergence of Markov chains and the analysis of iterative load balancing schemes. In: 39th Annual Symposium on Foundations of Computer Science, (FOCS), pp. 694–705 (1998)
10. Sauerwald, T., Sun, H.: Tight bounds for randomized load balancing on arbitrary network topologies. In: 53rd Annual IEEE Symposium on Foundations of Computer Science (FOCS), pp. 341–350 (2012)

Silent MST Approximation for Tiny Memory

Lélia Blin[1,2], Swan Dubois[3], and Laurent Feuilloley[4(✉)]

[1] Sorbonne Université, CNRS, LIP6, 75005 Paris, France
[2] Université Evry-Val d'Essone, 91000 Evry, France
[3] Sorbonne Université, CNRS, Inria, LIP6, 75005 Paris, France
[4] DII, Universidad de Chile, Santiago, Chile
feuilloley@dii.uchile.cl

Abstract. In this paper we show that approximation can help reduce the space used for self-stabilization. In the classic *state model*, where the nodes of a network communicate by reading the states of their neighbors, an important measure of efficiency is the space: the number of bits used at each node to encode the state. In this model, a classic requirement is that the algorithm has to be *silent*, that is, after stabilization the states should not change anymore. We design a silent self-stabilizing algorithm for the problem of minimum spanning tree, that has a trade-off between the quality of the solution and the space needed to compute it.

1 Introduction

1.1 Our Questions

Context. Self-stabilization is a technique to ensure fault-tolerance in distributed systems. It aims at designing systems that can recover from arbitrary faults. Silent self-stabilization consists in asking for the additional property that, once a correct configuration has been reached, the processors basically stop computing.

In the context of self-stabilization, the most studied measure of performance is the time to stabilize to a correct configuration. Another essential parameter is the space used by each processor. This parameter not only captures some notion of memory (and is actually also called *the memory*), but more remarkably, it captures the performance in terms of communication, as self-stabilizing algorithms communicate by reading the states of their neighbors (when they are described in the so-called *state model*, which is the most common model).

For silent self-stabilizing algorithms, this memory usage is tightly related to the space needed to locally certify that a configuration is correct. Such certifications, also called proofs, have been studied independently under the name of *proof-labeling schemes*. On the one hand, it is known that the space needed for the proof is a lower bound on the space required for silent stabilization. Indeed, after stabilization, a silent algorithm is only checking that the configuration is

---

Support by ANR ESTATE.

S. Devismes and N. Mittal (Eds.): SSS 2020, LNCS 12514, pp. 118–132, 2020.
https://doi.org/10.1007/978-3-030-64348-5_10

correct via reading its neighbors' states, which is exactly what a distributed proof is made for. On the other hand, it is proved in [7] that one can always design an algorithm matching this lower bound (up to an additive logarithmic factor), even in the most asynchronous setting. Thus in some sense, one can always achieve optimal space.

There are two issues to this situation. First the general technique of [7] is inherently exponential in time: it basically consists in looking for the distributed proof via an exhaustive search. Second, the space required for silent stabilization can be simply be too large for applications.

Approximation-Memory Trade-Off. The core of our paper is to give a solution for the second problem, which can be rephrased as: what can be done when we do not even have the space needed for a distributed proof? One technique is to consider non-silent algorithms, that keep changing their states. For example, [8] achieves $O(\log \log n)$ space for leader election on a ring, when the lower bound for silent stabilization is $\Omega(\log n)$ (where n is the number of nodes in the network). In this paper, we make the choice of keeping the silence property, but to be less demanding on the quality of the solution. More precisely we are aiming at a trade-off between the memory used and the quality of the solution produced, that is, we want to design approximation algorithms for optimization problems, such that the larger the memory allowed, the better the approximation ratio. To our knowledge this is the first time approximation is used to reduce memory usage for self-stabilization (although it has recently been proved fruitful in the more restricted context of proof-labeling schemes [10,16]).

Optimal Space in Polynomial Time. Now a second question, which follows from the exponential-time algorithm of [7], is: when we can afford the optimal space to compute an exact solution, can we get it in polynomial time? The answer is no in general. Consider for example the task of 3-coloring a 3-colorable graph. The distributed proof uses only constant space, because the colors are enough for local checkability. On the other hand, it is known that no algorithm can compute a 3-coloring in constant space. Indeed, in order to perform even a minimal symmetry breaking (such as having two nodes with two different outputs), an algorithm needs strictly more than constant space [2] (actually $\Omega(\log \log n)$ bits are necessary [5]). On the positive side, [6] shows that for various tree construction problems, one can match the optimal space bound and have polynomial-time stabilization. In particular, one can get down to $\Theta(\log^2 n)$ bits for minimum spanning tree, which is optimal when the edge weights are in a polynomial range. As we will see, we can improve on this, as a side result of our approximation algorithm.

1.2 Our Results

In this paper, we focus on the central problem of minimum spanning tree (MST). Our main result is an approximation-memory trade-off for this problem. The theorem below, and all the results of this paper hold under the classic assumption that the edge weights are in $[1, n]$, where n is the size of the network.[1]

Theorem 1. *There exists a silent self-stabilizing approximation algorithm for minimum spanning tree, that stabilizes in polynomial time and has a trade-off between memory and approximation. This trade-off goes from space $O(\log^2 n)$ for a minimum spanning tree to space $O(\log n)$ for a simple spanning tree.*

The precise trade-off has a complicated expression, thus we do not write explicitly here. It is given in Lemma 1. The two extreme values, $O(\log^2 n)$ for an MST and $O(\log n)$ for a simple spanning tree are optimal (see [22,23] for the lower bounds). We get a smooth trade-off between these extremes, with for example $O(\log n \log \log n)$ space for a 2-approximation.

One of the two ingredients to achieve this result is an exact algorithm for MST, which is self-stabilizing, silent and polynomial-time, and uses $O(\log n \cdot s)$ space,[2] where s is the number of bits used to encode an edge weight.[3]

Theorem 2. *There exists a silent self-stabilizing algorithm for (exact) minimum spanning tree, with $O(\log n \cdot s)$ memory, that stabilizes in polynomial time.*

It is known that an MST requires $\Omega(\log n \cdot s)$ space [22]. Therefore our algorithm improves on the state-of-the-art by proving that, even with a parametrization by s, MST is part of the set of problems that can be solved in optimal space *and* polynomial-time.

1.3   Our Techniques

Our algorithm has two main ingredients: the exact algorithm that we have already mentioned and a technique to transform the weights. The weight transformation changes the original weights into approximated weights that can be encoded in smaller space. Then we basically feed these approximated weights to the exact algorithm and get as a result an approximate solution. The better the approximation of the weight, the better the approximation of the final solution, but the larger the space used.

The weight transformation (Lemma 1) takes as input a weight in $[1, poly(n)]$, encoded on $\Theta(\log n)$ bits, and outputs a weight in a smaller range, hence using less bits of memory. The simplest form of the technique is the following: replace each weight by the position of its most significant bit. This way when we write weights in the memory, we use exponentially less space: s will be in $O(\log \log n)$ instead of $\Theta(\log n)$. Of course by this operation we loose some precision. Namely, we only have the information to recover a 2-approximation of each weight. Now

---

[1] Assuming the maximum to be n and not poly(n) allows to have cleaner proofs without additional constants, but the asymptotic results are also correct for polynomial weights.

[2] Here and everywhere in the paper, $\log n \cdot s$ should be read as $(\log n) \times s$.

[3] Note that as we assume the weights are polynomial in n, s is in $O(\log n)$.

if we feed these "new weights" to an exact algorithm for minimum spanning tree, then we will get a 2-approximation, and using much less space. We design an extension of this technique, that allows to get the whole trade-off between space and approximation. This extension is more complicated, but is still based on manipulations of the binary representation of the weights.

The rest of the paper consists in designing the exact self-stabilizing algorithm for minimum spanning tree in space $O(\log n \cdot s)$. This exact algorithm does not follow the usual design of silent self-stabilizing algorithms. A silent algorithm typically stores some key pieces of information, *while* it is building the output, *e.g.* while selecting the edges of the MST. These pieces of information form a certificate of correctness that allows, during and after stabilization, to check whether the construction is correct. And if the construction is correct then the output is correct. This is for example the way the $O(\log^2 n)$-space algorithm of [6] is designed, book-keeping the important information of a Boruvka-inspired algorithm. Unfortunately, it seems that this approach is difficult if not impossible to use when one wants to go below $\log^2 n$ space for minimum spanning tree. Instead, we use a two-phase approach: we first build a minimum spanning tree, and, once it is finished, we certify it. This paper is, as far as we know, the first occurrence of such a modular approach. The certification we use is the proof-labeling scheme of [22]. As a side result we answer a question of [22] where designing a self-stabilizing algorithm using this certification was left as an open problem.

1.4   Outline

We start in Sect. 2 with a description of the model. In Sect. 3, we describe the general structure of our algorithm. Then in Sect. 4, 5, and 7, we describe the different components of our algorithm. Section 6 is a high-level description of the certification of [22] that we use in Sect. 7.

2   Model

We consider that a network is represented by an undirected connected graph $G = (V, E)$ where V is a set of processors (or nodes), and E is a set of edges that represent communication channels. We denote the number of nodes by n. Every node v is given a unique identifier ID_v, and every edge has a weight. Both identifiers and weights are polynomial, that is, they are integers in $[1, poly(n)]$. The identifiers are all distinct, but the weights are not required to be distinct.

We want to compute an approximation of a minimum spanning tree. A k-approximation of an MST is a spanning tree whose weight is not larger than k times the weight of an MST. Note that in our definition of approximation, we do not relax the requirement of acyclicity, thus it is not the same kind of approximation as the one used in the literature about spanners.

State Model. Our algorithm is designed for the classic *state model* [13]. In this model, every node has a state, and communication between neighbors is modeled by direct reading of states instead of exchanges of messages. This state is the *mutable memory*, that is, it is the part of the memory that can be modified, and also the one that is counted when we consider space complexity. There is also the *non-mutable memory*, that contains for each node, its identifier and the weights of the adjacent edges, as well as the code of the algorithm. The tuple of all the states of the network is called the *configuration*, and the execution of an algorithm is therefore described by a sequence of configurations.

In the state model, an algorithm is usually described as a set of rules. A rule basically states that if the local view of a node satisfies some property, then the node can change its state to a specified new value. Here by "local view" we mean the state of a node, the states of its neighbors, the identifier of the node, and the weights of the adjacent edges. If there exists a rule, such that the property of the node's view is satisfied, then we say that the node is *enabled*. (Note that for a node deciding whether it is enabled or not could take some time and space. As in most models in network distributed computing, such local computation is not taken into account for the time and space complexity.) The asynchrony of the system is modeled by a scheduler who chooses, at each step, a non-empty subset of enabled nodes, that are allowed to apply a rule. We consider the harshest scheduler, the *distributed unfair scheduler*, which has no further constraint for the choices it makes. Other schedulers considered in the literature can have, for example, fairness constraints: they cannot activate always the same nodes. See [15] for a survey of the schedulers of the literature. The choice of the distributed unfair scheduler makes our algorithm the most robust possible.

To compute time complexities, we use the definition of *round* of [14]. This definition captures the execution rate of the slowest process in any execution. The first round of an execution ϵ, noted ϵ', is the minimal prefix of ϵ such that every node that was enabled in the initial configuration, either has taken a step or is not enabled anymore. Let ϵ'' be the suffix of ϵ such that $\epsilon = \epsilon'\epsilon''$. The second round of ϵ is the first round of ϵ'', and so on.

Distributed Proof, Proof-Labeling Schemes and Silent Stabilization. Given a property, for example 'a set of pointers defines a spanning tree', a *distributed proof* (or *distributed certificate*) is a labeling of the nodes that certifies that the property is satisfied. It is usually presented as a *proof-labeling scheme* [23]. In such a scheme, the first element is an oracle, called the *prover*, which provides each node with a label. The second element of a scheme is a *verification algorithm*. This algorithm, run at a node v, takes as input the view of v, including the labels of v and of its neighbors, and decides whether to *accept* or to *reject*. A scheme is correct for a property P, if (1) for any configuration satisfying P, there is a way for the prover to make all nodes accept, and (2) for any configuration *not* satisfying P, there is *no* way for the prover to make all nodes accept. The performance of a scheme is measured by the size of the labels in number of bits (all labels have the same size). The notion of distributed proof is tightly related to the concept of *silent self-stabilizing algorithm.* An algorithm is *self-stabilizing* if starting from an arbitrary configuration, it reaches a correct configuration after some finite time, called the *stabilization time,* and stays in correct configurations afterwards. Such an algorithm is *silent,* if the algorithm reaches a correct configuration and then stays silent: it does not change the states anymore, or in other words, no node is enabled. To be sure to be in a correct configuration, a silent self-stabilizing algorithm has to keep some certification of the correctness. This certification ensures that, if the configuration is not correct, then at least one node will detect it, be enabled, and start the recovery (for example launching a reset). This is basically the same as the notion of distributed proof above [7], except that the proof is not given by an oracle: it is built by the algorithm. We refer to [17,20] for surveys on distributed proofs.

3   General Description

Our algorithm is made of several components. Basically, we have several algorithms that will operate one after the other, in order to reach a configuration with a minimum spanning tree certified by a distributed proof. The algorithms are designed to work if they start from a clean situation (for example our first algorithm expects its variables to be empty) and we have a reset mechanism that will go back to such a situation if one of our algorithms detects a problem.

Two-Phase Approach. In order to reach a configuration where the nodes can safely stop updating their states, we need first to have a correct solution at hand, and second to allow the nodes to be sure that indeed the solution is correct. As said earlier, the classic way to do this is to keep in memory some key extra pieces of information gathered during the computation. As it seems that this approach is difficult to implement here, we use another way: we first build the solution alone, and then we build a certification (*i.e.* a distributed proof) of this solution.

The strategy of aiming for a distributed proof simplifies the design of the algorithm. Indeed, if something goes wrong in the computation because of the initial configuration, then we have to face two rather simple situations. In the first case, one of our algorithms detects the error, for example because there is an obvious inconsistency between the neighbors' states. In the second case, the problem is subtle enough to not be detected during the run of our algorithms, but then if the output is not correct, the distributed proof that is built cannot be correct either. In both cases the error is detected, and a reset is launched. In other words, either there is something obviously wrong that is caught on the fly, or there is something that is more subtle, and it is caught at the end.

The difficulties that remain are the same as for any self-stabilizing algorithm. First, one has to ensure that for any computation, starting from a clean configuration, we cannot end up in a configuration that is detected as incorrect. If this were to happen, then the scheduler could make the algorithm go through a reset infinitely often, and it would never stabilize. Second, we have to make sure that the algorithm does not get stuck in a position where no node can be activated.

The Components and How They Work Together. We have three main components: one algorithm that builds the minimum spanning tree (detailed in Sect. 5),
one algorithm that takes this tree and certifies it, thus allows the silent stabilization (detailed in Sect. 7), and a reset algorithm that can erase everything and
go back to a clean configuration. We describe the algorithm in a modular way
to ease the reading, but in the end our result is one algorithm. In particular, it
is important to have the three pieces working together. Note that such modular
design for a self-stabilizing algorithms is not new, even for unfair schedulers: for
example in the recent and celebrated coloring algorithm of [1] there is a first
algorithm reducing the number of colors very fast and then a second algorithm
to finish the job by eliminating the last extra-colors.

In our algorithm, the reset procedure is dominant, in the sense that if a
reset is launched it will basically overrule the other procedures. For this reset we
take a solution from the literature: Devismes and Johnen [12] recently proposed
a cooperative (that is, tolerating multiple simultaneous initiators) silent self-
stabilizing reset algorithm that satisfies our constraints in terms of scheduler,
stabilization time and space complexity. Then to articulate the two other pieces,
we simply use flags that indicate for each node which algorithm it is running.
If the algorithm are not run one after the other, then there will be a local
inconsistency between these flags, and the reset will be launched.

One last element about how the pieces work together is related to the weight
transformation. As said earlier, a key ingredient to get our approximation algorithm in small space (Theorem 1) is a transformation of the weights. This transformation consists in replacing each weight by a smaller weight. But, as we have
limited space, we cannot store the transformed weights of all the edges adjacent
to a node in its state (there can be $n - 1$ edges adjacent to a node). Instead,
every time a step is taken, the node recomputes the new weights to know which
rule applies.

4   Approximation-Memory Trade-Off

In this section, we show how we can replace the original weights with approximated weights to decrease the number of bits used to encode them, while preserving guarantees on the MST computed. The precise trade-off between the
space used for the new weights and the approximation is given in Lemma 1. As
the expression is not very elegant, this is the only place of the paper where we
write it explicitly. The trade-off for the whole algorithm can be derived from the
values of this lemma: simply multiply by $O(\log n)$.

Lemma 1. *There exists a transformation of the weights that allows for a trade-
off between space needed to encode the new weights, and the quality of the tree
one can compute from them. More precisely, for every integer k in the range
$[-\log \log n, \log n]$, we can get new weights with the following properties:*

- *for $k = \log n - 1$, approximation 1 and size $\log n + 1$,*
- *$k \in [0, \log n - 2]$, approximation $1 + \frac{1}{2^k}$ and space $k + \log(\log n - k + 1)) + 1$,*

– *for $k \in [-\log\log n, 0]$ approximation $2^{2^{-k}}$ and space $\log\left(\frac{\log n}{2^{-k}} + 1\right) + 1$.*

Before proving the lemma, let us restate and prove the three points of the trade-offs that we have already mentioned.

Corollary 1. *The construction of Lemma 1 gives in particular:*

– *Exact solution, with weights encoded on $O(\log n)$ bits.*
– *2-approximation, with weights encoded on $O(\log\log n)$ bits.*
– *A trivial guarantee (poly(n)-approximation), with weights encoded on a constant number of bits.*

The exact solution directly follows from the case $k = \log n - 1$. The 2-approximation corresponds to $k = 0$. The arbitrary approximation corresponds to $k = -\log\log n$.

The full proof of Lemma 1 is deferred to the full version [4], but we give a sketch of the argument now. The explanation is illustrated by Fig. 1.

The core idea of the weight transformation is to group the weights into buckets, that will be assigned the same new weight. The larger the buckets, the less bits needed to encode the group name, but the larger the rounding error on each weight. The basic idea is to use exponential bucketing. For example, with exponent 2, a weight w, will be in a bucket b, if $2^{b-1} < w \leq 2^b$. This way, every weight is at most doubled, and an MST computed on these new weights is at most twice as heavy as the MST with the original weights. The good thing is that now there are $O(\log n)$ different weights instead of n, which means it can be encoded on $O(\log\log n)$ bits instead of $O(\log n)$. Now for various reasons explained in the full version [4], we do not use this vanilla version of exponential bucketing with other bases for other approximation ratios. Instead we consider the series of the rounding values 1, 2, 4, 8, ..., $2^{\log n}$ given by the technique above, that we call *milestones*, and work on it. We remove some of the milestones to get a coarser approximation or we add new ones to get a more fine-grain approximation.

5   MST Construction Algorithm

In this section, we give a distributed algorithm for minimum spanning tree. This algorithm is not self-stabilizing; the self-stabilizing part will be taken care of in Sect. 7. As our main goal is to use small space, we do not optimize the time of this construction algorithm (except that we want it to be polynomial), and keep it as simple as possible.

Lemma 2. *There exists a distributed algorithm in space $O(s + \log n)$ that builds a minimum spanning tree in polynomial time.*

Proof. Our algorithm is a distributed version of Kruskal's algorithm. Remember that Kruskal's algorithm sorts the edges by increasing weight, and then adds the edges to the tree one by one, discarding any edge that would close a cycle. There

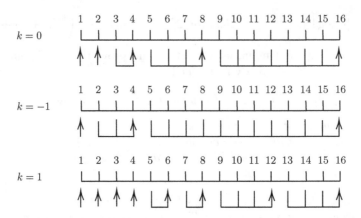

Fig. 1. Illustration of Lemma 1. For $k = 0$, for example the weight 2, stays as 2, but weight 3 is transformed into 4. For the case $k - 1$, we remove basically half milestones, and for $k = 1$ we basically double the number of milestones.

are several modifications to perform in order to make this algorithm work in our framework.

First, to consider the edges by increasing weights, we work in phases, each phase corresponding to a specific value that a weight can take. We have a phase for the possible edges of weight 1, then a phase for those of weight 2, etc.

Then in order to have these phases somehow synchronized on the whole network, and to avoid simultaneous additions of edges of the same weight (which could create cycles), we use a token. This token visits the nodes of the graph, and only the node with the token will be allowed to add edges to the tree. The token transports the information about the weight of the current phase. To make the token visit all the nodes we need to build a spanning tree beforehand, and we make the token traverse the tree. We use the same spanning tree construction and token circulation as in [3], inspired by the tree algorithm of [11] and the token algorithm of [24]. These algorithms ensure, under the distributed unfair scheduler, stabilization to a proper token circulation in $O(n)$ rounds, and they use only $O(\log n)$ bits of memory.

Finally, a node must be able to decide locally whether adding a specific edge would close a cycle or not. To do so every node will hold a *component name*. This name is the minimum identifier in the connected component of the node, in the current state of the tree. This way, a node can safely add an edge if its component name is different from the component name of the other endpoint. To maintain this name, we will need to perform a traversal of a part of the tree every time we add an edge. This is also known to be doable in our setting in polynomial time and logarithmic memory [9].

The space complexity is in $O(s + \log n)$, because we just need to store a constant number of IDs, weights and additional $O(\log n)$-size objects. The time complexity in terms of rounds is polynomial. Indeed the number of phases is polynomial, because we consider polynomial weights, and each phase lasts a polynomial number of rounds (because the primitives we use, token circulation and traversal of the tree, are known to be polynomial [3,9,11,24]).

Augmented MST. Actually, for the next steps (described in Sect. 7), we will need a bit more than the minimum spanning tree. We want to have for each node: an orientation to an arbitrary root, the number of nodes that are descendants of this node (including the node itself), and the total number of nodes in the graph. These are easy to compute by simple traversals.

6   Distributed Proof of MST

The second part of our algorithm, that makes it self-stabilizing, consists in labeling the nodes with a distributed proof. This labeling certifies the correctness of the minimum spanning tree. It comes from a proof-labeling scheme described in [22]. It is necessary here to describe the scheme of [22] with some details (even though it is not our work) in order to allow the reader to understand the next section without reading the full version in [22]. For an intuitive but more in depth description of the main ideas behind the scheme, we refer to [18].

Lemma 3 ([22]). *There exists a distributed proof of size $O(\log n \cdot s)$ for minimum spanning trees.*

Proof (Sketch of the scheme and of the proof). The labeling is actually in two parts. The first part certifies the acyclicity of the tree. It is well known that acyclicity can be certified with $O(\log n)$ bits [23], for example with each node storing its number of descendants, thus we focus on the second part.

The second part of the scheme has a recursive shape. Let us first describe the top-most level of it. The prover chooses a node to be the *center* of the tree, and orients the tree towards this node such that it becomes a root. This node has some x subtrees, that is, if we remove this center from the graph, there would be x trees. The prover gives a distinct number to each subtree from 1 to x. Every node (except the center) is labeled with the number of its subtree. Also every node is given the maximum weight on its path to the center along the edges of the tree. It is rather easy to check that the correctness of these pieces of information can be checked locally.

To show why this is useful, consider a node u of a subtree A, that is adjacent in the graph (but not in the tree) to a node v of a subtree B. Thanks to the subtree numbers in their labels, the nodes u and v know that they do not belong to the same subtree. We claim that they can check whether adding (u, v) to the tree could result in a smaller weight tree (which would contradict the fact that the selected edges form an MST). First, remember that adding (u, v) would lead to a lighter tree, if and only if, the path from u to v (in the tree) has an edge that is heavier than (u, v) (and thus that could be replaced by (u, v)). Therefore, u and v only have to look for such an edge. The path from u to v must go through the root, because these nodes are in different subtrees. Thus the maximum weight on this path is either the maximum weight from u to the root, or from v to the root. As the nodes are given these maximum weights on the paths to the root in their labels, they can check whether (u, v) is lighter or not than the heaviest edge on the path.

To allow the nodes to test for all the non-selected edges, we need to handle adjacent nodes that *are* in the same subtree. We do so by choosing recursively a center in each subtree, and providing the same information as for the top level. That is, every node will have information regarding the first center, its second center (*i.e.* the center of its subtree of the first center), its third center (*i.e.* the center of its subtree of the second center) etc. until the level where it is itself a center. Thanks to this recursive structure, for every pair of nodes, there is a level of the recursion for which they are assigned to two distinct subtrees (or to a subtree and its root), and the checking can be done in the same way as for the top-most level.

To be sure to use only $O(\log n \cdot s)$ space, for all this information about all the centers of a node, we need the above scheme to have two properties. First, we need the centers to be placed in a balanced manner. Precisely, we want the center of a (sub)tree T to be at a node such that every subtree has size at most $|T|/2$. (Such a node always exists and can be computed in a simple way, as described in the next section.) This balance for the centers implies that there is at most $O(\log n)$ centers per node. Second, we need that for each node, the concatenation of all its subtree numbers is not too long. To get this, it has been proved (see [21]) that it is enough that, for each center, the subtrees are numbered by decreasing number of nodes (the largest subtree gets number 1, the second largest gets number 2, etc.).

Also, remember that we need to have an orientation towards the center, for each of the $O(\log n)$ centers a node has. This takes $O(\log^2 n)$ bits if we use the ID of the parent as a pointer. Instead, we have only one orientation of the full tree encoded by IDs, and the other orientations are encoded with respect to this original orientation. Concretely, every node will store whether the center of the current level is in the direction of its parent in the original orientation, or if it is in the direction of one of its children. Surprisingly, this is enough to describe and check each orientation, and it takes only constant number of bits per level. (See [22] or [18], to see for example why not knowing the precise child-parent relation is not problematic.)

7   Certification Labeling Algorithm

In this section, we describe an algorithm that builds the labeling of Sect. 6. Remember that thanks to Sect. 5, we start with a tree that has an orientation toward a root and whose nodes are labeled with the number of nodes in their subtrees. Thus the first part of the labeling of Sect. 5 is already present.

Lemma 4. *Given a tree with an orientation to a root and subtree sizes, there exists an algorithm that builds the labeling of Sect. 6 in space $O(\log n \cdot s)$ and polynomial time.*

Proof. We first describe the algorithm, and then highlight some key properties. The algorithm takes as input a tree, and certifies it as a minimum spanning tree. The construction follows the recursive description of the labeling of Sect. 6.

Before any computation, we do a copy of the pointers, subtree number, etc. We will modify the copies, but we need to keep the original labels. Let us first describe the computation of a center. (This description is for the first center, we explain later how to adapt it to the other phases of the labeling.) Basically, we will move the root of the tree until it satisfies the property of a proper center. (Remember that this property is that the center separates the tree into parts that have size at most half the size of the full tree.) Our first candidate for a center is then the root of the tree. Thanks to the subtree sizes that every node holds, the root can detect whether it is in a central position or not. Indeed, as the subtree size of the root is the number of nodes in the full tree, the root can easily check whether its neighbors have subtree sizes at most half this number. If the root is not in a central position, then we transfer the root to the neighbor having the largest subtree size. This transfer of the root implies several computations:

- the old root designates the new root
- the old root orients its pointer to the new root
- the new root, erase its pointer and takes the root label
- the old root takes as new subtree size its old subtree size, minus the old subtree size of the new root
- the new root takes as subtree size the old subtree size of the old root.

Thanks to this step, we are in the same position as before (that is, we have a tree with a correct orientation, and correct subtree sizes), but in addition, the root is in a more central position, in the sense that the largest subtree size is smaller than before the transfer. After $O(n)$ such moves, the root is in a central position.

Once the center is validated, we have to label each node with: the orientation to the center, the maximum weight on the path to the center, and the subtree number. There is a difficulty here. Remember that only the center can decide which subtree gets which number, because these numbers depend on the relative sizes of all the subtrees. Thus the center has to announce the subtree numbers *e.g.* "the subtree with root ID_v has number 3". It is not possible to announce all these numbers at once. Indeed, the list of these numbers can have length of order n, therefore we cannot write the whole list in the state of the center. Instead the center will announce a first pair identifier-number, then wait for the node with this identifier to confirm this information, and then go to the second etc. Once the root of a subtree receives its subtree number, it broadcasts this information to all its subtree using a snap-stabilizing (*i.e.*, self-stabilizing with a stabilization time of 0 rounds) Propagation of Information with Feedback algorithm. Bui *et al.* [9] provide such an algorithm with constant space requirement and polynomial completion time (in rounds) on trees, that meets our requirements. In the same wave, the maximum weight to the center is computed by keeping and updating the maximum weight seen so far, while descending in the tree. The orientation is even easier to store: just copy the current orientation.

Once the broadcast waves have come back to the root of the subtree, every node has, in its label, all the information it needs to store for this phase. And then we can start immediately a new recursive call, looking for the next center. Indeed we have a tree, with a root, with a proper orientation, and with correct subtree numbers.

Now, let us highlight some key points of the algorithm.

– Once a center has finished launching the computations in each subtree, it becomes silent, and will not change its state, except if there is a reset.
– As soon as two adjacent nodes (in the graph) become centers, they can start checking their labelings to see if the distributed proof makes sense, at least for this edge (and launch a reset if it is not the case).
– Once the root has launched the computation in the subtrees, these subtrees will run independent computations (except for the cycle checking mentioned in previous item). This means that the scheduler can delay the computation in one subtree, by making a series of recursive calls in the other subtrees, but at some point these recursive calls will end up by having all nodes as centers, thus disabled, and the scheduler will have to enable the remaining nodes.
– Suppose that we are in a subtree, looking for the center for the second recursive call. The main center is already chosen, and thanks to the pieces of information stored, the nodes can check that it exists. But they cannot check whether it was placed in a central position, because we are reusing the subtree size variables. Thus if we start from an initial configuration that corresponds to this situation, we are not following the correct construction described in Sect. 6. This means that we could end up with a larger memory than what we claimed. What we do is that we control the size used. If we end up using more than $\alpha \cdot \log n \cdot s$ bits, for a constant α large enough to allow correct computations, then we launch a reset. Note that it may actually be the case that the centers are not perfectly placed, but that their positions are good enough to ensure that the memory size does not cross the $\alpha \cdot \log n \cdot s$ limit; in this case we do not detect it, and the outcome is correct.

When the computations ends on all nodes, every node is labeled as specified in Sect. 6, and the algorithm becomes silent.

8 Conclusion

This paper presents the first self-stabilizing algorithm using approximation to reduce memory usage. A step in this construction is the design of a polynomial-time silent self-stabilizing algorithm for MST construction using $O(\log n \cdot s)$ bits of memory. This later algorithm uses an unusual two-phase approach: building and then certifying the solution. We believe that this modular approach is the key to go down to complexity $O(\log n \cdot s)$, and we think that it is an interesting problem to formally prove this intuition.[4]

---

[4] Some elements indicating that at least the classic techniques cannot avoid a modular approach can be found in [19].

References

1. Barenboim, L., Elkin, M., Goldenberg, U.: Locally-iterative distributed ($\Delta+$ 1): - coloring below Szegedy-Vishwanathan barrier, and applications to self-stabilization and to restricted-bandwidth models. In: Newport, C., Keidar, I. (eds.) Proceedings of the 2018 ACM Symposium on Principles of Distributed Computing, PODC 2018, Egham, United Kingdom, 23–27 July 2018, pp. 437–446. ACM (2018)
2. Beauquier, J., Gradinariu, M., Johnen, C.: Memory space requirements for self-stabilizing leader election protocols. In: 18th Annual ACM Symposium on Principles of Distributed Computing, PODC 1999, pp. 199–207 (1999). https://doi.org/10.1145/301308.301358
3. Blin, L., Boubekeur, F., Dubois, S.: A self-stabilizing memory efficient algorithm for the minimum diameter spanning tree under an omnipotent daemon. J. Parallel Distrib. Comput. **117**, 50–62 (2018). https://doi.org/10.1016/j.jpdc.2018.02.007
4. Blin, L., Dubois, S., Feuilloley, L.: Silent MST approximation for tiny memory. CoRR abs/1905.08565 (2019). http://arxiv.org/abs/1905.08565
5. Blin, L., Feuilloley, L., Bouder, G.L.: Brief announcement: memory lower bounds for self-stabilization. In: 33rd International Symposium on Distributed Computing, DISC 2019, Budapest, Hungary, 14–18 October 2019, vol. 146, pp. 37:1–37:3 (2019). https://doi.org/10.4230/LIPIcs.DISC.2019.37
6. Blin, L., Fraigniaud, P.: Space-optimal time-efficient silent self-stabilizing constructions of constrained spanning trees. In: 35th IEEE International Conference on Distributed Computing Systems, ICDCS 2015, pp. 589–598 (2015). https://doi.org/10.1109/ICDCS.2015.66
7. Blin, L., Fraigniaud, P., Patt-Shamir, B.: On proof-labeling schemes versus silent self-stabilizing algorithms. In: Felber, P., Garg, V. (eds.) SSS 2014. LNCS, vol. 8756, pp. 18–32. Springer, Cham (2014). https://doi.org/10.1007/978-3-319-11764-5_2
8. Blin, L., Tixeuil, S.: Compact deterministic self-stabilizing leader election on a ring: the exponential advantage of being talkative. Distrib. Comput. **31**(2), 139–166 (2017). https://doi.org/10.1007/s00446-017-0294-2
9. Bui, A., Datta, A.K., Petit, F., Villain, V.: Snap-stabilization and PIF in tree networks. Distrib. Comput. **20**(1), 3–19 (2007). https://doi.org/10.1007/s00446-007-0030-4
10. Censor-Hillel, K., Paz, A., Perry, M.: Approximate proof-labeling schemes. Theoret. Comput. Sci. **811**, 112–124 (2020). https://doi.org/10.1016/j.tcs.2018.08.020
11. Datta, A.K., Larmore, L.L., Vemula, P.: An o(n)-time self-stabilizing leader election algorithm. J. Parallel Distrib. Comput. **71**(11), 1532–1544 (2011). https://doi.org/10.1016/j.jpdc.2011.05.008
12. Devismes, S., Johnen, C.: Self-stabilizing distributed cooperative reset. In: 39th IEEE International Conference on Distributed Computing Systems, ICDCS 2019, pp. 379–389 (2019). https://doi.org/10.1109/ICDCS.2019.00045
13. Dolev, S.: Self-Stabilization. MIT Press, Cambridge (2000)
14. Dolev, S., Israeli, A., Moran, S.: Resource bounds for self-stabilizing message-driven protocols. SIAM J. Comput. **26**(1), 273–290 (1997). https://doi.org/10.1137/S0097539792235074
15. Dubois, S., Tixeuil, S.: A taxonomy of daemons in self-stabilization (2011). arXiv: 1110.0334

16. Emek, Y., Gil, Y.: Twenty-two new approximate proof labeling schemes. In: 34th International Symposium on Distributed Computing, DISC 2020, Virtual Conference, 12–16 October 2020, vol. 179, pp. 20:1–20:14 (2020). https://doi.org/10.4230/LIPIcs.DISC.2020.20

17. Feuilloley, L.: Introduction to local certification. CoRR abs/1910.12747 (2019). http://arxiv.org/abs/1910.12747

18. Feuilloley, L.: Note on distributed certification of minimum spanning trees (2019). arXiv: 1909.07251

19. Feuilloley, L.: Can we always build and certify at the same time? (Maybe not). Discrete notes (2020). https://discrete-notes.github.io/build-certify-3

20. Feuilloley, L., Fraigniaud, P.: Survey of distributed decision. Bull. EATCS 119 (2016). bulletin.eatcs.org, arXiv: 1606.04434

21. Gavoille, C., Katz, M., Katz, N.A., Paul, C., Peleg, D.: Approximate distance labeling schemes. In: auf der Heide, F.M. (ed.) ESA 2001. LNCS, vol. 2161, pp. 476–487. Springer, Heidelberg (2001). https://doi.org/10.1007/3-540-44676-1_40

22. Korman, A., Kutten, S.: Distributed verification of minimum spanning trees. Distrib. Comput. **20**(4), 253–266 (2007). https://doi.org/10.1007/s00446-007-0025-1

23. Korman, A., Kutten, S., Peleg, D.: Proof labeling schemes. Distrib. Comput. **22**(4), 215–233 (2010). https://doi.org/10.1007/s00446-010-0095-3

24. Petit, F., Villain, V.: Time and space optimality of distributed depth-first token circulation algorithms. In: Distributed Data & Structures 2, Records of the 2nd International Meeting (WDAS 1999), pp. 91–106 (1999)

A Privacy-Preserving Collaborative Caching Approach in Information-Centric Networking

Andrew Jones$^{(\boxtimes)}$ (ID) and Robert Simon

George Mason University, Fairfax, VA 22030, USA
{ajones93,simon}@gmu.edu

Abstract. It has been established that in-network caching in an Information-Centric Network (ICN) environment significantly reduces required bandwidth and content retrieval delay, and reduces load on content producers. However, malicious actors masquerading as legitimate consumers can probe cache contents and use the resultant data to map content objects to, and thereby violate the privacy of, the consumer(s) who requested them. Existing mitigation approaches suffer a direct trade-off between privacy and utility; the two are diametrically opposed, and prioritizing either rapidly degrades its counterpart. This paper presents a *collaborative caching* approach with provable privacy and utility guarantees that instead monotonically increase as a function of one another, growing in tandem. Our proposed scheme preserves all true cache hits to utilize in-network caching as efficiently as possible. We have evaluated our method against a number of other in ICN caching policies for a variety of workloads and topologies. Our results show that our technique delivers high cache hit ratios and minimizes interest satisfaction delay while offering provable privacy guarantees.

Keywords: Secure information centric networking · Provable privacy · Distributed system security

1 Introduction

Information-Centric Networking (ICN) encompasses a paradigm shift from the point-to-point, address-based IP protocol which comprises the "thin waist" of today's internet. ICN eschews this existing model in favor of an architecture in which content is treated as a first-class citizen and is named, addressable, and routable [6]. At a high level, entities within an ICN are content producers, content consumers and routers. ICN development is motivated by modern internet usage patterns resembling those of a content distribution network (CDN). IP was designed to address the needs of a network of hosts intercommunicating via relatively equally-weighted full duplex conversations. However, many of the hosts in today's internet operate almost exclusively as consumers, requesting

© Springer Nature Switzerland AG 2020
S. Devismes and N. Mittal (Eds.): SSS 2020, LNCS 12514, pp. 133–150, 2020.
https://doi.org/10.1007/978-3-030-64348-5_11

content from those who produce it. ICN is the product of an attempt to design an internet architecture better suited to this model of communication.

An important feature of proposed ICN architectures is the utilization of in-network content caching at routers. However, if implemented in a naive fashion, ICN content caching is susceptible to attacks against consumer privacy. In this context consumer privacy is informally defined by asserting that a legitimate consumer (Alice) wishes to hide the fact that she has requested a content object $\mathcal{O}$. Suppose a malicious user (Darth) connects to same first-hop router $\mathcal{R}$ to which Alice is connected, and wants to determine if, indeed, Alice has requested $\mathcal{O}$. As described in greater detail in Sect. 2.1, Darth issues a request for $\mathcal{O}$. Darth also has determined the expected time $T_\mathcal{R}$ to satisfy content requests from $\mathcal{R}$. If the time to receive $\mathcal{O}$ is approximately equal to $T_\mathcal{R}$ then Darth can conclude that Alice has previously requested $\mathcal{O}$. Note that this attack still works if users besides Alice are connected to $\mathcal{R}$.

Defenses against the attack just described include TOR-like mechanisms or introducing artificial delays to request response. Both of these approaches introduce performance penalties. The contribution of our paper is to introduce a *collaborative caching* policy designed to defeat consumer privacy attacks without introducing significant performance penalties. We focus on domain-clustering ICN deployments and show how to serve a content request from an in-network cache in such a way as to hide from Darth information about Alice's content requests. We also show that our scheme produces a provable privacy bound, in the sense of providing (ϵ, δ)-probabilistic indistinguishability, a standard measurement used to quantify the utility of privacy protocols [8,16].

2   Background and Related Work

Over the last decade or so there have been several proposed Information Centric Networking architectures, such as the European PURSUIT project [5]. Our work is motivated by research in Content Centric Networking proposals and work in the ongoing Named Data Networking (NDN) project [10,19].

2.1   NDN Overview

Content retrieval in NDN[1] does not necessitate a persistent end-to-end connection between the entity which produced it and that which is requesting it. Rather, network endpoints fall into one or both of the following categories: **consumers**, which issue *interests* for the data they wish to retrieve, and **producers**, which dispatch *content* packets to *satisfy* received interests. Notably, a host in the network can be both a producer and consumer. *Pure* consumers or producers are those which perform solely the functions of consumers or producers, respectively. A pure consumer has no addressable namespace and no private/public key pair for signing and authenticating its (nonexistent) content. Content packets in NDN

---

[1] After this section we revert to the abbreviation ICN.

are only forwarded to consumers which specifically request them via interests. A noteworthy security implication of these policies is that pure consumers are not addressable entities in ICN. Two data structures are present in each router in an NDN network: A **Pending Interest Table** (PIT), which records each of the interests which have arrived at the router and the corresponding interfaces on which they were received, and a **Forwarding Interest Base** (FIB), which contains a mapping of content name prefixes to outgoing interfaces.

A router is not required to but may additionally possess a *content store* (CS). A router can opportunistically cache content in its CS, upon receiving that content in response to a previously forwarded interest. The router can then serve that content from its CS in response to future interests received for the same content. This has the benefit of potentially greatly reducing data retrieval delays, as an interest and its corresponding content may not need to traverse the entire path between a consumer and producer. Any content received by a router which does not match an entry in the router's PIT is discarded. The focus of our work is to ensure consumer privacy in the face of timed probing attacks against content stores. In NDN the content that satisfies an interest is always forwarded along the reverse path of the interest which requested it. The determination of this reverse path in the absence of a source address is accomplished by per-interest state recorded at each router hop in the form of an entry in the PIT. Upon receipt of an interest for the same content as another interest already in its PIT, a router will simply add the interface on which the new interest was received to the existing entry in its PIT and discard the remaining information in the new interest without forwarding it. Corresponding content is returned along all necessary interfaces whilst avoiding duplication. Producers and/or other routers are not inundated with multiple interests forwarded by a given router requesting the same content. The satisfaction of a single interest by a producer may serve the content in question to many consumers whose interests were collapsed into that received by the producer.

2.2   Related Work

There has recently been significant interest in ICN cache privacy issues. ANDaNA [4] and AC3N [25] are applications of the onion routing and ephemeral circuits of TOR to ICN. Though effective, these approaches increase latency, decrease available bandwidth compared to vanilla NDN, and- due to ephemeral encryption- prohibit any useful in-network caching.

A proposed mitigation to the cache privacy attack which incorporates a randomized content satisfaction delay to mask cache hits is presented in [2]. A router R can introduce an artificial delay before responding to an interest. In doing so, R prevents an adversary $\mathcal{A}$ from determining whether or not a given piece of content C is in its content store (CS), denoted CS_R. This work also establishes privacy bounds. As with all approaches that introduce artificial delays, performance can become an issue. Somewhat similar to [2] is the work presented in [18] and [17] which uses privacy preserving delays at edge routers. The work described in [1] uses the concept of "Betweenness Centrality-based" caching that caches content at nodes with a higher betweenness centrality value to put consumers in larger anonymity sets. Unlike this work, we focus on providing consumers with a uniform anonymity level, and we provide a computable privacy bound. Additional related efforts include a namespace-based privacy [14] approach, and a Long Short Term Memory detection approach to detect a timing attack [27]. The work described in [26] details an edge-based access control mechanism for ICNs. Our work differs from the above papers because we focus on protecting individual consumers without sacrificing performance, and because we offer a provable privacy bound. We also note that our methods are not vulnerable to attacks that exploit hop limit and scope fields in the NDN packet header [2].

We note that there is recent interest in using domain clustering methods, in conjunction with hash routing, to support large ICNs [23]. Our approach relies on the use of clustering.

3   Collaborative Caching Algorithm

As illustrated in Fig. 1a, our proposed caching scheme divides the network into clusters of routers, each of which will operate as a distributed aggregate in-network cache. This abstraction is transparent to producers, consumers, and other clusters. The cluster to which a router will belong is determined by the partitioning around medoids algorithm [12]. Upon the arrival of a content packet at a router on the edge of a cluster, a router is chosen uniformly at random from the members of the cluster (including the specific router which actually received the content packet) as the designated, or "authoritative", router at which to cache the content. The content is then multicast to both the designated router cache for later use and to the next-hop router on the path back to the consumer which originally issued the interest for the content.

When a router on the edge of a cluster encounters an interest, that interest is forwarded to the authoritative router cache pertaining to the content requested by the interest, if one has already been determined. If the requested content is in the content store of a router in the cluster, it is returned to the consumer which issued the interest. If not, a single interest for the entire cluster is propagated upstream toward the appropriate producer by the cluster router closest to that producer. This process is illustrated in Fig. 1b and detailed in Algorithm 1.

We now describe our system and adversary model. Let Σ^* and Γ denote the universes of all content names (composed of some finite alphabet Σ) and content objects, respectively, using notation in common with [2]. Let G represent a cluster of collaborating routers according to our proposed caching model, and U represent the set of all consumers downstream from G. $S : (\Gamma, U) \to \mathbb{N}$ represents, for a given content item $C \in \Gamma$ in the cache of any router in G, the number of times C has been forwarded by G to $u \in U$. Note that we use the definition of $\mathbb{N}$ from ISO 8000-2 [9], where $0 \in \mathbb{N}$. $S(C, u) = 0$ for all content not in any router cache in G, and for all content for which u has not issued an interest.

We allow consumers to determine whether specific content has been forwarded by G via probing attacks. As in [2], this is modeled by a function

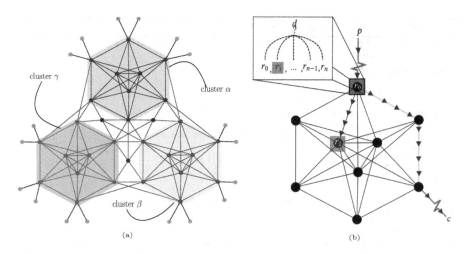

Fig. 1. (a) A network divided into router clusters α, β, and γ, the routers belonging to which lie within the pink, cyan, and orange shaded regions, respectively. Producers are denoted by blue nodes, consumers by green nodes, and routers by black nodes. Black solid edges represent intra-cluster connections, whereas red edges are inter-cluster or cluster-to-producer connections and consumer-to-router connections are indicated with dashed black edges. (b) Caching process within a cluster. A content packet d, requested by consumer c, has just arrived at router r_0 from producer p. r_1 is selected uniformly at random from all cluster routers to cache d. Blue arrowheads indicate the multicast flow of d to router r_1 and along the (green) shortest path to c. (Color figure online)

$Q_S : \Sigma^* \rightarrow \{0,1\}$. We let $N_C \in \Sigma^*$ denote the name associated with content $C \in \Gamma$. In network state S,

$$Q_S(N_C, G) = \begin{cases} 1, & \text{if cached content } C \text{ in } G \text{ matches the input name } N_C \\ 0, & \text{otherwise} \end{cases}$$

(1)

Each invocation of $Q_S(N_C, G)$ by a given consumer u causes S to transition to S' such that:

1. $S'(C, u) = S(C, u) + 1$
2. $\forall C' \in \Gamma \setminus \{C\}, S'(C', u) = S(C', u)$
3. $\forall C'' \in \Gamma, u' \in U \setminus \{u\}, S'(C'', u') = S(C'', u')$

The attack we are concerned with operates as follows [7]: A malicious consumer, $\mathcal{A}$, is connected to an edge router $\mathcal{R}$, the only other consumer connected to which is u. $\mathcal{A}$ determines the round-trip time to $\mathcal{R}$ by issuing two identical interests with the same random content name and observing the content return delay. $\mathcal{A}$ then issues an interest for some content C and measures that content retrieval delay. If that content (C) is returned with a delay approximately equal to the round trip time (RTT) from $\mathcal{A}$ to $\mathcal{R}$, $\mathcal{A}$ concludes that u recently requested C, as the interest must have been satisfied at the first hop router.

Algorithm 1: Collaborative-Caching

Input: Interest I from consumer N, requesting content C produced by P, Collaborating router cluster G

Output: C, $x = \begin{cases} 1, & \text{if collaborative cache hit} \\ 0, & \text{otherwise} \end{cases}$

1 $CS_{loc} :=$ local content store of router R receiving interest;
2 **if** $C \notin CS_{loc}$ **then**
3   **if** C *in* CS_g *for some* $g \in G$ **then**
4   Route I to authoritative router for C;
5   Return $(C, 1)$ when C returned from authoritative router;
6   **else**
7   Decrement HopLimit;
8   **if** *HopLimit = 0* **then**
9   Return $(NULL, 0)$;
10   **end**
11   Forward I to router R_E on edge of G and onward toward P;
12   **while** C *has not arrived at* R_E *from* P **do**
13   Wait;
14   **end**
15   Determine authoritative router $g_C \in G$ with $\Pr = \frac{1}{|G|}$ (uniformly random);
16 R_E: (Mulicast) send C to g_C for caching and return $(C, 0)$ on shortest path to N;
17   **end**
18 **else**
19   Return $(C, 1)$;
20 **end**

---

4   Provable Privacy

4.1   Quantifying "Privacy"

We derive our understanding of cache privacy from the concept of (ϵ, δ)-probabilistic indistinguishability, which we define using a definition motivated by that provided in [2].

Definition 1 (ϵ, δ)-**probabilistic indistinguishability.** *Two distributions* $\mathcal{D}_1$ *and* $\mathcal{D}_2$ *are* (ϵ, δ)-*probabilistically indistinguishable if we can divide the output space* $\Omega = Range(\mathcal{D}_1) \cup Range(\mathcal{D}_2)$ *into* Ω_1 *and* Ω_2 *such that, letting* $\mathcal{Q}_1$ *and* $\mathcal{Q}_2$ *be random variables with probability distributions* $\mathcal{D}_1$ *and* $\mathcal{D}_2$ *respectively,*

1. for all $O \in \Omega_1$, $e^{-\epsilon} \leq \frac{\Pr[\mathcal{Q}_1 = O]}{\Pr[\mathcal{Q}_2 = O]} \leq e^{\epsilon}$
2. $\Pr[\mathcal{Q}_1 \in \Omega_2] + \Pr[\mathcal{Q}_2 \in \Omega_2] \leq \delta$

The similarity of distributions $\mathcal{D}_1$ and $\mathcal{D}_2$ is directly proportional to the magnitude of both ϵ and δ. Minimizing the upper bound on both ϵ and δ is therefore desirable when seeking to prove that two distributions are indistinguishable. Suppose we observe a network in two measurable states, represented by $\mathcal{D}_1$ and $\mathcal{D}_2$, respectively. Intuitively, this definition merely implies that, when ϵ and δ are both small, those states are quite similar. This similarity makes it difficult to distinguish between the distributions. If the two distributions were to respectively represent states of the network in which, on the one hand, consumer u had not requested content C, and, on the other, it had, then the difficulty of distinguishing between the two distributions would be directly related to the difficulty of defeating cache privacy. This is the difficulty of mounting a successful attack which can identify the source of an interest based on cached content.

Our definition of (k, ϵ, δ)-privacy is a modified version of that presented in [2], adapted to suit a collaborative caching model.

Definition 2 *(k, ϵ, δ)-privacy. For all names $n \in \Sigma^*$, subset of content $M \subset \Gamma$, and pairs of states S_0, S_1 such that $S_0(\gamma, u') = S_1(\gamma, u')$ for all $\gamma \in \Gamma \setminus M$ and for all $u' \in U$, and $S_0(C, u) = 0$ and $0 < S_1(C, u) \leq k$ for all $C \in M$ (i.e., S_0 and S_1 differ only on content objects in M) and for consumer u downstream from router cluster G; $Q_{S_0}(n, G)$ and $Q_{S_1}(n, G)$ are (ϵ, δ)-probabilistically indistinguishable.*

Notably, the above definition does not prohibit S_0 and S_1 from differing in terms of the number or distribution of requests made for content C by routers other than u. We allow, but do not require, $S_0(C, u') \neq S_1(C, u'), \forall u' \in U \setminus \{u\}$, and $S_0(C, u')$ and $S_1(C, u')$ could each be zero or positive for any given router $u' \in U \setminus \{u\}$ (as long as $S_1(C, u') \geq S_0(C, u')$).

4.2   Provable Privacy Guarantee

Our approach, like those which employ artificial delay, ultimately serves to prohibit an adversary from learning if a given consumer has issued an interest for a particular piece of content. However, the two methodologies are divergent with respect to the manner in which this is accomplished. *Random-Caching* [2] seeks to conceal the existence of any particular piece of content in a router's cache, assuming that cognizance of the content's presence there would allow an adversary $\mathcal{A}$ to correctly infer that a specific consumer u requested that content. *Collaborative-Caching* (Algorithm 1) decouples the existence of a content item in a router's cache from the implication that a consumer directly downstream from that router issued an interest for that content. We allow an adversary to successfully determine that a consumer downstream from a collaborating router cluster has issued an interest for some specific content- and even the exact router in the cluster at which that content is cached- without revealing the precise identity of the consumer from which the interest originated. We achieve this by maintaining an anonymity set of a specified size for every router downstream from a collaborating router cluster; the size of a router's anonymity set is no longer determined by the number of other consumers which share its first-hop router.

Let $m_{(v,i)}$ denote the number of interests issued by consumer $v \in U$ for all content in state S_i. Let $Q_0(C, r_0)$ and $Q_1(C, r_1)$ denote the output x of Algorithm 1 in states S_0 and S_1, respectively, with C where r_i denotes the set of expected values of the number of interests for C issued by consumers- other than u- downstream from the collaborating router cluster G in each state S_i. That is, $r_i = \{\mathbb{E}(S_i(C, v)), \forall v \in U \setminus \{u\}\}$ where U denotes the set of all consumers downstream from G. Note that we use zero-based array indexing when referring to elements of r in the subsequent formulae.

Theorem 1 *If all cached content is statistically independent, and consumers issue interests for specific content with uniformly random probability, Collaborative-Caching is*

$$\left(\sum_{v=0}^{|U|-1} m_{(v,0)}, \ln |r_0|, \left(1 - \frac{1}{|\Gamma|}\right)^{\sum_{v=0}^{|r_0|-1} m_{(v,0)}} \right) - private.$$

Proof. Per Definition 2, $S_0(C, u) = 0$ and $S_1(C, u) = n$, where $1 \leq n \leq k$. $Q_0^t(C, r_0)$ and $Q_1^t(C, r_1)$ denote the sequence of outputs produced by Algorithm 1 when executed t consecutive times with C, in states S_0 and S_1 respectively. As noted in [2], the presumed statistical independence of content simplifies this analysis by allowing us to focus on the difference between S_0 and S_1 only as it relates to C- whether or not other content has been requested and by whom is irrelevant. The following probabilistic analyses therefore assume content object independence and leverage the idea that separate requests for content are statistically independent events. Let $\mathcal{Q}_0^t$ and $\mathcal{Q}_1^t$ denote two random variables describing $Q_0^t(C, r_0)$ and $Q_1^t(C, r_1)$ respectively when consumers request content uniformly at random. Each entry in the zero-indexed set r_i will therefore be: $r_i[v] = m_{(v,i)} \cdot \frac{1}{|\Gamma|}$ where $m_{(v,i)}$ denotes the total number of requests (for all content) which have been made by downstream consumer v in state S_i.

We show that, assuming consumers are equally likely to request or not request C, $\mathcal{Q}_0^t$ and $\mathcal{Q}_1^t$ are (and consequently *Collaborative-Caching* is)

$$\left(\ln |r_0|, \left(1 - \frac{1}{|\Gamma|}\right)^{\sum_{v=0}^{|r_0|-1} m_{(v,0)}} \right) - probabilistically\ indistinguishable$$

for any C (a corollary of our assumption that content is statistically independent). Note that in our adversarial model, $\mathcal{A}$ has no knowledge of the likelihood that any particular consumer would be interested in a given piece of content. As such, $\mathcal{A}$ possesses no information from which it can extrapolate that the probability distribution of a given consumer's requests is anything but uniform.

The output x (defined in Algorithm 1) of $\mathcal{Q}_1^t$ will be $\{1\}^t$ because, in state S_1, u has already issued at least one interest for C and it is therefore cached at some router in the router cluster. The output of $\mathcal{Q}_0^t$ will be:

$$\mathcal{Q}_0^t = \begin{cases} \{1\}^t, & \text{if } \exists v \in U \setminus \{u\} \text{ s.t. } S_0(C, v) \geq 1 \\ 0||\{1\}^{t-1}, & \text{otherwise} \end{cases}$$

That is to say, $\mathcal{Q}_0^t$ will be either $\{1\}^t$ (a sequence of t ones), if a consumer other than u has already issued at least one interest for C in S_0, or $0||\{1\}^{t-1}$, if no consumer other than u has issued an interest for C. We partition the output space $\Omega = Range(\mathcal{Q}_0^t) \cup Range(\mathcal{Q}_1^t)$ into Ω_1 and Ω_2, for all t and C, as follows:

- $\Omega_1 = Range(\mathcal{Q}_0^t) \setminus Range(\mathcal{Q}_1^t)$: If no consumer downstream from G other than u has issued an interest for C, then the first interest issued will result in a cache miss (in S_0, u must not have issued an interest for C yet either). However, this cannot occur in S_1, as u would have already requested C and it would be in G's collaborative cache. Therefore, $\nexists r_1$ such that $Q_0^t(C, r_0) = Q_1^t(C, r_1)$.
- $\Omega_2 = Range(\mathcal{Q}_0^t) \cap Range(\mathcal{Q}_1^t)$: Either some consumer other than u has requested C in S_0, or we are in S_1 so u has issued an interest for C (but may not be the only consumer to have done so). Either way, the output will be t cache hits: $\{1\}^t$. Thus, $Q_0^t(C, r_0) = Q_1^t(C, r_1)$.

Note that $\Omega_1 \cup \Omega_2 = \Omega$, as there are no possible outputs of $\mathcal{Q}_1^t$ that are not possible outputs of $\mathcal{Q}_0^t$ (whereas the converse is true).

A series of t ones is the only output $O \in \Omega_2$. Therefore, for all $O \in \Omega_2$, $\Pr[\mathcal{Q}_1^t = O] = 1$. For all $O \in \Omega_2$,

$$
\Pr[\mathcal{Q}_0^t = O] = \Pr[\exists v \in U \setminus \{u\} \text{ s.t. } S_0(C, v) \geq 1]
$$

$$
= \sum_{v=0}^{|r_0|-1} \Pr[S_0(C, v) \geq 1] = \sum_{v=0}^{|r_0|-1} (1 - \Pr[S_0(C, v) = 0])
$$

$$
= \sum_{v=0}^{|r_0|-1} \left(1 - \left(1 - \frac{1}{|\Gamma|} \right)^{m_{(v,0)}} \right) = \sum_{v=0}^{|r_0|-1} 1 - \sum_{v=0}^{|r_0|-1} \left(1 - \frac{1}{|\Gamma|} \right)^{m_{(v,0)}}
$$

$$
= |r_0| - \sum_{v=0}^{|r_0|-1} \left(1 - \frac{1}{|\Gamma|} \right)^{m_{(v,0)}}
\tag{2}
$$

Substituting these values into clause 1 of Definition 1, we obtain

$$
\forall O \in \Omega_2, \frac{\Pr[\mathcal{Q}_1^t = O]}{\Pr[\mathcal{Q}_0^t = O]} = \frac{1}{|r_0| - \sum\limits_{v=0}^{|r_0|-1} \left(1 - \frac{1}{|\Gamma|} \right)^{m_{(v,0)}}}
\tag{3}
$$

To circumvent the issue of division by zero, we assume there is at least a single piece of content in the network and each consumer downstream from the collaborating router cluster has issued at least one interest (for at least one piece of content). We then derive the value of ϵ as defined in Definition 1:

$$|\Gamma| \geq 1, m_{(v,0)} \geq 1 \; \forall v \in U \setminus \{u\} \Rightarrow \forall v \in U \setminus \{u\}, \; 1 > \left(1 - \frac{1}{|\Gamma|}\right)^{m_{(v,0)}}$$

$$\Rightarrow \sum_{v=0}^{|r_0|-1} 1 > \sum_{v=0}^{|r_0|-1} \left(1 - \frac{1}{|\Gamma|}\right)^{m_{(v,0)}}$$

$$\Rightarrow |r_0| > \sum_{v=0}^{|r_0|-1} \left(1 - \frac{1}{|\Gamma|}\right)^{m_{(v,0)}} \Rightarrow |r_0| - \sum_{v=0}^{|r_0|-1} \left(1 - \frac{1}{|\Gamma|}\right)^{m_{(v,0)}} \geq 1 \qquad (4)$$

$$\Rightarrow \frac{1}{|r_0| - \sum_{v=0}^{|r_0|-1}\left(1 - \frac{1}{|\Gamma|}\right)^{m_{(v,0)}}} = \frac{\Pr[\mathcal{Q}_1^t = O]}{\Pr[\mathcal{Q}_0^t = O]} \leq 1$$

Having determined an upper bound on $\frac{\Pr[\mathcal{Q}_1^t=O]}{\Pr[\mathcal{Q}_0^t=O]}$, we now calculate a lower bound on the same value in terms of ϵ.

$$\frac{1}{e^\epsilon} \leq \frac{1}{|r_0| - \sum_{v=0}^{|r_0|-1}\left(1 - \frac{1}{|\Gamma|}\right)^{m_{(v,0)}}} = \frac{\Pr[\mathcal{Q}_1^t = O]}{\Pr[\mathcal{Q}_0^t = O]} \Rightarrow e^\epsilon \geq |r_0| - \sum_{v=0}^{|r_0|-1}\left(1 - \frac{1}{|\Gamma|}\right)^{m_{(v,0)}}$$

$$(5)$$

Under the reasonable assumption that the total number of requests made by any given consumer downstream from a collaborating router cluster will grow faster than the amount of content in the network, we find $\lim_{m_{(v,0)} \to \infty} \left(1 - \frac{1}{|\Gamma|}\right)^{m_{(v,0)}} = 0$, meaning this value approaches 0 as the number of interests issued in a given network state increase- a natural consequence of typical network traffic. Substituting this limit into the RHS of our inequality to allow us to calculate a concrete value for ϵ, we arrive at:

$$e^\epsilon \geq |r_0| - \sum_{v=0}^{|r_0|-1} \Rightarrow e^\epsilon \geq |r_0| \Rightarrow \ln e^\epsilon \geq \ln |r_0| \Rightarrow \epsilon \geq \ln |r_0| \qquad (6)$$

Combining the upper and lower bounds computed on $\frac{\Pr[\mathcal{Q}_1^t=O]}{\Pr[\mathcal{Q}_0^t=O]}$, we conclude:

$$e^{-\ln |r_0|} \leq \frac{\Pr[\mathcal{Q}_1^t = O]}{\Pr[\mathcal{Q}_0^t = O]} \leq 1 \leq e^{\ln |r_0|} \therefore \epsilon = \ln |r_0| \qquad (7)$$

We now derive the value of δ as defined in clause 2 of Definition 1. A zero followed by $t - 1$ ones is the only output $O \in \Omega_1$. If $O \in \Omega_1$,

$$\delta = \Pr[\mathcal{Q}_0^t = O] + \Pr[\mathcal{Q}_1^t = O] = \Pr[\mathcal{Q}_0^t = O] + 0 = \Pr[S_0(C, v) = 0], \forall v \in U \setminus \{u\}$$

$$= \prod_{v=0}^{|r_0|-1} \Pr[S_0(C, v) = 0] = \prod_{v=0}^{|r_0|-1}\left(1 - \frac{1}{|\Gamma|}\right)^{m_{(v,0)}}$$

$$(8)$$

$$= \prod_{v=0}^{|r_0|-1}\left(1 - m_{(v,0)} \cdot \frac{1}{|\Gamma|}\right) = \left(1 - \frac{1}{|\Gamma|}\right)^{\sum_{v=0}^{|r_0|-1} m_{(v,0)}}$$

Finally, we determine an appropriate value for k as defined in Definition 2. As previously stated, S_0 and S_1 differ only in requests made for content object C. Therefore, $M = \{C\}$, where M is defined as in Definition 2. The only stipulation which must be satisfied by the chosen value of k is therefore that $0 < S_1(C, u) \leq k$. $S_1(C, u)$ is defined to be the cumulative number of requests made by router u for content C in state S_1. The maximum possible number of such requests is the cumulative number of requests for all content made by all consumers collectively in state S_1 (in the case where only consumer u issues any interests and every one of those interests is for content C). Therefore, we let $k = \sum_{v=0}^{|U|-1} m_{(v,0)}$.

Though the values of ϵ and δ we have proven may at first appear too complex to be meaningful, it is illustrative to consider them in the context of a network as time (and with it network traffic) progresses. The value of ϵ we have found grows logarithmically with respect to the number of consumers downstream from a router cluster. Complementing the slow logarithmic growth of ϵ is the observation that, compared to the amount of content and requests in a network, the number of consumers attached to a router cluster could reasonably be expected to grow quite slowly (or even remain relatively stagnant). We can conclude something even more concrete about the value of δ in this context. Assuming, as before, that the number of requests issued by consumers in a network grows at a rate faster than the number of unique content objects in the network, we see that:

$$\lim_{m_{(v,0)} \to \infty} \delta = \lim_{m_{(v,0)} \to \infty} \left(1 - \frac{1}{|\Gamma|}\right)^{\sum_{v=0}^{|r_0|-1} m_{(v,0)}} = 0$$

Note that the binary representation of the output of $Q_i(C)$ deliberately abstracts away any details of the *delay* associated with interest satisfaction beyond the granularity of a cache hit or miss. Any attack which leverages a timing side-channel to determine the precise router within a cluster at which content is stored, even if successful, reveals no more information about individual consumers' requests than the knowledge that the content is cached *anywhere* in the cluster. Because the router at which content is to be cached is chosen uniformly at random, knowledge of the content in a particular router's cache in no way leaks any information about which specific consumer downstream from the cluster requested that content.

4.3 Quantifying Utility

Definition 3 Utility [2]. *Let $\mathcal{H}(\rho)$ denote the random variable describing the distribution of the number of cache hits depending on the total number of requests ρ ($\rho \geq 1$). The utility function $u : \mathbb{N} \to \mathbb{R}+$ of a cache management scheme is defined as: $u(\rho) = \frac{1}{\rho}\mathbb{E}(\mathcal{H}(\rho))$*

Intuitively, we define utility as the expected number of cache hits as a fraction of total interests issued. Using notation and assumptions from Sect. 4.2, let G denote the set of all clusters in the network (cluster count $n = |G|$) and $r_{g,0}$ denote r_0 for a given cluster $g \in G$.

Theorem 2. *The utility $u(\rho) = \frac{1}{\rho}\mathbb{E}(\mathcal{H}(\rho))$ of Collaborative-Caching is:*

$$\frac{1}{\rho} \cdot \frac{\left(\sum\limits_{g=0}^{n} \left(|r_{g,0}| - \sum\limits_{v=0}^{|r_{g,0}|-1} \left(1 - \frac{1}{|\Gamma|} \right)^{m_{(v,0)}} \right) \right)}{n}$$

Proof. If we consider the scale of a single collaborating router cluster, intuitively, the probability of a cache hit for a given interest is directly related to the probability that any consumer (including the one that issued the interest in question) downstream from the cluster has already requested the same content for which the interest was issued. Using the same notation as in our proof of Theorem 4.1, and again making the assumption that the content requested in consumers' interests is uniformly distributed, we denote this probability as the following:

$$\Pr[\exists v \in U \text{ s.t. } S_0(C,v) \geq 1] = \sum_{v=0}^{|r_0|-1} \left(1 - \left(1 - \frac{1}{|\Gamma|} \right)^{m_{(v,0)}} \right) = |r_0| - \sum_{v=0}^{|r_0|-1} \left(1 - \frac{1}{|\Gamma|} \right)^{m_{(v,0)}} \tag{9}$$

for some network state S_0. Now consider the wider scope of all clusters in the network, letting $n = |G|$ denote the total number of clusters, where G is the

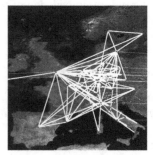

(b) WIDE Topology. 11 producers, 6 consumers, 13 routers acting as in-network caches. From the Internet Topology Zoo[13]. Figure from http://www.topology-zoo.org/dataset.html

(a) GEANT Topology. 13 producers, 8 consumers, 32 routers acting as in-network caches. From the Internet Topology Zoo[13]. Figure from http://www.topology-zoo.org/dataset.html

(c) GARR Topology. 13 producers, 21 consumers, 27 routers acting as in-network caches. From the Internet Topology Zoo[13]. Figure from http://www.topology-zoo.org/dataset.html

(d) TISCALI topology. 44 producers, 36 consumers, 160 routers acting as in-network caches. Parsed from the Rocketfuel dataset[24]. Figure from https://research.cs.washington.edu/networking/rocketfuel/interactive/

Fig. 2. Network topologies used in simulations.

set of all clusters in the network, and letting $r_{g,0}$ denote r_0 for a given cluster $g \in G$. Averaging the above probability over all clusters yields, as the value of $\mathbb{E}(\mathcal{H}(\rho))$:

$$\frac{\sum_{g=0}^{n-1} \left(|r_{g,0}| - \sum_{v=0}^{|r_{g,0}|-1} \left(1 - \frac{1}{|\Gamma|} \right)^{m_{(v,0)}} \right)}{n}$$

which we then multiply by $\frac{1}{\rho}$ to derive the expected cache hit ratio in Theorem 2.

5   Simulation Results

Having defined *Collaborative Caching*'s utility as a function of network traffic and topology, we now establish the practical implications of those bounds in a variety of simulated environments, with the intent of precisely quantifying the utility penalty one might expect to suffer in exchange for the proven privacy guarantees established in Sect. 4.2 and evaluating the impact of cache eviction on utility. Of particular interest is the comparative performance of multicast *Hash Routing* [21], a scheme expressly designed for improving ICN caching performance and which, when benchmarked against *Collaborative Caching*, should produce telling results regarding the trade-off between privacy and utility. Using the Icarus ICN caching performance simulation framework [22], experiments encompassing a variety of network topologies (detailed in Fig. 2), traffic characteristics, and caching schemes were performed. Constant factors in all experiments included the existence of $3x10^5$ unique content objects, $3x10^5$ "warmup" requests (issued prior to the beginning of performance measurements, to populate in-network caches), $6x10^5$ measured requests (used to compute results), an aggregate request rate of 1 per second, uniform content distribution amongst producers, and uniform cache space allocation amongst all in-network caches. Each experiment was parameterized by a unique combination of: cache eviction policy $p \in$ {Least Recently Used (LRU), Practical/In-Cache Least Frequently Used (LFU)}, traffic including requests characterized by a Zipf distribution with coefficient $\alpha \in \{0.6, 0.8, 1.0\}$, total network cache size $n \in \{0.002, 0.008, 0.02\}$ as a fraction of content objects in the network, topology $t \in$ {GEANT, WIDE, GARR, TISCALI}, and caching strategy $s \in$ {No Caching, Leave Copy Everywhere [11], Cache Less For More [3], ProbCache [20], Leave Copy Down [15], Random Choice [22], Random Bernoulli [22], Multicast Hash Routing [21], Collaborative Caching}. For all topologies, "unclustered" variants of multicast *Hash Routing* and *Collaborative Caching* (wherein all cache nodes form one large cluster, implying a total cluster count of 1) were tested, whereas cluster counts of 2, 4, and 8 were also used in experiments involving "clustered" variants of multicast *Hash Routing* and *Collaborative Caching* on smaller topologies (WIDE and GARR), as opposed to cluster counts of 5, 10, and 20 for those same experiments on larger topologies (TISCALI and GEANT). Exhaustive simulations were conducted, including all possible experiments parameterized by each element of the

Table 1. Comparing the interest satisfaction latency and cache hit ratio observed in experiments pitting *Collaborative Caching* against a variety of other caching schemes using several network topologies. Zipf coefficient $\alpha = 0.8$. Cache size as a fraction of total content objects in network $= 0.008$. Reported values over ten trials per experiment indicated as "<mean> $\pm$ <error>", where (mean $-$ error, mean $+$ error) denotes a confidence interval of 99%. Data grouped into columns by cache eviction policy (Least Recently Used (LRU) vs. Practical/In-Cache Least Frequently Used (LFU)) and evaluated performance metric (cache hit ratio vs. interest satisfaction latency in milliseconds).

Caching	Topology	LRU Cache Hit Ratio	LRU Latency (ms)	LFU Cache Hit Ratio	LFU Latency (ms)
No Caching	GEANT	0.0000 ± 0.0000	87.1733 ± 0.0411	0.0000 ± 0.0000	87.2018 ± 0.0449
	WIDE	0.0000 ± 0.0000	78.2623 ± 0.0388	0.0000 ± 0.0000	78.2523 ± 0.0578
	GARR	0.0000 ± 0.0000	81.5429 ± 0.0243	0.0000 ± 0.0000	81.5514 ± 0.0261
	TISCALI	0.0000 ± 0.0000	91.9094 ± 0.0418	0.0000 ± 0.0000	91.9487 ± 0.0360
Leave Copy Everywhere	GEANT	0.1340 ± 0.0013	77.2756 ± 0.0875	0.2490 ± 0.0009	68.0778 ± 0.0531
	WIDE	0.1131 ± 0.0025	70.0952 ± 0.1585	0.2175 ± 0.0015	62.4335 ± 0.1063
	GARR	0.0877 ± 0.0023	75.1306 ± 0.1537	0.1931 ± 0.0014	67.2137 ± 0.1026
	TISCALI	0.0629 ± 0.0029	87.2501 ± 0.2061	0.1592 ± 0.0014	79.3937 ± 0.0980
Cache Less for More	GEANT	0.0976 ± 0.0012	79.3849 ± 0.0976	0.1455 ± 0.0021	75.4987 ± 0.1557
	WIDE	0.1175 ± 0.0012	69.6373 ± 0.0795	0.1851 ± 0.0016	64.7108 ± 0.1153
	GARR	0.1257 ± 0.0016	72.0542 ± 0.1068	0.1748 ± 0.0017	68.5384 ± 0.1129
	TISCALI	0.0810 ± 0.0015	85.2013 ± 0.1149	0.1142 ± 0.0031	82.9270 ± 0.2286
ProbCache	GEANT	0.1901 ± 0.0012	73.1032 ± 0.0810	0.2097 ± 0.0009	70.8426 ± 0.0762
	WIDE	0.1600 ± 0.0021	66.7402 ± 0.1302	0.2065 ± 0.0015	63.2143 ± 0.0988
	GARR	0.1346 ± 0.0012	71.8143 ± 0.0769	0.1787 ± 0.0010	68.2259 ± 0.0675
	TISCALI	0.0966 ± 0.0023	84.7233 ± 0.1597	0.1254 ± 0.0019	81.7880 ± 0.1281
Leave Copy Down	GEANT	0.1901 ± 0.0006	72.2230 ± 0.0482	0.2321 ± 0.0010	69.3720 ± 0.0750
	WIDE	0.1552 ± 0.0012	66.9467 ± 0.0728	0.2179 ± 0.0018	62.4464 ± 0.1200
	GARR	0.1420 ± 0.0012	70.8519 ± 0.0873	0.1865 ± 0.0020	67.7530 ± 0.1326
	TISCALI	0.1111 ± 0.0011	82.9278 ± 0.0867	0.1492 ± 0.0026	80.4490 ± 0.1715
Random Choice	GEANT	0.1714 ± 0.0008	74.2851 ± 0.0523	0.2451 ± 0.0012	68.3600 ± 0.0832
	WIDE	0.1396 ± 0.0021	68.1645 ± 0.1348	0.2176 ± 0.0014	62.4283 ± 0.0900
	GARR	0.1136 ± 0.0013	73.1932 ± 0.0904	0.1896 ± 0.0017	67.4475 ± 0.1115
	TISCALI	0.0848 ± 0.0030	85.4584 ± 0.2096	0.1544 ± 0.0018	79.7452 ± 0.1239
Random Bernoulli	GEANT	0.1691 ± 0.0008	74.5642 ± 0.0509	0.2462 ± 0.0007	68.2988 ± 0.0494
	WIDE	0.1408 ± 0.0022	68.0703 ± 0.1325	0.2171 ± 0.0010	62.4687 ± 0.0634
	GARR	0.1136 ± 0.0022	73.2183 ± 0.1399	0.1908 ± 0.0013	67.3758 ± 0.0870
	TISCALI	0.0840 ± 0.0034	85.6602 ± 0.2316	0.1573 ± 0.0016	79.5238 ± 0.1101
Multicast Hash Routing	GEANT (1 Cluster)	0.2024 ± 0.0004	77.1330 ± 0.0360	0.2980 ± 0.0007	69.9174 ± 0.0533
	GEANT (5 Clusters)	0.1498 ± 0.0011	81.0665 ± 0.2781	0.2614 ± 0.0024	71.8809 ± 0.5363
	GEANT (10 Clusters)	0.1445 ± 0.0031	80.3193 ± 0.6959	0.2581 ± 0.0031	71.4754 ± 0.5042
	GEANT (20 Clusters)	0.1327 ± 0.0030	80.3586 ± 0.4330	0.2520 ± 0.0034	70.5184 ± 0.4933
	WIDE (1 Cluster)	0.2022 ± 0.0003	69.4218 ± 0.0231	0.2992 ± 0.0009	62.5137 ± 0.0660
	WIDE (2 Clusters)	0.1772 ± 0.0027	70.7592 ± 0.5882	0.2753 ± 0.0017	63.9659 ± 0.5566
	WIDE (4 Clusters)	0.1529 ± 0.0037	71.8960 ± 0.8385	0.2594 ± 0.0058	64.2794 ± 0.5646
	WIDE (8 Clusters)	0.1412 ± 0.0039	70.5638 ± 0.5580	0.2391 ± 0.0035	62.7147 ± 0.4347
	GARR (1 Cluster)	0.2022 ± 0.0005	72.1170 ± 0.0419	0.2985 ± 0.0007	65.1865 ± 0.0504
	GARR (2 Clusters)	0.1678 ± 0.0016	76.0041 ± 0.1719	0.2834 ± 0.0014	67.3761 ± 0.1233
	GARR (4 Clusters)	0.1540 ± 0.0010	75.9113 ± 0.2279	0.2633 ± 0.0010	67.5829 ± 0.1285
	GARR (8 Clusters)	0.1435 ± 0.0036	76.0570 ± 0.4793	0.2488 ± 0.0022	67.9702 ± 0.4740
	TISCALI (1 Cluster)	0.2023 ± 0.0005	85.9880 ± 0.0348	0.2986 ± 0.0009	78.6860 ± 0.0671
	TISCALI (5 Clusters)	0.1447 ± 0.0027	91.9963 ± 0.3347	0.2515 ± 0.0033	83.1105 ± 0.2408
	TISCALI (10 Clusters)	0.1361 ± 0.0043	91.6054 ± 0.4447	0.2393 ± 0.0023	82.6559 ± 0.3442
	TISCALI (20 Clusters)	0.1221 ± 0.0043	92.1700 ± 1.1268	0.2261 ± 0.0049	83.7201 ± 0.4924
Collaborative Caching	GEANT (1 Cluster)	0.2019 ± 0.0003	77.1064 ± 0.0229	0.2983 ± 0.0006	69.8392 ± 0.0500
	GEANT (5 Clusters)	0.1501 ± 0.0021	81.1400 ± 0.4658	0.2619 ± 0.0014	71.9911 ± 0.4231
	GEANT (10 Clusters)	0.1424 ± 0.0022	80.4245 ± 0.3510	0.2557 ± 0.0030	71.6417 ± 0.4440
	GEANT (20 Clusters)	0.1357 ± 0.0028	80.2378 ± 0.4199	0.2543 ± 0.0030	70.2752 ± 0.3869
	WIDE (1 Cluster)	0.2021 ± 0.0005	69.3865 ± 0.0365	0.2989 ± 0.0008	62.4855 ± 0.0626
	WIDE (2 Clusters)	0.1755 ± 0.0033	71.0600 ± 0.7987	0.2756 ± 0.0022	63.7720 ± 0.6622
	WIDE (4 Clusters)	0.1593 ± 0.0032	71.5101 ± 0.5026	0.2565 ± 0.0077	64.2052 ± 0.8650
	WIDE (8 Clusters)	0.1390 ± 0.0048	71.0124 ± 0.6516	0.2449 ± 0.0068	63.1633 ± 0.8622
	GARR (1 Cluster)	0.2019 ± 0.0004	72.1890 ± 0.0332	0.2985 ± 0.0006	65.2583 ± 0.0396
	GARR (2 Clusters)	0.1695 ± 0.0013	75.9091 ± 0.1358	0.2829 ± 0.0015	67.5114 ± 0.1462
	GARR (4 Clusters)	0.1544 ± 0.0016	75.7915 ± 0.3114	0.2627 ± 0.0010	67.5725 ± 0.1946
	GARR (8 Clusters)	0.1440 ± 0.0030	76.1438 ± 0.4875	0.2504 ± 0.0023	67.9309 ± 0.3139
	TISCALI (1 Cluster)	0.2003 ± 0.0003	86.4305 ± 0.0235	0.2973 ± 0.0009	79.0758 ± 0.0690
	TISCALI (5 Clusters)	0.1444 ± 0.0026	92.0837 ± 0.2891	0.2501 ± 0.0016	83.2046 ± 0.3384
	TISCALI (10 Clusters)	0.1331 ± 0.0031	92.3790 ± 0.3812	0.2384 ± 0.0025	83.4196 ± 0.6007
	TISCALI (20 Clusters)	0.1216 ± 0.0036	92.4759 ± 0.2060	0.2230 ± 0.0049	84.5297 ± 0.8395

Cartesian product of all aforementioned parameter sets (10 trials for each of 1,080 unique experiments).

The data produced by the experiments with a Zipf coefficient of 0.8 and total network cache size of 0.008 (as a fraction of content objects in the network) proved to be representative of trends in the larger collected results as a whole, and is therefore provided in Table 1 as a focused subset thereof. Predictably, the performance of all schemes (with the exception of "No Caching") improved as aggregate cache size and the Zipf coefficient α (indicating the relative similarity/overlap of content requests) increased. *Collaborative Caching* consistently performed on par with *Hash Routing* regardless of cluster count, in some cases out-performing it relative to both latency and cache hit ratio metrics, and occasionally trailing *Hash Routing*'s performance by a very thin margin. Unclustered *Collaborative Caching* and unclustered *Hash Routing* achieved notably higher cache hit ratios than other schemes for each topology, and were often among the schemes with the lowest reported interest satisfaction latencies, as well. Interestingly, as cluster count decreased (and cluster size consequently increased), both *Collaborative Caching* and *Hash Routing* performed more favorably (lower latencies and higher cache hit ratios).

This trend is likely the result of the focus of our chosen simulation framework (namely, the measurement of caching performance). Our scheme has the potential to increase utility and privacy simultaneously. As cluster size increases, the likelihood that a given interest intercepted by the cluster corresponds to content cached in the cluster must monotonically increase, regardless of the distribution of content and interests. Also, the number of connected consumers must monotonically increase, increasing the size of the anonymity set of which downstream consumers are a part. The downside of increased cluster size is the overhead incurred by coordination and communication within the cluster, and simulating the resulting link saturation and congestion is not a problem Icarus claims to accurately emulate. We supplement these empirical observations with a theoretical calculation of *Collaborative Caching*'s utility as a function of network characteristics in Theorem 2.

6   Conclusions

We set out to demonstrate a caching scheme for ICN which would provide provable privacy guarantees and attack vector resilience for network consumers with negligible performance degradation. We have shown that, in a variegated pool of simulated environments, the interest satisfaction latencies and cache hit ratios afforded by our caching scheme are comparable to, and occasionally better than, those observed when schemes solely designed for improving cache utility are used. However, unlike those alternative methods, *Collaborative Caching* is able to accomplish this whilst preserving consumer privacy.

References

1. Abani, N., Braun, T., Gerla, M.: Betweenness centrality and cache privacy in information-centric networks. In: Proceedings of the 5th ACM Conference on Information-Centric Networking, ICN: 518, pp. 106–116. Association for Computing Machinery, New York (2018). https://doi.org/10.1145/3267955.3267964
2. Acs, G., Conti, M., Gasti, P., Ghali, C., Tsudik, G., Wood, C.A.: Privacy-aware caching in information-centric networking. IEEE Trans. Dependable Secure Comput. **16**(2), 313–328 (2019). https://doi.org/10.1109/TDSC.2017.2679711
3. Chai, W.K., He, D., Psaras, I., Pavlou, G.: Cache "less for more" in information-centric networks (extended version). Comput. Commun. **36**(7), 758–770 (2013). https://doi.org/10.1016/j.comcom.2013.01.007, https://doi.org/10.1016/j.comcom.2013.01.007
4. DiBenedetto, S., Gasti, P., Tsudik, G., Uzun, E.: Andana: anonymous named data networking application. In: 19th Annual Network and Distributed System Security Symposium, NDSS 2012, San Diego, California, USA, 5–8 February, 2012 (2012), https://www.ndss-symposium.org/ndss2012/andana-anonymous-named-data-networking-application
5. Fotiou, N., Nikander, P., Trossen, D., Polyzos, G.C.: Developing information networking further: from PSIRP to PURSUIT. In: Tomkos, I., Bouras, C.J., Ellinas, G., Demestichas, P., Sinha, P. (eds.) BROADNETS 2010. LNICST, vol. 66, pp. 1–13. Springer, Heidelberg (2012). https://doi.org/10.1007/978-3-642-30376-0_1
6. Gasti, P., Tsudik, G.: Content-centric and named-data networking security: the good, the bad and the rest. In: 2018 IEEE International Symposium on Local and Metropolitan Area Networks (LANMAN), pp. 1–6, June 2018. https://doi.org/10.1109/LANMAN.2018.8475052
7. Ghali, C., Tsudik, G., Wood, C.A.: When encryption is not enough: privacy attacks in content-centric networking. In: Proceedings of the 4th ACM Conference on Information-Centric Networking, ICN 2017, pp. 1–10. ACM, New York (2017). https://doi.org/10.1145/3125719.3125723, http://doi.acm.org/10.1145/3125719.3125723
8. Gotz, M., Machanavajjhala, A., Wang, G., Xiao, X., Gehrke, J.: Publishing search logs–a comparative study of privacy guarantees. IEEE Trans. Knowl. Data Eng. **24**(3), 520–532 (2012)
9. ISO: ISO 80000-2: Quantities and units – Part 2: Mathematical signs and symbols to be used in the natural sciences and technology. International Organization for Standardization, Geneva, Switzerland, December 2009. https://www.iso.org/standard/31887.html
10. Jacobson, V., Smetters, D.K., Thornton, J.D., Plass, M.F., Briggs, N.H., Braynard, R.L.: Networking named content. In: Proceedings of the 5th International Conference on Emerging Networking Experiments and Technologies, CoNEXT Õ 2009, pp. 1–12. Association for Computing Machinery, New York (2009). https://doi.org/10.1145/1658939.1658941
11. Jacobson, V., Smetters, D.K., Thornton, J.D., Plass, M.F., Briggs, N.H., Braynard, R.L.: Networking named content. In: Proceedings of the 5th International Conference on Emerging Networking Experiments and Technologies, CoNEXT 2009, pp. 1–12. Association for Computing Machinery, New York (2009). https://doi.org/10.1145/1658939.1658941

12. Kaufman, L., Rousseeuw, P.J.: Partitioning Around Medoids (Program PAM), chap. 2, pp. 68–125. John Wiley & Sons, Ltd. (2008). https://doi.org/10.1002/9780470316801.ch2, https://onlinelibrary.wiley.com/doi/abs/10.1002/9780470316801.ch2

13. Knight, S., Nguyen, H.X., Falkner, N., Bowden, R., Roughan, M.: The internet topology zoo. IEEE J. Sel. Areas Commun. **29**(9), 1765–1775 (2011)

14. Kumar, N., Aleem, A., Singh, A.K., Srivastava, S.: NBP: namespace-based privacy to counter timing-based attack in named data networking. J. Network Comput. Appl. **144**, 155–170 (2019). https://doi.org/10.1016/j.jnca.2019.07.004, http://www.sciencedirect.com/science/article/pii/S1084804519302280

15. Laoutaris, N., Che, H., Stavrakakis, I.: The LCD interconnection of LRU caches and its analysis. Perform. Eval. **63**(7), 609–634 (2006). https://doi.org/10.1016/j.peva.2005.05.003

16. Machanavajjhala, A., Kifer, D., Abowd, J., Gehrke, J., Vilhuber, L.: Privacy: theory meets practice on the map. In: 2008 IEEE 24th International Conference on Data Engineering, pp. 277–286 (2008)

17. Mohaisen, A., Mekky, H., Zhang, X., Xie, H., Kim, Y.: Timing attacks on access privacy in information centric networks and countermeasures. IEEE Trans. Dependable Secure Comput. **12**(6), 675–687 (2015)

18. Mohaisen, A., Zhang, X., Schuchard, M., Xie, H., Kim, Y.: Protecting access privacy of cached contents in information centric networks. In: Proceedings of the 8th ACM SIGSAC Symposium on Information, Computer and Communications Security, ASIA CCS: 513, pp. 173–178. Association for Computing Machinery, New York (2013). https://doi.org/10.1145/2484313.2484335, https://doi.org/10.1145/2484313.2484335

19. NDN Project Homepage (2020). https://named-data.net/. Accessed 4 June 2020

20. Psaras, I., Chai, W.K., Pavlou, G.: Probabilistic in-network caching for information-centric networks. In: Proceedings of the Second Edition of the ICN Workshop on Information-Centric Networking, ICN 2012, pp. 55–60. Association for Computing Machinery, New York (2012). https://doi.org/10.1145/2342488.2342501, https://doi.org/10.1145/2342488.2342501

21. Saino, L., Psaras, I., Pavlou, G.: Hash-routing schemes for information centric networking. In: Proceedings of the 3rd ACM SIGCOMM Workshop on Information-Centric Networking, ICN 2013, pp. 27–32. Association for Computing Machinery, New York (2013). https://doi.org/10.1145/2491224.2491232

22. Saino, L., Psaras, I., Pavlou, G.: Icarus: a caching simulator for information centric networking (ICN). In: Proceedings of the 7th International ICST Conference on Simulation Tools and Techniques (SIMUTOOLS 2014). ICST, ICST, Brussels, Belgium (2014)

23. Sourlas, V., Psaras, I., Saino, L., Pavlou, G.: Efficient hash-routing and domain clustering techniques for information-centric networks. Comput. Networks **103**, 67–83 (2016). https://doi.org/10.1016/j.comnet.2016.04.001, http://www.sciencedirect.com/science/article/pii/S1389128616300998

24. Spring, N., Mahajan, R., Wetherall, D.: Measuring ISP topologies with rocketfuel. SIGCOMM Comput. Commun. Rev. **32**(4), 133–145 (2002). https://doi.org/10.1145/964725.633039

25. Tsudik, G., Uzun, E., Wood, C.A.: Ac3n: anonymous communication in content-centric networking. In: 2016 13th IEEE Annual Consumer Communications Networking Conference (CCNC), pp. 988–991, January 2016. https://doi.org/10.1109/CCNC.2016.7444924

26. Xue, K., et al.: A secure, efficient, and accountable edge-based access control framework for information centric networks. IEEE/ACM Trans. Networking **27**(3), 1220–1233 (2019)
27. Yao, L., Jiang, B., Deng, J., Obaidat, M.S.: LSTM-based detection for timing attacks in named data network. In: 2019 IEEE Global Communications Conference (GLOBECOM), pp. 1–6 (2019)

Affine Tasks for k-Test-and-Set

Petr Kuznetsov[1] and Thibault Rieutord[2(✉)]

[1] LTCI, Télécom Paris, Institut Polytechnique Paris, Palaiseau, France
[2] Université Paris-Saclay, CEA, List, 91120 Palaiseau, France
thibault.rieutord@cea.fr

Abstract. The paper proposes a surprisingly simple characterization of task computability of the wait-free shared-memory model in which processes, in addition to read-write registers, have access to k-*test-and-set* objects. Our characterization is expressed in the form of an *affine task*: a subcomplex of some iteration of the standard chromatic subdivision. This appears to be the first topological characterization of a model in which processes communicate via long-lived objects beyond read-write registers.

Keywords: Affine tasks · Distributed computability · Test-and-set

1 Introduction

One of the central challenges in the theory of distributed computing is determining *relative computability* of its numerous models, parameterized by types of failures they expose (crash, omission, Byzantine), synchrony hypotheses they assume (asynchronous, partially synchronous, synchronous), and communication primitives they employ (message-passing, read-write registers, powerful shared objects). Starting from the seminal work by Herlihy and Shavit [18], task computability of multiple models of computation have been characterized using the language of combinatorial topology [12,17,23,26]. More precisely, given a task T and a model of computation M, we can equate the question of whether T is solvable in M with the existence of a specific continuous map between a transformed *input complex* of T and the *output complex* of T, *carried by* T, i.e., preserving the task specification. M entirely determines the way the input complex is transformed.

For example, to characterize task computability in the *wait-free* read-write model [15], we can simply consider a *subdivision* of the input complex [18]. In particular, we can choose this subdivision to be a number of iterations of the *standard chromatic* subdivision (denoted Chr, Fig. 1). The complex captures one round of *immediate snapshot* (IS) [3].

Task computability in the t-resilient read-write model has been characterized [26] via a specific task $\mathcal{R}_{t-res}$. The task is defined for n processes as a restriction of the *double* immediate snapshot task: the output complex of the task is a sub-complex consisting of *all* simplices of the second iteration of the

© Springer Nature Switzerland AG 2020
S. Devismes and N. Mittal (Eds.): SSS 2020, LNCS 12514, pp. 151–166, 2020.
https://doi.org/10.1007/978-3-030-64348-5_12

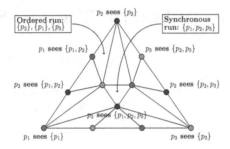

Fig. 1. Chr(**s**), the standard chromatic subdivision of a 2-simplex, the output complex of the 3-process *IS* task.

standard chromatic subdivision of the task's input complex, except the simplices adjacent to the $(n-t-1)$-skeleton of the input complex. Intuitively the output complex of $\mathcal{R}_{t-res}$ contains all of 2-round *IS* runs in which every process "sees" at least $n-t-1$ other processes. Figure 2 depicts the output complex of $\mathcal{R}_{1-res}$, the affine task for the 3-process 1-resilient model.

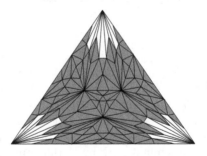

Fig. 2. $\mathcal{R}_{1-res}$, the affine task of 1-resilience (in blue). (Color figure online)

Complex Chr and $\mathcal{R}_{t-res}$ are called *affine tasks* [11,12] for the wait-free model and *t*-resilient model, respectively. More generally, an affine task $\mathcal{R}_M$ for a model M is defined as a subcomplex of a finite number of iteration of the standard chromatic subdivision, such that a task T is solvable in M if and only if there exists a continuous map from a finite number of iterations of $\mathcal{R}_M$ on the input complex of T to the output complex of T, carried by T.

Recently, affine tasks for a large class of *fair adversarial* models have been characterized via affine tasks [23]. An adversarial [8] shared-memory is defined via a collection $\mathcal{A}$ of process subsets, called *live sets*. A run is in the corresponding *adversarial $\mathcal{A}$-model* if the set of processes taking infinitely many steps in it is a live set of $\mathcal{A}$. Computability of adversarial models has been characterized for several special cases: wait-freedom [18], *t*-resilience [26], *k*-concurrency [11] and, finally, fair adversaries [23].

These characterizations mostly focused on *read-write* shared memory. The only exception is the work by Gafni et al. [11], where the processes could, additionally, access k-set consensus objects. Via simulations, this model was shown to be equivalent to the model of k-*concurrency* that assumes up to k processes are allowed to be concurrently active. However, to the best of our knowledge, no direct topological characterization has been proposed for models in which processes communicate via powerful shared objects, other than read-write registers. In this paper, we complement earlier results with a simple characterization of a model in which processes, in addition to read-write registers, can access k-test-and-set objects, for a fixed natural k. A k-test-and-set object is accessed by a single operation that eventually returns either 1 or 0, so that at least 1 and at most k of the participating processes obtain 1.

It has been observed that computability of a model in distributed computing is tightly coupled with its ability to solve *set consensus*. The *agreement function* [20,25] of a model specifies the "best" level of set consensus, i.e., the minimal guaranteed number of distinct output values, for all sets of *participating* (proposing inputs and expecting outputs) and *active* (taking steps) processes. Depending on their agreement functions, models can be classified in (1) *symmetric* models, where agreement functions only depend on the set cardinalities, (2) *fair*, where agreement functions do not depend on the active sets, (3) *local*, where agreement functions do not depend on the participating sets, and (4) *regular*, where agreement function increase with increasing participation. In Fig. 3, we depict interrelations between these and other shared-memory models.

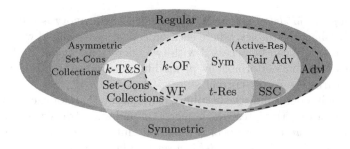

Fig. 3. A classification of shared-memory models based on agreement functions.

In this paper, we define an affine task $\mathcal{R}_{k-T\&S}$ capturing the task computability of the wait-free read-write shared memory models equipped with k-test-and-set objects. Our characterization can be put as the following generalization of the ACT [18]:

A task $T = (\mathcal{I}, \mathcal{O}, \Delta)$ is solvable in a wait-free shared memory models enhanced with k-test-and-set objects if and only if there exists a natural number ℓ and a simplicial map $\phi : \mathcal{R}^{\ell}_{k-T\&S}(\mathcal{I}) \to \mathcal{O}$ carried by Δ.

Task $\mathcal{R}_{k-T\&S}$ (see Fig. 5) can be defined as a subcomplex of Chr, just a single iteration of the standard chromatic subdivision. In contrast, affine tasks of other models, such as k-concurrency or t-resilience ($1 < k < n$ and $0 < t < n - 1$) require at least two iterations [7,11,26].

We believe that the results can be extended to all "practical" restrictions of the wait-free model which may result in a complete computability theory for distributed computing shared-memory models. Affine tasks may also lead to decidable characterization of relative task computability, as has been recently shown for 2-process models [22].

Roadmap. Section 2 reviews our model definitions. Section 3 defines our affine task $\mathcal{R}_{k-T\&S}$. In Sect. 4, we show that $\mathcal{R}^*_{k-T\&S}$ can be simulated in the wait-free shared memory model enhanced with k-test-and-set objects. In Sect. 5, we show that, reciprocally, any task solvable in the wait-free shared memory model enhanced with k-test-and-set objects can be solved in $\mathcal{R}^*_{k-T\&S}$. Section 6 reviews related work and concludes the paper. Missing proofs can be found in the full version of the paper [21].

2   Preliminaries

We assume a system of n asynchronous processes, $\Pi = \{p_1, \ldots, p_n\}$. Two models of communication are considered: (1) *atomic snapshots* [1] and (2) *iterated immediate snapshots* [4,18].

Atomic Snapshot Models. The atomic-snapshot (AS) memory is represented as a vector of n shared variables, where each process p_i is associated with the position i. The memory can be accessed with two operations: *update* and *snapshot*. An *update* operation performed by p_i modifies the value at position i and a *snapshot* returns the vector current state.

A *protocol* is a deterministic distributed automaton that, for each process and each its local state, stipulates which operation and state transition the process may perform. A *run* of a protocol is a possibly infinite sequence of alternating states and operations. An AS *model* is a set of infinite runs.

In an infinite run of the AS model, a process that takes only finitely many steps is called *faulty*, otherwise it is called *correct*. We assume that in its first step, a process shares its initial state using the *update* operation. If a process completed this first step in a given run, it is said to be *participating*, and the set of participating processes is called the *participating set*.

Iterated Immediate Snapshot Model. In the iterated immediate snapshot (IIS) model, processes proceed through an infinite sequence of independent memories $M_1, M_2, \ldots$. Each memory M_r is always accessed by a process with a single *WriteSnapshot* operation [3]: the operation performed by p_i takes a value v_{ir} and returns a set V_{ir} of submitted values (w.l.o.g, values of different processes are distinct), satisfying the following properties (See Fig. 4 for IS examples):

- self-inclusion: $v_{ir} \in V_{ir}$;
- containment: $(V_{ir} \subseteq V_{jr}) \vee (V_{jr} \subseteq V_{ir})$;
- immediacy: $v_{ir} \in V_{jr} \Rightarrow V_{ir} \subseteq V_{jr}$.

Fig. 4. Examples of valid sets of IS outputs. On the left, we have the "ordered" execution in which every process outputs a distinct set of inputs (blue - only itself, red - blue and itself, and green - all three). On the right, we have the "synchronous" execution in which all three processes output all the inputs. (Color figure online)

In the IIS communication model, we assume that processes run the *full-information* protocol, in which, the first value each process writes is its *initial state*. For each $r > 1$, the outcome of the WriteSnapshot operation on memory M_{r-1} is submitted as the input value for the WriteSnapshot operation on M_r. There are no failures in the IIS model, all processes go through infinitely many *IS* instances.

Note that the wait-free AS model and the IIS model are equivalent as regards task solvability [3,16].

Tasks. In this paper, we focus on distributed *tasks* [18]. A process invokes a task with an *input* value and the task returns an *output* value, so that the inputs and the outputs across the processes respect the task specification. Formally, a *task* is defined through a set $\mathcal{I}$ of input vectors (one input value for each process), a set $\mathcal{O}$ of output vectors (one output value for each process), and a total relation $\Delta : \mathcal{I} \mapsto 2^{\mathcal{O}}$ that associates each input vector with a set of possible output vectors. We require that Δ is a *carrier* map: $\forall \rho, \sigma \in \mathcal{I}, \rho \subseteq \sigma : \Delta(\rho) \subseteq \Delta(\sigma)$. An input $\perp$ denotes a *non-participating* process and an output value $\perp$ denotes an *undecided* process. Check [16] for more details on the definition.

In the *k-set consensus* task [6], input values are in a set of values V ($|V| \geq k + 1$), output values are in V, and for each input vector I and output vector O, $(I, O) \in \Delta$ if the set of non-$\perp$ values in O is a subset of values in I of size at most k. The case of 1-set consensus is called *consensus* [9].

A protocol solves a task $T = (\mathcal{I}, \mathcal{O}, \Delta)$ in a model M, if it ensures that in every run of M in which processes start with an input vector $I \in \mathcal{I}$, there is a finite prefix R of the run in which: (1) decided values form a vector $O \in \mathcal{O}$ such that $(I, O) \in \Delta$, and (2) all correct processes decide. Hence, in the IIS model, all processes must decide.

k-Test-and-Set Model. For an integer $k \geq 1$, a *k-test-and-set* object exports one operation, *apply*(), that may be only accessed once by each process, takes no parameters, and returns a boolean value. It guarantees that at most k processes will get 1 as output and that not all processes accessing it obtained 0. In the special case of $k = 1$ the object is simply called *test-and-set*.

The *k-test-and-set model* is then simply defined as the wait-free model with, additionally, access to any number of k-test-and-set objects. Hence processes run a full-information protocol on an AS memory without any restrictions on the set of possible runs, but processes may proceed, between any operations on the AS memory, to operations on any number of k-test-and-set objects.

Simplicial Complexes. We use the standard language of *simplicial complexes* [16, 27] to give a combinatorial representation of the IIS model. A *simplicial complex* is defined as a set of *vertices* and an inclusion-closed set of vertex subsets, called *simplices*. The *dimension* of a simplex σ is the number of vertices of σ minus one, and any subset of σ is one of its *faces*. We denote by **s** the *standard $(n-1)$-simplex*: a fixed set of n vertices and all its subsets.

Given a complex K and a simplex $\sigma \in K$, σ is a *facet* of K, denoted $facet(\sigma, K)$, if σ is not a face of any strictly larger simplex in K. Let $facets(K) = \{\sigma \in K, facet(\sigma, K)\}$. A simplicial complex is *pure* (of dimension m) if all its facets have dimension m.

A map α from the vertices of a complex K to the vertices of a complex L is *simplicial* if each simplex in K is mapped to a simplex in L. A simplicial map $\alpha : K \to L$ is *rigid* if for all $\sigma \in K$, $|\sigma| = |\alpha(\sigma)|$.

A simplicial complex is *chromatic* if it is equipped with a *coloring function*— a rigid simplicial map χ from its vertices to **s**, in one-to-one correspondence with n *colors*. In our setting, colors correspond to processes identifiers.

Standard Chromatic Subdivision and IIS. The *standard chromatic subdivision* [18] of a complex K, $\mathrm{Chr}\,K$ ($\mathrm{Chr}\,\mathbf{s}$ is depicted in Fig. 1), is a complex where vertices of $\mathrm{Chr}\,K$ are couples (c, σ), where c is a color and σ is a face of K containing a vertex of color c. Simplices of $\mathrm{Chr}\,K$ are the sets of vertices $(c_1, \sigma_1), \ldots, (c_m, \sigma_m)$ associated with distinct colors (i.e., $\forall i, j, c_i \neq c_j$) such that the σ_i satisfies the containment and immediacy properties of *IS*.

Every simplex σ has a *geometric realization* $|\sigma|$. It is obtained by representing the vertices of σ as an affinely independent set of points in a Euclidean space and then taking the convex hull of them. A geometric realization of a complex K, denoted by $|K|$, is the union of geometric realizations of its simplices, properly "glued" along their faces [16].

It has been shown that Chr is a *subdivision* [19], i.e., informally, $|\mathrm{Chr}\,\mathbf{s}|$ is homeomorphic to $|\mathbf{s}|$. If we *iterate* this subdivision m times, each time applying Chr to all simplices, we obtain the mth chromatic subdivision, $\mathrm{Chr}^m\mathbf{s}$. $\mathrm{Chr}^m\mathbf{s}$ precisely captures the runs of the m-round IIS model, IS^m [4,18].

Given a complex K and a subdivision of it $Sub(K)$, the carrier of a simplex $\sigma \in Sub(K)$ in K, $carrier(\sigma, K)$, is the smallest simplex $\rho \in K$ such that the geometric realization of σ, $|\sigma|$, is contained in $|\rho|$: $|\sigma| \subseteq |\rho|$. The carrier of a vertex $(p, \sigma) \in \mathrm{Chr}\,\mathbf{s}$ is σ. In the matching *IS* task, the carrier corresponds to the snapshot returned by p, i.e., the set of processes *seen* by p. The carrier of a simplex $\rho \in \mathrm{Chr}\,K$ is simply the union (or, due to inclusion, the maximum) of the carriers of vertices in ρ. Given a simplex $\sigma \in \text{Chr}^2\mathbf{s}$, $carrier(\sigma, \mathbf{s})$ is equal to $carrier(carrier(\sigma, \text{Chr}\,\mathbf{s}), \mathbf{s})$. $carrier(\sigma, \text{Chr}\,\mathbf{s})$ corresponds to the set of all snapshots seen by processes in $\chi(\sigma)$. Hence, $carrier(\sigma, \mathbf{s})$ corresponds to the union of all these snapshots. Intuitively, it results in the set of all processes *seen* by processes in $\chi(\sigma)$ through the two successive immediate snapshots instances.

Simplex Agreement and Affine Tasks. In the *simplex agreement task*, processes start on vertices of some complex K, forming a simplex $\sigma \in K$, and must output vertices of some subdivision of K, $Sub(K)$, so that outputs form a simplex ρ of $Sub(K)$ respecting carrier inclusion, i.e., $carrier(\rho, K) \subseteq \sigma$. In the simplex agreement tasks considered in the characterization of wait-free task computability [4,18], K is the standard simplex $\mathbf{s}$ and the subdivision is usually iterations of Chr.

An *affine task* is a generalization of the simplex agreement task, where $\mathbf{s}$ is fixed as the input complex and where the output complex is a pure non-empty sub-complex of some iteration of the standard chromatic subdivision, $\text{Chr}^\ell\mathbf{s}$. Formally, let L be a pure non-empty sub-complex of $\text{Chr}^\ell\mathbf{s}$ for some $\ell \in \mathbb{N}$. The affine task associated with L is then defined as $(\mathbf{s}, L, \Delta)$, where, for every face $\sigma \subseteq \mathbf{s}$, $\Delta(\sigma) = L \cap \text{Chr}^\ell(\sigma)$. Note that $L \cap \text{Chr}^\ell(\mathbf{t})$ can be empty, in which case the set of participating processes must increase before processes may produce outputs. Note that, since an affine task is characterized by its output complex, with a slight abuse of notation, we use L for both the affine task $(\mathbf{s}, L, \Delta)$ and its output complex.

By running m iterations of this task, we obtain L^m, a sub-complex of $\text{Chr}^{\ell m}\mathbf{s}$, corresponding to a subset of $IS^{\ell m}$ runs (each of the m iterations includes ℓ IS rounds). The affine model associated with L, denoted L^*, corresponds to the set of infinite runs of the IIS model where every prefix restricted to a multiple of ℓ IS rounds belongs to the subset of $IS^{\ell m}$ runs associated with L^m.

3   Affine Task for k-Test-and-Set

In this section, we define affine task $\mathcal{R}_{k-T\&S}$ capturing computability of the k-test-and-set model. The task is defined as a simple subcomplex of a single iteration of the standards chromatic subdivision.

The intuition is the following. Test-and-set solves, in a straightforward manner, perfect renaming [2]. It also provides adaptive solutions to renaming in which *names* of the processes reflect the order in which they access the task. Hence a process obtaining the name j can see all values shared previously by processes that receive a smaller name i with $i < j$. It can also be used to solve immediate snapshot [3] in a way that every process obtains a distinct rank j and observes the inputs of all processes with smaller positions: the process of level j sees precisely j inputs, its own plus those of the processes with strictly lower ranks. Such an immediate snapshot execution essentially impose *total order* on the processes.

We can naturally generalize this observation to k-test-and-set objects. Indeed, consider a partial order in which every process is associated with a rank so that at most k processes share the same rank. Similarly, a process can observe the values previously shared by processes obtaining a lower or equal rank. In an immediate snapshot, this is equivalent to having at most k processes sharing the same output. It leads to a definition of k-*ordered* executions or corresponding simplices of Chr: among any set of $k+1$ processes, at least two have different ranks. Note that total order executions corresponding to 1-test-and-set are 1-ordered.

$\mathcal{R}_{k-T\&S}$ captures the set of k-ordered executions. Formally, $\mathcal{R}_{k-T\&S}$ is the set of simplices of Chr $\mathbf{s}$, the standard chromatic subdivision, in which at most k vertices share the same carrier:

Definition 1. $\mathcal{R}_{k-T\&S}$ *is equal to:*

$$\sigma \in Chr(\mathbf{s}) : \forall \sigma' \subseteq \sigma, (\forall v, v' \in \sigma', carrier(v) = carrier(v')) \implies |\sigma'| \le k.$$

To be an affine task, $\mathcal{R}_{k-T\&S}$ needs to be a pure sub-complex of Chr $\mathbf{s}$ of the same dimension:

Property 1. $\mathcal{R}_{k-T\&S}$ is an affine task.

Proof. The fact that $\mathcal{R}_{k-T\&S}$ is a sub-complex of Chr $\mathbf{s}$ is trivial. Indeed, the definition is inclusion-closed. Consider any simplex $\sigma \in \mathcal{R}_{k-T\&S}$ and any face of it $\sigma' \subseteq \sigma$. Any face of σ' is a face of σ and hence satisfies the condition of having at most k vertices sharing the same carrier.

Showing that $\mathcal{R}_{k-T\&S} \subseteq$ Chr $\mathbf{s}$ is pure and of the same dimension as $\mathbf{s}$ is less trivial. For this, we need to show that any simplex $\sigma \in \mathcal{R}_{k-T\&S}$ is a face of a simplex $\sigma' \in \mathcal{R}_{k-T\&S}$ of dimension equal to $dim(\mathbf{s})$. Note that by transitivity, it is sufficient to show that any simplex of $\mathcal{R}_{k-T\&S}$ of a strictly smaller dimension than $dim(\mathbf{s})$ is the face of a strictly larger simplex of $\mathcal{R}_{k-T\&S}$.

Consider a simplex $\sigma \in \mathcal{R}_{k-T\&S}$ and any color c from Π such that $c \notin \chi(\sigma)$ and let $v \in \mathbf{s}$ be the vertex of $\mathbf{s}$ of color c. Two cases may happen: either c is a color of the carriers every vertex of in σ, or else, there exists a vertex in σ with the largest carrier $\mathbf{t}$ such that $c \notin \chi(\mathbf{t})$. In the former case, we can add vertex $(c, \{v\})$ to σ. In the latter case, we can add the vertex $(c, \{v\} \cup \mathbf{t})$ to σ. It is easy to check that the new simplex still verifies the immediacy, self-inclusion, and containment properties and, thus, belongs to Chr $\mathbf{s}$. Moreover, the carrier of the new vertex is distinct (shared by no vertex in σ), and hence the new simplex belongs to $\mathcal{R}_{k-T\&S}$. Indeed, any vertex $v \in \sigma$ such that $\chi(v) \in \chi(\mathbf{t})$ has a carrier that is a face of $\mathbf{t}$ due to the immediacy property. Hence, as long as there are missing colors, we can find a larger simplex in $\mathcal{R}_{k-T\&S}$ including σ as a face. Hence, $\mathcal{R}_{k-T\&S} \subseteq$ Chr $\mathbf{s}$ is indeed a pure complex of dimension of $\mathbf{s}$.   □

The affine tasks corresponding to 3-process models of 1-test-and-set and 2-test-and-set are depicted in Fig. 5. Note that affine tasks' facets are displayed in blue and, thus, the faces of blue simplices also belong to the affine task.

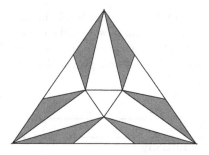 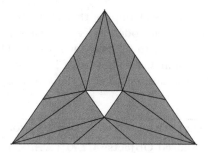

Fig. 5. 3-process affine tasks $\mathcal{R}_{1-T\&S}$ and $\mathcal{R}_{2-T\&S}$ with their facets displayed in blue. (Color figure online)

4   Solving $\mathcal{R}_{k-T\&S}$ in the k-Test-and-Set Model

Solving $\mathcal{R}_{k-T\&S}$ using k-test-and-set objects and read-write registers is rather straightforward. The idea, originally suggested in [14,24], consists in using a level-based immediate snapshot implementation that additionally uses access to k-test-and-set objects.

Recall that the level-based implementation of an immediate snapshot [3] operates as follows. Starting with level $\ell = n$, every process (1) writes its input and ℓ in the memory array; (2) takes a snapshot of the array; (3) if the snapshot contains ℓ values associated with levels $\ell' \leq \ell$, the process returns the snapshot consisting of these ℓ values; otherwise, the process proceeds to level $\ell - 1$ (see Fig. 4).

The modification proposed in [14,24] to solve $\mathcal{R}_{k-T\&S}$, consists in modifying step (3) of this implementation as follows: (3') if the snapshot contains ℓ values associated with a level $\ell' \leq \ell$ then processes accesses the k-test-and-set object number ℓ. If the k-test-and-set object returns *true*, then the process terminates with the snapshot consisting of these ℓ values; otherwise (in both else conditions), the process proceeds to level $\ell - 1$. The formal description is depicted in Algorithm 1, Figure 6 illustrates the general procedure.

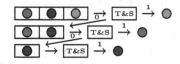

Fig. 6. Ordered IS algorithm with test-and-set.

The proof of the following theorem is delegated to the full version of the paper [21].

Theorem 1. *Algorithm 1 solves $\mathcal{R}_{k-T\&S}$.*

This result implies that every task solvable in $\mathcal{R}^*_{k-T\&S}$ is solvable in the k-test-and-set model. Indeed, a task solvable in $\mathcal{R}^*_{k-T\&S}$ implies a solution ϕ from $\mathcal{R}^m_{k-T\&S}(\mathcal{I})$ to $\mathcal{O}$, for some given m. One can thus simply iterate the solution of Algorithm 1 m times using its input and return the solution provided by ϕ to obtain a task solution in the k-test-and-set model.

Algorithm 1: Solving $\mathcal{R}_{k-T\&S}$ for process p_i.

---

1 **Shared Objects:** $MEM[1\ldots n] \in Val \times \mathbb{N})$, **initially** $(\perp, \perp)$;
2 **Init:** $level = n + 1$, $exit = true$, $value = InputValue$, $snap = \emptyset$;

3 **Do**
4 | **Do**
5 | | $level = level - 1$;
6 | | $MEM[i].Update(value, level)$;
7 | | $snap = MEM.Snapshot()$;
8 | **While** $|\{(v, \ell) \in snap, \ell \le level\}| \ne level$;
9 | $exit = k\text{-}Test\&Set[level]$;
10 **While** $\neg exit$;
11 **Return** $\{(v, \ell) \in snap, \ell \le level\}$;

---

5   Simulating k-Test-and-Set Model in $\mathcal{R}^*_{k-T\&S}$

Simulating the k-test-and-set model using iterations of $\mathcal{R}_{k-T\&S}$ ($\mathcal{R}^*_{k-T\&S}$) is slightly less straightforward. Simulating shared memory is well known for iterations of the standard chromatic subdivision, hence, for a subset of such runs as well. A standard simulation described in [13] (for completeness described in the full version of the paper [21]) ensures progress to the non-terminated processes with infinitely often the smallest view. This provides lock-freedom, one process makes progress and will eventually return with a task output. Hence, this is enough to ensure that eventually all processes can obtain a task output. Our goal is to ensure that the same set of "fast" processes make progress with their k-test-and-set operations as well.

As not all processes may participate in a k-test-and-set operation, processes with the smallest view need to progress independently. But processes with a larger view could later participate in a round in which they have the smallest view. Hence, to ensure that not more than k processes return 1, slow processes must preemptively fail the test-and-set operations. Unfortunately, this is not possible, as a process can only identify processes with smaller views and not precisely those that have the smallest view. Indeed, we could have a process preemptively fail without anyone returning 1. We resolve this issue by simulating k-set-consensus operations among sets of $k + 1$ processes and showing then that it is sufficient to simulate n-process k-test-and-set operations.

Let us first show how a $k+1$-process k-test-and-set can be used to implement an n-process k-test-and-set. Then, we will show how k-set consensus among $k+1$ processes can be used to implement k-test-and-set among $k+1$ processes. Lastly, we are going to show how to simulate operations of k-set consensus among $k+1$ processes in $\mathcal{R}^*_{k-T\&S}$.

From $k+1$-Process Test-and-Set to n-Process k-Test-and-Set. In the solution, depicted in Algorithm 2, every process participates in k-test-and-set operations among any possible subset of $k+1$ processes it belongs to. The processes iterate on over these sets of $k+1$ processes in the same deterministic order and return 0 as soon as they obtain 0 from a k-test-and-set operation. If a process manages to obtain 1 from all k-test-and-set operations, it returns 1.

Algorithm 2: n-process $k - T\&S$ using $k+1$-process k-$T\&S$ for p_i.

---

1   **forall** $S \subseteq \Pi, |S| = k+1$ **do**
2   **if** $i \in S$ **then**
3   **if** $k - T\&S[S].apply()=0$ **then** **Return** 0 ;
4   **Return** 1;

Theorem 2. *Algorithm 2 solves n-process k-test-and-set.*

Proof. Let us first show that at most k process may return 1. It is straightforward. Indeed, assume that there are $k+1$ processes returning 1. They must have all accessed the same k-test-and-set operation corresponding to their set of $k+1$ processes. But at most k of them may have obtained 1 from it, the remaining ones must have therefore returned from the protocol with 0—a contradiction.

Let us now show that not all participating processes may return 0. Indeed, consider the last set of $k+1$ processes in the sequence for which the associated k-test-and-set object has been accessed. Not all processes accessing this object can return 0. Hence, they must either access another k-test-and-set object afterward, which is not the case by assumption or return with 1 or crash. Therefore, not all participating processes can return 0.   □

From $(k+1)$-Process k-set Consensus to k-Test-and-Set. Using k-set consensus among $k+1$ processes to solve k-test-and-set operations among $k+1$ processes is straightforward. Processes can access a k-set consensus operation with their identifier. Then they write their output to the shared-memory and take a snapshot. If a process sees that some process obtained its identifier as output, it returns 1, and otherwise, it returns 0. See Algorithm 3 for a formal description.

Algorithm 3: $k + 1$-process $k - T\&S$ using $k + 1$-process k-set-consensus
for p_i.

---

1 $res = k$-set-consensus$[i]$;
2 $MEM.update(res)$;
3 $snap = MEM.snapshot()$;
4 **if** $i \in snap$ **then**
5 | **Return** 1;
6 **else**
7 | **Return** 0;

Theorem 3. *Algorithm 3 solves $k + 1$-process k-test-and-set.*

Proof. A process can return with 1 only if its identifier was returned to some process, hence at most k process can obtain 1. Assume now that all participating processes terminate and consider the process for which its identifier was first written to the shared memory. This process must see its identifier in its snapshot and return 1, hence not all participating processes may return 0.   □

Solving k-Set-Consensus Among $k+1$ processes in $\mathcal{R}^*_{k-T\&S}$. The advantage of k-set consensus operations compared to k-test-and-set operations is that processes can participate as soon as they see some process participating. Indeed, since it is a colorless task, processes can adopt inputs from any other process. Hence, to solve k-set consensus operations among $k + 1$ processes, processes maintain a decision estimate for all k-set consensus operations and share them in all iterations of $\mathcal{R}^*_{k-T\&S}$. When a process initiates a new operation for which it has no decision estimate yet, it simply adds a decision estimate corresponding to its input value. Moreover, when a process sees a process participating in a new operation, it adopts its decision estimate.

Now, at the end of each iteration of $\mathcal{R}_{k-T\&S}$, processes look at the decision estimate for all operations. If a process sees all $k + 1$ potential participants of an operation, then it replaces its decision estimate by the decision estimate of the process with the next identifier (going back to the first to form a loop when there are none higher). For a process to terminate, it must see that all potential participants share a decision estimate for the same round. If it happens, processes return their potentially updated decision estimate as k-set consensus output.

Note that once terminated, processes use a special input value $\bot$. When a process competes with a terminated process for a k-set consensus operation, then it directly returns with its proposal.

Correctness of the Simulation of k-set Consensus Among $k+1$ Processes. Let us first show that simulated operations respect the specification of k-set consensus among $k + 1$ processes before showing that sufficient progress is also guaranteed.

Lemma 1. *The simulation satisfies the safety properties of k-set consensus among $k + 1$ processes in $\mathcal{R}^*_{k-T\&S}$.*

Proof. Processes return their decision estimate which is initially set to their input or adopted from other processes decision estimates. Hence validity is satisfied.

Now consider the first iteration of the affine task after which the first process returns with an output. In this iteration, all processes with the smallest view shared a decision estimate. Hence, all processes adopted a decision estimate at the end of the round. If at the end of the round there are less than k distinct decision estimates, then the agreement property will be ensured as the number of distinct decision estimates in later rounds is a subset of this one.

To see that there are at most k distinct decision estimates at the end of this first iteration in which a process decides, consider the processes which see the $k + 1$ potential participants. These processes adopt the decision estimate of the next process (relatively to identifier ranks). But in $\mathcal{R}_{k-T\&S}$, at most k vertices may share the same carrier. Hence a process seeing all participants must adopt the decision estimate of a process not seeing all of them. But this process does not change its decision estimate. Thus, two processes share the same decision estimate. The number of distinct decision estimates is, therefore, smaller than or equal to k and hence at most k distinct outputs may be returned.   □

Lemma 2. *The simulation of k-set consensus among $k + 1$ in $\mathcal{R}^*_{k-T\&S}$ provides progress to processes having infinitely often the smallest view among non-terminated processes.*

Proof. Processes participate in an operation as soon as they see another process participating. In particular if a process with the smallest view among non-terminated participates in some iteration, all processes participate in the next iteration (terminated processes are always participating). But if all processes observed in some iteration are participating, then processes return at the end of the round. Hence, a process with a k-set consensus operation terminates at most one round after obtaining the smallest view among non-terminated processes. □

Equivalence Between the k-Test-and-Set Model and $\mathcal{R}^*_{k-T\&S}$. Both the AS memory and k-set consensus among $k + 1$ provides progress to the non-terminated processes with the smallest view infinitely often. Hence, some process will eventually output and terminate as long as they are non-terminated processes. Thus, all processes eventually produces valid task outputs.

We can conclude with the equivalence of the two classes of models. Indeed, this simulation and Algorithm 1 can be used to simulate the affine model $\mathcal{R}^*_{k-T\&S}$ in the k-test-and-set model and reciprocally. Therefore:

Theorem 4. *A task is solvable in the k-test-and-set model if and only if it is solvable in the affine model $\mathcal{R}^*_{k-T\&S}$.*

Thus, we get the following generalization of the asynchronous computability theorem for the k-test-and-set model:

Theorem 5. *Task $T = (\mathcal{I}, \mathcal{O}, \Delta)$ is solvable in the k-test-and-set model if and only if there exist $\ell \in \mathbb{N}$ and a simplicial map $\delta : (\mathcal{R}_{k-T\&S})^\ell(\mathcal{I}) \to \mathcal{O}$ carried by Δ.*

6   Related Work and Concluding Remarks

Herlihy and Shavit [18] proposed a characterization of wait-free task computability through the existence of a simplicial map from a subdivision of the input complex of a task $\mathcal{I}$ to its output complex $\mathcal{O}$. (The reader is referred to [16] for a thorough discussion of the use of combinatorial topology in distributed computability.) Herlihy and Rajsbaum [17] studied colorless task computability in the particular case of *superset-closed* adversaries. They show that the protocol complex of a superset-closed adversary with *minimal core size* c is $(c-2)$-connected. This result, obtained via an iterative application of the Nerve lemma, gives a combinatorial characterization of superset-closed adversaries. The characterization only applies to colorless tasks, and it does not allow us to express the adversary in an affine way.

Gafni et al. [12] introduced the notion of an affine task and characterized task computability in *iterated* adversarial models via infinite subdivisions of input complexes, assuming a limited notion of solvability that only guarantees outputs to "fast" processes [5,10] (i.e., "seen" by every other process infinitely often). The liveness property defined in this paper for iterated models guarantees outputs for *every* process, which allowed us to establish a task-computability equivalence with conventional non-iterated models.

Affine tasks have been defined for the *read-write* models of wait-free [18], t-resilience [26], k-concurrency [11] and, finally, for the general class of fair adversaries [23], encompassing all these models. In this paper, we complement the characterization of [23] with a model in which processes can communicate via k-test-and-set objects, in addition to read-write registers.

This paper proposes a new affine characterization of the wait-free shared memory model enhanced with k-test-and-set objects. Just as the wait-free characterization [18] implies that the *IS* task captures the wait-free model, our characterization equates any such model with a (compact) affine task embedded in the standard chromatic subdivision.

Interestingly, unlike [26], we cannot rely on the *shellability* [16] (and, thus, link-connectivity) of the affine task. Link-connectivity of a simplicial complex $\mathcal{C}$ allows us to work in the *point set* of its geometrical embedding $|\mathcal{C}|$ and use continuous maps (as opposed to simplicial maps that maintain the simplicial structure). For example, the existence of a continuous map from $|\mathcal{R}_{\mathcal{A}_{t-res}}|$ to any $|\mathcal{R}^k_{\mathcal{A}_{t-res}}|$ implies that $\mathcal{R}_{\mathcal{A}_{t-res}}$ indeed captures the general task computability of $\mathcal{A}_{t-res}$ [26]. In general, however, the existence of a continuous map onto $\mathcal{C}$ only allows us to converge on a *single* vertex [16]. If $\mathcal{C}$ is not link-connected, converging on one vertex allows us to compute an output only for a single process and not more. Unfortunately, only very special adversaries, such as $\mathcal{A}_{t-res}$, have link-connected counterparts (see, e.g.., the affine task corresponding to 1-test-and-set in Fig. 5). Instead of relying on link-connectivity, this paper takes an explicit algorithmic way of showing that iterations of $\mathcal{R}_{k-T\&S}$ simulate the wait-free shared memory model enhanced with k-test-and-set objects. An interesting question is to which extent point-set topology and continuous maps can be applied in affine characterizations.

References

1. Afek, Y., et al.: Atomic snapshots of shared memory. J. ACM **40**(4), 873–890 (1993)
2. Anderson, J.H., Moir, M.: Using local-spin k-exclusion algorithms to improve wait-free object implementations. Distrib. Comput. **11**(1), 1–20 (1997)
3. Borowsky, E., Gafni, E.: Immediate atomic snapshots and fast renaming. In: PODC, pp. 41–51 (1993)
4. Borowsky, E., Gafni, E.: A simple algorithmically reasoned characterization of wait-free computation. In: PODC, pp.189–198 (1997)
5. Bouzid, Z., Gafni, E., Kuznetsov, P.: Strong equivalence relations for iterated models. In: OPODIS, pp. 139–154 (2014)
6. Chaudhuri, S.: Agreement is harder than consensus: set consensus problems in totally asynchronous systems. In: PODC, pp. 311–324 (1990)
7. Delporte, C., Fauconnier, H., Rajsbaum, S., Raynal, M.: t-resilient immediate snapshot is impossible. In SIROCCO, pp. 177–191 (2016)
8. Delporte-Gallet, C., Fauconnier, H., Guerraoui, R., Tielmann, A.: The disagreement power of an adversary. Distrib. Comput. **24**(3–4), 137–147 (2011)
9. Fischer, M.J., Lynch, N.A., Paterson, M.S.: Impossibility of distributed consensus with one faulty process. J. ACM **32**(2), 374–382 (1985)
10. Gafni, E.: Round-by-round fault detectors: unifying synchrony and asynchrony. In: PODC, pp. 143–152 (1998)
11. Gafni, E., He, Y., Kuznetsov, P., Rieutord, T.: Read-write memory and k-Set consensus as an affine task. In: OPODIS, pp. 6:1–6:17 (2017)
12. Gafni, E., Kuznetsov, P., Manolescu, C.: A generalized asynchronous computability theorem. In: PODC, pp. 222–231 (2014)
13. Gafni, E., Rajsbaum, S.: Distributed programming with tasks. In: OPODIS, pp. 205–218 (2010)
14. Gafni, E., Raynal, M., Travers, C.: Test & set, adaptive renaming and set agreement: a guided visit to asynchronous computability. In: SRDS, pp. 93–102 (2007)
15. Herlihy, M.: Wait-free synchronization. ACM Trans. Program. Lang. Syst. **13**(1), 123–149 (1991)
16. Herlihy, M., Kozlov, D.N., Rajsbaum, S.: Distributed Computing Through Combinatorial Topology. Morgan Kaufmann (2014)
17. Herlihy, M., Rajsbaum, S.: Simulations and reductions for colorless tasks. In: PODC, pp. 253–260 (2012)
18. Herlihy, M., Shavit, N.: The topological structure of asynchronous computability. J. ACM **46**(2), 858–923 (1999)
19. Kozlov, D.N.: Chromatic subdivision of a simplicial complex. Homol. Homotopy Appl. **14**(1), 1–13 (2012)
20. Kuznetsov, P., Rieutord, T.: Agreement functions for distributed computing models. In: NETYS, pp. 175–190 (2017)
21. Kuznetsov, P., Rieutord, T.: Affine Tasks for k-Test-and-Set. working paper or preprint, June 2018
22. Kuznetsov, P., Rieutord, T.: Brief announcement: on decidability of 2-process affine models. In: DISC, pp. 54:1–54:3 (2020)
23. Kuznetsov, P., Rieutord, T., He, Y.: An asynchronous computability theorem for fair adversaries. In: PODC, pp. 387–396 (2018)

24. Mostefaoui, A., Raynal, M., Travers, C.: Exploring Gafni's reduction land: from Ω^k to wait-free adaptive $(2p - \lceil \frac{p}{k} \rceil)$-renaming via k-set agreement. In: Dolev, S. (ed.) DISC 2006. LNCS, vol. 4167, pp. 1–15. Springer, Heidelberg (2006). https://doi.org/10.1007/11864219_1
25. Rieutord, T.: Combinatorial characterization of asynchronous distributed computability. Ph.D. thesis, Université Paris Saclay (COmUE), October 2018
26. Saraph, V., Herlihy, M., Gafni, E.: Asynchronous computability theorems for t-resilient systems. In: Gavoille, C., Ilcinkas, D. (eds.) DISC 2016. LNCS, vol. 9888, pp. 428–441. Springer, Heidelberg (2016). https://doi.org/10.1007/978-3-662-53426-7_31
27. Spanier, E.H.: Algebraic Topology. McGraw-Hill Book Co., New York (1966)

A Combinatorial Characterization of Self-stabilizing Population Protocols

Shaan Mathur[(✉)] and Rafail Ostrovsky[(✉)]

University of California, Los Angeles, USA
{shaan,rafail}@cs.ucla.edu

Abstract. We characterize self-stabilizing functions in population protocols for complete interaction graphs. In particular, we investigate self-stabilization in systems of n finite state agents in which a malicious scheduler selects an arbitrary sequence of pairwise interactions under a global fairness condition. We show a necessary and sufficient condition for self-stabilization. Specifically we show that functions without certain set-theoretic conditions are impossible to compute in a self-stabilizing manner. Our main contribution is in the converse, where we construct a self-stabilizing protocol for all other functions that meet this characterization. Our positive construction uses Dickson's Lemma to develop the notion of the root set, a concept that turns out to fundamentally characterize self-stabilization in this model. We believe it may lend to characterizing self-stabilization in more general models as well.

Keywords: Population protocols · Self-stabilization · Anonymous · Finite-state · Chemical reaction networks

1 Introduction

The population protocol computational model assumes a system of n identical finite state transducers (which we call *agents*) in which pairwise interactions between agents induce their respective state transitions. Each agent is provided a starting input and starting state, and an adversarial scheduler decides at each time step which two agents are to interact. In order to make the behavior of the scheduler precise, the scheduler is allowed to act arbitrarily so long as the *global fairness condition* (defined in the paper defining the model by Angluin, Aspnes, Diamadi, Fischer, and Peralta [1]) is satisfied: if a configuration of agent states appears infinitely often, then any configuration that can follow (say, after an interaction) must also appear infinitely often. We stress that agents individually do not have unique identifiers and a bound on n in not known. We call all n agents jointly a *population*. When an agent interacts with another agent, both agents change states as a function of each agent's input and state tuple. Each agent outputs some symbol at every time step as a function of their current state, and once every agent agrees on some common output for all subsequent time steps, we say the protocol has *converged* to that symbol. We say a protocol

© Springer Nature Switzerland AG 2020
S. Devismes and N. Mittal (Eds.): SSS 2020, LNCS 12514, pp. 167–182, 2020.
https://doi.org/10.1007/978-3-030-64348-5_13

computes (or decides) some function f if distributing the input symbols of input x and running the protocol causes the population (i.e. all agents) to converge to $f(x)$. In the general model an accompanying *interaction graph* restricting which agents can ever interact may be provided as a constraint on the scheduler. In this paper, we will be considering the original, basic model introduced in [1], where we deal with complete interaction graphs and inputs that do not change with time. For the population protocol model on complete graphs, the characterization of computable predicates (not necessarily with self-stabilization) has been studied by Angluin, Aspnes, Eisenstat, and Ruppert, who proved it to be equivalent to the set of semilinear predicates [2].

It is desirable to have population protocols that can handle transient faults: specifically, we would like it to be the case that no matter what multiset of states the agents are initialized with, the protocol will eventually converge to the correct output. For this reason, such a protocol is called *self-stabilizing*, being able to converge after experiencing any adversarial fault that erroneously changes an agent state. Since fault-tolerance is a desirable property of any distributed system, we aim to determine exactly what computable functions in this model admit self-stabilizing solutions, and which do not. We prove the following main theorem for population protocols on complete interaction graphs.

Theorem (Main). *Let $f : \mathcal{X} \to Y$ be a function where $\mathcal{X}$ is any set of finite multisets on a finite alphabet. Then f has a self-stabilizing protocol $\iff$ (For any $A, B \in \mathcal{X}, A \subseteq B \implies f(A) = f(B)$).*

We remark on the definition of $\mathcal{X}$: Firstly, it is a set of multisets; thus $A \subseteq B$ refers to multiset inclusion, not set inclusion (e.g. $\{a, a, b\} \subseteq \{a, a, b, b\}$ but Moreover, the domain $\mathcal{X}$ can be *any* set of multisets; that is, the domain does not necessarily contain all possible nonempty finite multisets on a finite alphabet, but could be a subset of it. If $\mathcal{X}$ is all possible nonempty finite multisets on a finite alphabet, then the theorem states f is a constant function: the output of f on the singleton multisets is the output of f on the union of all singletons; since f agrees on all singletons, it agrees on all larger multisets. Thus self-stabilizing decision problems, where all possible inputs are included, will be a constant function. In contrast, self-stabilizing promise problems, where only a subset of all possible inputs are included, may be non-constant.

Our Techniques: The technique that we use to show when self-stabilization is not possible follows from the work of Angluin, Aspnes, Fischer, and Jiang in [3]. Informally, self-stabilizing functions must not allow subpopulations to "re-converge" to a different answer; so if $A \subseteq B$ but $f(A) \neq f(B)$, running a protocol with input B from any configuration could lead to the subpopulation with input A converging on erroneous output $f(A)$. The converse, that functions where subsets lead to the same output are self-stabilizing, is an unstudied problem that involves a technically intricate construction. Tools from partial order theory (Dickson's Lemma) are used to observe that the domain $\mathcal{X}$ will have a finite set of minimal elements under the $\subseteq$ partial order; these minimal elements completely determine the output of f, and so our protocol computes f

by identifying which of these minimal elements is present in the population. A function is *self-stabilizing* if it admits a self-stabilizing protocol under the basic model. A nontrivial corollary of our main theorem is that for a fixed (possibly infinite) domain $\mathcal{X}$ and a finite output alphabet Y, there are only finitely many self-stabilizing functions $f : \mathcal{X} \to Y$.

Corollary (Self-Stabilizing Functions are Rare). *Fix some set, $\mathcal{X}$, of finite multisets on a finite alphabet. Fix a finite output alphabet Y. There are only finitely many self-stabilizing functions of the form $f : \mathcal{X} \to Y$.*

This follows from the fact that the outputs of finitely many minimal elements in the domain fully characterize a self-stabilizing function. This validates the intuition that the set of all self-stabilizing functions is very limited in comparison to the set of computable functions. We note that our general self-stabilizing protocol is not efficient, and often specific problems have much faster self-stabilizing protocols. Unsurprisingly the functions we list in Sect. 1.3 admit faster solutions than our protocol.

1.1   Related Work

The population protocol model was first introduced by Angluin, Aspnes, Diamadi, Fischer, and Peralta in [1] to represent a system of mobile finite-state sensors. Dijkstra was the first to formalize the notion of self-stabilization within a distributed system, although the models and problems he discussed imposed different constraints than that of population protocols, such as distinguishing agents with unique identifiers [13]. Self-stabilizing population protocols were first formalized in the work of Angluin, Aspnes, Fischer, and Jiang [3]. Their work generated self-stabilizing, constant-space protocols for problems including round-robin token circulation, leader election in rings, and 2-hop coloring in degree-bounded graphs. Moreover their work established a crucial method of impossibility result. Call a class of graphs *simple*, if there does not exist a graph in the class which contains two disjoint subgraphs that are also in the class. Example of this include the class of all rings or the class of all connected degree-d regular graphs. Angluin et al.'s work demonstrated that leader election in *non-simple* classes of graphs are impossible. Our paper's impossibility result follows from Angluin et al.'s technique, as the class of complete graphs that we work with is non-simple. Other impossibility results come from Cai, Izumi, and Wada [10] using *closed sets*, which are sets of states such that a transition on any two of the states results in a state within the set; impossibility in leader election is demonstrated by identifying a closed set excluding the leader state.

Many self-stabilizing population protocol constructions besides those from [3] tend to give the model additional properties to achieve self-stabilization. Beauquier, Burman, Clement, and Kutten introduce intercommunication speeds amongst agents, captured by the *cover time*; they also add a distinguished, non-mobile agent with unlimited resources called a *base station* [7]. Under this model, Beauquier, Burman, and Kutten design an *automatic tranformer* that takes a population protocol algorithm solving some static problem and transforms it into a self-stabilizing algorithm [8]. Izumi, Kinpara, Izumi, and Wada also use this model to create efficient protocols for performing a self-stabilizing count of the number of agents in the network, where there is a known upper bound P on the number of sensors. Their protocol converges under global fairness with $3 \cdot \lceil \frac{P}{2} \rceil$ agent states [16]. Fischer and Jiang introduce an *eventual leader detector* oracle into the model that allows self-stabilizing leader election in complete graphs and rings. The complete graph protocol works with local and global fairness conditions, while the ring protocol requires global fairness [14]. Beauquier, Blanchard, and Burman extend this work by presenting self-stabilizing leader election in arbitrary graphs when a composition of eventual leader detectors is introduced into the model [6]. Knowledge of the number of agents allows Burman, Doty, Nowak, Severson, and Xu to develop several efficient self-stabilizing protocols for leader election; with no silence or space constraints they achieve optimal expected parallel time of $\mathcal{O}(\log n)$ [9]. Loosely-stabilizing protocols relax self-stabilization to allow more tractable solutions, such as leader election protocols with polylogarithmic convergence time by Sudo, Ooshita, Kakugawa, Masuzawa, Datta, and Larmore [19]. Self-stabilizing Leader election (under additional symmetry-breaking assumptions) is possible without unique identifiers, as discussed in [5,18], but the communication happens over a fixed graph - unlike population protocols where interaction of agents is arbitrary and interaction pattern is controlled by an adversarial scheduler. Population self-stabilizing protocols are also related to biological systems self-stabilization, see [15] for further discussion.

In this work, we do not extend the basic model with any extra abilities. We demonstrate a universal self-stabilizing population protocol for any function $f : \mathcal{X} \to Y$ where for any $A, B \in \mathcal{X}$, $A \subseteq B$ implies that $f(A) = f(B)$. We do this by using a result from partial order theory by Dickson [12] that states that any set of finite dimensional vectors of natural numbers have finitely many minimal elements under the pointwise partial order.

1.2   The Number of Self-stabilizing Functions Depends on the Number of Minimal Elements

In Definition 7 we define the *root set*. Formally, let $\mathcal{X}$ be a (possibly infinite) set of finite multisets over a finite alphabet (e.g. $\mathcal{X} = \{\{a\}, \{a, a\}, \dots\}$ over alphabet $\Sigma = \{a\}$). The root set is some subset $\mathcal{R} \subseteq \mathcal{X}$ with the following property: for any element of the domain $A \in \mathcal{X}$, there is some element $R \in \mathcal{R}$ such that $R \subseteq A$. We call R a root. Section 2.1 uses Dickson's Lemma to prove that there exists a unique, finite, and minimally sized root set $\mathcal{R}$. We note the following corollary to the main theorem: to determine the output of any $A \in \mathcal{X}$ for a self-stabilizing function $f : \mathcal{X} \to Y$, it suffices to identify some root $R \subseteq A$ since $f(R) = f(A)$. In fact the entire output of f is determined by $f(R)$ for each $R \in \mathcal{R}$. The number of possible outputs is then upper bounded by the size of the smallest root set, which is an interesting fact in of its own right.

Corollary. *Let $\mathcal{X}$ be a set of finite multisets over a finite alphabet, let Y be a finite output alphabet, and let $\mathcal{R}$ be the minimally sized root set of $\mathcal{X}$. Then $|im(f)| \leq |\mathcal{R}|$.*

Fix some domain $\mathcal{X}$ and finite output alphabet Y, and let $\mathcal{R}$ be the minimally sized root set of $\mathcal{X}$. The number of self-stabilizing functions $f : \mathcal{X} \rightarrow Y$ are precisely the number of ways to assign an output to each root in $\mathcal{R}$, of which there are at most $N = |Y|^{|\mathcal{R}|}$ ways. In fact, the number of functions is exactly N if and only if every $A \in \mathcal{X}$ has a unique root $R \subseteq A$. This is true because if two roots R and R' are both subsets of A, then they must be assigned the same output $f(R) = f(R') = f(A)$; otherwise if every A has a unique root, then there is no overlap and each root has $|Y|$ choices for its output.

Corollary. *Let $\mathcal{X}$ be a set of finite multisets over a finite alphabet, let Y be a finite output alphabet, and let $\mathcal{R}$ be the minimally sized root set of $\mathcal{X}$. The number of self-stabilizing functions $f : \mathcal{X} \rightarrow Y$ is finite. Specifically there are at most $|Y|^{|\mathcal{R}|}$ self-stabilizing functions. There are exactly this number of self-stabilizing functions if and only if every $A \in \mathcal{X}$ has a unique root $R \in \mathcal{R}$.*

1.3   Nontrivial Examples of Self-stabilizing Functions

If we can restrict the domain $\mathcal{X}$ to exclude some inputs, then it can become easier to generate problems that admit self-stabilizing solutions.

- In **Chemical Reaction Networks (CRN)**, it can be desirable to compute boolean circuits. In CRNs one prevalent technique to compute some boolean function $g : \{0,1\}^n \rightarrow \{0,1\}^m$ is to have n different species, each with 2 sub-species each [11]. That is, we have molecules $s_1^0, s_1^1, s_2^0, s_2^1, \ldots, s_n^0, s_n^1$, where molecule s_i^j signifies that the ith bit has value $j \in \{0,1\}$. All species will appear in the input, but only one sub-species per species will appear (i.e. s_i^0 and s_i^1 will not both appear, but one of them will). This way of formulating the input allows for self-stabilizing computation of boolean functions! This is because if $A \subseteq B$ for some input multisets, A and B, of molecules, they will include the same sub-species and hence have the same output $g(A) = g(B)$.
- Generalizing the former example, suppose we have an input alphabet of n symbols, but only $k < n$ of these symbols will ever show up in the population (though a given symbol could occur multiple times). Then we can compute *any* self-stabilizing function of the k present members.

  For instance, suppose we have k different classes of finite-state mobile agents, $a_1, \ldots, a_k$. There are a nonzero number of agents in each class a_i, and agents within the same class are running a (possibly not self-stabilizing) population protocol. Eventually every agent in class a_i will converge and be outputting the same common value o_i. These outputs $o_1, \ldots, o_k$ will then be the input to our self-stabilizing population protocol. The agents might perform some sort of protocol composition where a tuple of states (q, s) are used for each agent; q would correspond to the agent state in the first protocol, and then s would correspond to the agent state in the self-stabilizing protocol. If the first protocol was also self-stabilizing (it doesn't have to be), then the entire protocol composition would be self-stabilizing as well.

As an example, we could have each class of agents a_i compute some boolean circuit C_i in a self-stabilizing way. Eventually each agent of class a_i will be outputting some m character string over the set $\{0_i, 1_i\}$. Notice that even both agents in class a_1 and in class a_2 intend to output 0110, the former agents will output $0_1 1_1 1_1 0_1$ and the latter agents will output $0_2 1_2 1_2 0_2$. Once all of these boolean circuit computations are done, all the agents across classes will calculate the majority output in a self-stabilizing way (and they would be able to do so since the outputs $o_1, \ldots, o_k$ are all distinguishable).

- Any computable function in which the number of agents is a fixed constant k admits a self-stabilizing solution (there can't be any subsets, so the condition is vacuously true). For instance, distribute bits amongst exactly k agents; we can create a self-stabilizing protocol to output 1 if any permutation of those k bits represents a prime number in binary.

1.4   The Basic Model

There are different formalizations of the basic population protocol model. We adopt the basic one first introduced by Angluin et al. [1], except where we impose that any two agents are allowed to interact.

A **population protocol** is a tuple $\mathcal{P} = (Q, \Sigma, Y, I, O, \delta)$ where Q is the finite set of agent states; Σ is a finite set of input symbols; Y is a finite set of output symbols; $I : \Sigma \to Q$ is an input function; $O : Q \to Y$ is the output function; and $\delta : (Q \times \Sigma) \times (Q \times \Sigma) \to (Q \times Q)$ is the transition function.

Note that population protocols are independent over the size of the system; rather, first a population protocol is specified and then it is run on some set of agents V. At the beginning of execution, an input assignment $\alpha : V \to \Sigma$ is provided, providing each agent an input symbol (we will observe that in complete graphs, we can view input assignments as merely a multiset of inputs). Since our model focuses on the computation of functions, we will enforce that the input does not change (i.e. the input is hardwired into every agent). If an agent is assigned input symbol $\sigma \in \Sigma$, it will determine its starting state via the input function as $I(\sigma)$ (note that in the basic model the input determines the starting states, but in self-stabilization we do not consider starting states). At each time step, a scheduler selects (subject to a global fairness condition) an agent pair (u, v) for interaction; semantically the scheduler is selecting agents u and v to interact, where u is called the *initiator* and v is called the *responder*. Agents u and v will then state transition via δ; letting $q_u, q_v \in Q$ and $\sigma_u, \sigma_v \in \Sigma$ be the states and inputs for u and v respectively, the new respective states will be the output of $\delta((q_u, \sigma_u), (q_v, \sigma_v))$.

We use the notion of a *configuration* to describe the collective agent states.

Definition 1. *Configuration. Let V be the set of agents and Q be a set of states. A configuration of a system is a function $C : V \to Q$ mapping every agent to its current state.*

When running the protocol we will go through a sequence of configurations. If C and C' are configurations under some population protocol such that C' can follow from a single agent interaction in C, we write $C \rightarrow C'$. If a series of interactions takes us from C to C', we write $C \overset{*}{\rightarrow} C'$.

Definition 2. *Execution. An execution of a population protocol is a sequence of configurations* $\mathcal{C} = C_1 C_2 \ldots$ *where for all i, $C_i \rightarrow C_{i+1}$.*

The scheduler is subject to a global fairness condition, which states that if a configuration can follow from an infinitely occurring configuration, then it must also occur infinitely often.

Definition 3. *Global Fairness Condition [1]. Let C and C' be configurations such that $C \rightarrow C'$. If C appears infinitely often during an execution, then C' appears infinitely often during that execution.*

At each time step every agent outputs some symbol from Y via output function O. If all agents output the same symbol and continue to do so for each time step afterwards, we say the protocol's output is that symbol (interactions may continue, but the output of the agents remain that symbol). Note that when computing functions, the protocol should output the same symbol when the same inputs are provided, irrespective of the globally fair scheduler's behavior. When the population has determined the final output, we say it has *converged*.

Definition 4. *Convergence [1]. A population is said to have converged to an output $y \in Y$ during a population protocol's execution if the current configuration C is such that each agent's output is y and $C \rightarrow C'$ implies that C' has each agent with the same output y.*

When agents u and v interact, their state transition is a function of both agent's current state and respective inputs; since this is all the information they receive, an agent does not learn the identity of the agent it interacts with, but merely the state and input of that agent. In this sense, agents with the same input that are in the same state are *indistinguishable* from one another. Furthermore, population protocols are independent of the number of agents, making it impossible to design the state set to give every agent a unique identifier, so agents are truly *anonymous*.

As noted earlier, our model accepts input via an input assignment $\alpha : V \rightarrow \Sigma$, where V is the set of agents and Σ is our finite input alphabet. It is useful to note, though, that we can actually view our input instead as some finite multiset A over alphabet Σ, with $|A| = |V|$ [4]. Though we omit the proof, the idea follows from the fact that any two agents can interact, so which agent gets what input symbol is less important than what input symbols are provided to the system in the first place. We use the notation $m_A(\sigma)$ to denote the *multiplicity*, the number of occurrences, of element σ in multiset A.

Definition 5. *Population Protocol Functions. Let $f : \mathcal{X} \rightarrow Y$ be a function where $\mathcal{X}$ is a set of multisets over the finite alphabet Σ. A population protocol $\mathcal{P}$ computes f if and only if for any $A \in \mathcal{X}$, all executions of $\mathcal{P}$ with input A converge to $f(A)$.*

In this paper when we say $\mathcal{P}$ is a population protocol (or simply protocol), we mean that it computes some function f. When we don't care about whether a population protocol computes some function, we will refer to it as a *sub-protocol*. This will be useful jargon in our protocol composition in Sect. 2.3.

We say that a population protocol computing a function is *self-stabilizing* when it can begin in any configuration and eventually converge (to the same output). Such a function is called a *self-stabilizing* function.

Definition 6. *Self-Stabilizing Protocol. Let $\mathcal{P} = (Q, \Sigma, Y, I, O, \delta)$ be a population protocol computing some function $f : \mathcal{X} \to Y$. $\mathcal{P}$ is called self-stabilizing if and only if for any input multiset $A \in \mathcal{X}$, any set of agents V of cardinality $|A|$, and any starting configuration $C : V \to Q$, we have that any execution of $\mathcal{P}$ converges to $f(A)$.*

2   Constructing a Self-stabilizing Protocol for the Basic Model

We aim to show that not only does Theorem 1 specify *necessary* conditions for computing self-stabilizing functions (shown in the full version of this paper [17]), but they are also *sufficient* for self-stabilization. We do this by generating a self-stabilizing protocol for computing all functions of the form $f : \mathcal{X} \to Y$ where for all $A, B \in \mathcal{X}, A \subseteq B \implies f(A) = f(B)$. To do this, we must introduce a new notion known as the root set of a set of multisets.

2.1   The Root Set

The inputs for our agents are represented by a multiset of inputs, an element of domain $\mathcal{X}$. We are interested in a kind of subset $\mathcal{R} \subseteq \mathcal{X}$ such that all multisets in $\mathcal{X}$ are a superset of some multiset in $\mathcal{R}$. This section aims to show that all sets of multisets $\mathcal{X}$ actually have a *finite* $\mathcal{R}$.

Definition 7. *Root Set and its Roots.   Let $\mathcal{X}$ be a set of finite multisets. A subset $\mathcal{R} \subseteq \mathcal{X}$ is called a root set of $\mathcal{X}$ if and only if for all $A \in \mathcal{X}$, there exists $R \in \mathcal{R}$ such that $R \subseteq A$. We call a multiset $R \in \mathcal{R}$ a root of A.*

Notice that $\mathcal{X}$ is always trivially its own root set. However, $\mathcal{X}$ can be infinitely large; for example the set of all nonempty finite multisets on alphabet $\Sigma = \{a\}$ is $\mathcal{X} = \{\{a\}, \{a, a\}, \ldots\}$. However, we are primarily interested in the existence of a *finite* root set over our function f's domain. Dickson's Lemma [12] provides us what we need, though an alternative proof is in our full paper [17].

First, consider a finite multiset over an alphabet, Σ, of n elements. An equivalent representation of a multiset is as a vector of multiplicities in $\mathbb{N}^n$. For instance, the multiset $\{a, a, b\}$ on ordered alphabet $\Sigma = \{a, b, c\}$ would be represented by $(2, 1, 0)$. Hence a set of finite multisets on an alphabet of size n could be considered a subset $S \subseteq \mathbb{N}^n$. Consider two vectors $\boldsymbol{n}, \boldsymbol{m} \in \mathbb{N}^n$, and denote n_i and m_i as the ith component in the corresponding vectors. Define the pointwise partial order $n \leqslant m \iff n_i \leqslant m_i$ for all i. A minimal element of a subset $S \subseteq \mathbb{N}^n$ is an element that has no smaller element with respect to this partial order. Now we can state Dickson's Lemma.

Lemma 1. *Dickson's Lemma. In every subset $S \neq \varnothing$ of $\mathbb{N}^n$, there is at least one but no more than a finite number of elements that are minimal elements of S for the pointwise partial order.*

This is equivalent to the existence of a finite root set.

Corollary 1. *Let $\mathcal{X}$ be a set of finite multisets over a finite alphabet. $\mathcal{X}$ has a finite root set. In other words, there exists a finite subset $\mathcal{R} \subseteq \mathcal{X}$ such that for any $A \in \mathcal{X}$, there exists $R \in \mathcal{R}$ such that $R \subseteq A$.*

A interesting corollary is that the root set of minimal size is unique, making it legitimate to speak of *the* minimally-sized root set. See the full version of this paper for a proof [17].

Corollary 2. *Let $\mathcal{X}$ be a set of finite multisets over a finite alphabet. The minimal length root set $\mathcal{R}$ of $\mathcal{X}$ is unique.*

From order theory, we note that the minimal root set is a strong downwards antichain. We call a subset A of a poset P a strong downwards antichain if A is an antichain (a subset of P with incomparable elements) and no two distinct elements of A have a smaller element in P. The proof is in the full version [17].

Corollary 3. *Let $\mathcal{X}$ be a set of finite multisets over a finite alphabet. The minimal length root set $\mathcal{R}$ of $\mathcal{X}$ is a strong downwards antichain with respect to the subset partial ordering.*

Now suppose we have a function $f : \mathcal{X} \rightarrow Y$ over multisets such that for $A, B \in \mathcal{X}, A \subseteq B \implies f(A) = f(B)$. We have that there exists a finite, minimal length root set $\mathcal{R} \subseteq \mathcal{X}$. If the input to the population is A, it suffices to identify the root $R \in \mathcal{R}$ such that $R \subseteq A$, since the output would be $f(R) = f(A)$.

2.2   Self-stabilizing Population Protocol Construction

Before we formally specify a universal self-stabilizing population protocol, it is much more helpful to first understand how it works at a high level. As a reminder, we will be working with functions of the form $f : \mathcal{X} \rightarrow Y$, where $\mathcal{X}$ is some set of finite multisets on a finite alphabet and Y is the finite output alphabet. The function f is also assumed to satisfy the following property:

$$\forall A, B \in \mathcal{X}, A \subseteq B \implies f(A) = f(B).$$

We need to determine how to design our protocol to compute f in a self-stabilizing manner. Since $\mathcal{X}$ is a set of finite multisets, it has a finite root set; let $\mathcal{R} \subseteq \mathcal{X}$ be the minimally sized, finite root set. $\mathcal{R}$ will be given some arbitrary fixed ordering so that we can index into it. Loosely speaking, we can compute f

by having each agent iterate through every root in the finite root set to see if a given root is a subset of the population's input A.

As an agent iterates through the root set, how can it tell if the current root is indeed a subset of the population's input multiset, A? Suppose an agent has guessed that root $R_i \subseteq A$. Consider some other root R_j where $f(R_i) \neq f(R_j)$. Naturally there are some symbols that occur more often in R_j than in R_i, and vice versa. Consider a symbol σ that occurs n times in R_j and m times in R_i, where $n > m$. Suppose agents start counting how many other agents they see with input σ, and manage to identify there are at least n of them. Now further suppose that this is the case for all such σ; that is, every σ occurring more often in R_j has at least that many inputs distributed in the population. Then we now know that $f(R_i) \neq f(A)$ by way of contradiction. For any symbol σ occurring n times in R_j and m times in R_i, we have two cases. If $n > m$, then we know that input multiset A has at least n instances of σ. If $n \leq m$, since $R_i \subseteq A$ by our hypothesis then we also know that there are at least n instances of σ. Therefore we simultaneously have that $R_j \subseteq A$ and $R_i \subseteq A$, which is a contradiction since $f(A) = f(R_j) \neq f(R_i)$! Therefore our initial assumption was wrong and R_i is not a subset of the input multiset A, and so we should increment our index i to guess root R_{i+1}. Note that we could also change our index i to become j, and the authors believe this would be a faster protocol; however since this is a paper about computability, we leave this optimization as an observation.

On the other hand if R_i really is a subset of A, then the existence of such a R_j would be impossible. If our counters are all initialized to 0, then no agent counter would ever increment high enough to identify such a R_j. However since the protocol may start in an arbitrary configuration, we have to make sure to reset counters whenever an agent increments and guesses a new root.

To formalize this idea, we need convenient notation. We define $MORE_{i,j}$ to be the set of symbols occurring more often in R_j than in R_i when $f(R_i) \neq f(R_j)$.

$$MORE_{i,j} = \begin{cases} \{\sigma \mid m_{R_i}(\sigma) < m_{R_j}(\sigma)\} & f(R_i) \neq f(R_j) \\ \varnothing & \text{otherwise} \end{cases}.$$

Each agent will have a table indexed by i and j, where each entry is a binary string of length $|MORE_{i,j}|$. The kth bit of this string is set to 1 when the agent counts that the kth symbol of $MORE_{i,j}$ occurs in the population as often as it does in R_j; otherwise it is 0. Once any entry in this table becomes a (nonempty) bit string of all 1's, then the agent tries a new root. To keep track of these counts, each agent will also maintain a nonnegative integer $count$; when two agents with the same $count$ and same input symbol σ meet, the responder will increment its value. To keep the states finite, the count is bounded above by the maximum multiplicity that occurs in the root set. It's important to note that the fact that the root set is finite is crucial to keeping the number of states here finite.

2.3   Sub-protocols for Protocol Composition

The self-stabilizing population protocol we construct will be a protocol composition $A \times B \times C$, where the input is given to A, the input to B is the output of

A, the input to C is the output of B, and the composition output is the output of C. With the previous discussion as our design motivation, we decompose our protocol into three distinct sub-protocols:

1. *SymbolCount*. This sub-protocol implements a simple modular counting mechanism, where agents with the same input symbol σ compare their counts. If two agents with the same count meet, then the responder increments its count (modulo the maximum multiplicity of any symbol in the root set, M, to keep things finite). If there are at least $k < M$ agents with the same symbol, then some agent must eventually have their count *at least* $k - 1$ by the Pigeonhole Principle, irrespective of initial configuration. The converse is only true if all the counts are initialized to 0, which is problematic if we are designing a self-stabilizing protocol that can initialize in any configuration. We will circumvent this by having the larger protocol composition reset this counter whenever moving on to the next root.

2. *WrongOutput?*. This sub-protocol maintains a table indexed by $i, j \in \{0, 1, \ldots, |\mathcal{R}| - 1\}$, where $\mathcal{R}$ is the minimally sized root set. Each entry will be a binary string of length $|MORE_{i,j}|$, where $MORE_{i,j}$ is the set of all input symbols occurring more often in R_j than in R_i. If the kth symbol of $MORE_{i,j}$ occurs in the population at least as often as it does in R_j, then the corresponding bit in the binary string is set to 1. Notice that this table is only useful when it is initialized with all entries as binary strings of all 0's. Again, we will circumvent this by having the larger protocol composition reset the table whenever moving on to the next root.

3. *RootOutput*. This sub-protocol uses an index $root \in \{0, 1, \ldots, |\mathcal{R}| - 1\}$. Adopting the notation from the previous bullet and letting $root = i$, this protocol increments $root$ if there is a j such that the (i, j) entry in the table has a binary string of all 1's. The protocol composition will use the incrementing of $root$ to signal that the other sub-protocol states should reset.

We now list the three sub-protocols below.

Definition 8. *SymbolCount. Let $f : \mathcal{X} \to Y$ be a function over finite multisets on a finite alphabet Σ, and let $\mathcal{R}$ be a finite and minimally sized root set of $\mathcal{X}$. Each agent has a state called* count, *where* count $\in \{0, 1, \ldots, M\}$ *and M is the maximum multiplicity of any symbol in the root set. Let $M = \max_{R \in \mathcal{R}, \sigma \in \Sigma} m_R(\sigma)$. Each agent takes as input some σ from some input multiset $A \in \mathcal{X}$. When an agent meets another agent with the same σ and same* count, *one of them will increment their* count *modulo M. This guarantees that if there are n agents with symbol σ, then eventually one agent will have* count $\geq n - 1$.

$$(\text{count}, \sigma), (\text{count}, \sigma) \to (\text{count}, \text{count} + 1 \mod M)$$

An agent in state count *with input σ outputs* (count, σ).

Note that even though output is a function of just the state, we can formally allow *SymbolCount* agents to output their input symbol as well, since we could take our current states Q and define a new set of states via the cross product

$Q' = Q \times \Sigma$. Then transitions on this could be defined in a similar way, where we would also have to make agents transition into a state that reflects their own input symbol (which can happen after every agent interacts once).

For $WrongOutput?$, our transitions need to satisfy two properties.

1. First, it is natural to have agents share their tables with each other to share their collected information about the counts of the inputs. Whenever an agent meets another agent, they bitwise OR their tables.
2. Fix some ordering on $MORE_{i,j}$ and denote the kth symbol by σ_k. Consider the bitstring entry at index i and j. By our previous discussion we want the kth bit to be 1 if the number of occurrences of σ_k in the population is at least its multiplicity in $R_j \in \mathcal{R}$.

   This can be enforced by setting this bit to 1 when the *count* of some agent with input σ_k is sufficiently high enough; then the first property will ensure this bit is set for the other agents by bitwise OR'ing tables. This can formally be accomplished by bitwise OR'ing the agent's table with an indicator table of the same dimension. Let i and j be the indices into the table, k be the index into the bitstring entry, σ_k be the kth symbol of $MORE_{i,j}$, and *count* and σ be the agent's inputs. Define

$$INDICATOR_{i,j,k} = \begin{cases} 1 & \text{if } \sigma = \sigma_k \text{ and } count \geqslant m_{R_j}(\sigma) - 1 \\ 0 & \text{otherwise} \end{cases}.$$

We will denote this table as $INDICATOR(count, \sigma)$ to be explicit about the agent's inputs *count* and σ. Note that the *count* is 0-indexed, which is why we subtract by 1. Also, since agents can start in arbitrary states, every agent has to transition once to have the table updated with $INDICATOR$.

Definition 9. *WrongOutput? Let $f : \mathcal{X} \to Y$ be a function over finite multisets on a finite alphabet, and let $\mathcal{R}$ be a finite and minimally sized root set of $\mathcal{X}$. Each agent has a state called HAS-MORE, a table indexed by i and j where each entry is a binary string of length $|MORE_{i,j}|$ (as described in previous discussion).*

$$HAS\text{-}MORE_{i,j} \in \{0,1\}^{|MORE_{i,j}|}.$$

Each agent takes as input (from SymbolCount) some count $\in \{0,1,\ldots,M\}$, where M is the maximum multiplicity of any symbol in the root set, and some σ from the input multiset $A \in \mathcal{X}$.

Denote bitwise OR with symbol $\vee$. Our transition rule is

$$(HAS\text{-}MORE^1, (\text{count}, \sigma)), (HAS\text{-}MORE^2, (\text{count}', \sigma'))$$

$$\downarrow$$

$$(HAS\text{-}MORE^3 \vee INDICATOR(\text{count}, \sigma),$$

$$HAS\text{-}MORE^3 \vee INDICATOR(\text{count}', \sigma'))$$

where $HAS\text{-}MORE^3 = HAS\text{-}MORE^1 \vee HAS\text{-}MORE^2$.

An agent outputs their HAS-MORE table.

RootOutput will have an integer *root* that is an index into the root set. An agent with input *HAS-MORE* increments its *root* modulo $|\mathcal{R}|$ if there is a j such that $HAS\text{-}MORE_{root,j}$ is all 1's. Of course, this would mean that the state *root* keeps cycling every time an agent with such an input interacts. When we define the overall protocol composition after, this will be resolved by resetting the previous two sub-protocols' states when *root* increments. Notice that agents don't care about the states of the other agents in an interaction; instead the behavior depends on how sub-protocol *WrongOutput?* changes its output over time. Notice that we don't allow protocols to have changing inputs, but our 3 protocol composition allows two of the sub-protocols to have a changing input as the states of the other sub-protocols change.

Definition 10. *RootOutput. Let $f : \mathcal{X} \to Y$ be a function over finite multisets on a finite alphabet, and let $\mathcal{R}$ be a finite and minimally sized root set of $\mathcal{X}$. Each agent has a state called root, where root $\in \{0, 1, \ldots, |\mathcal{R}| - 1\}$. Each agent takes as input HAS-MORE, a table indexed by i and j where each entry is a binary string of length $|MORE_{i,j}|$ (as described in previous discussion).*

$$HAS\text{-}MORE_{i,j} \in \{0,1\}^{|MORE_{i,j}|}.$$

Our transition rules are:

$$((root_1, HAS\text{-}MORE^1), (root_2, HAS\text{-}MORE^2)) \to (root_1', root_2')$$

$$where \; i' = \begin{cases} i+1 \mod |\mathcal{R}| & if \; \exists j \; s.t. \; HAS\text{-}MORE_{i,j} = 1^{|MORE_{i,j}|} \\ i & otherwise \end{cases}.$$

An agent outputs $f(R_{\text{root}})$.

Putting these three sub-protocols together, we get the *SS-Protocol*, a self-stabilizing population protocol for f.

Definition 11. *SS-Protocol. Let $f : \mathcal{X} \to Y$ be a function over finite multisets on a finite alphabet, and let $\mathcal{R}$ be a finite and minimally sized root set of $\mathcal{X}$. Define SS-Protocol as protocol composition*

$$SymbolCount \times WrongOutput? \times RootOutput,$$

where we define protocol composition in the beginning of Sect. 2.3. We additionally modify this composition's transition function so that whenever an agent's RootOutput state root gets incremented, then

- *The count state from SymbolCount becomes 0.*
- *The HAS-MORE state from WrongOutput? becomes a table of all 0's.*

When this modified transition occurs, we say the agent has been reset and is in a reset state (note there are multiple reset states as we allow root to be arbitrary).

Say the input to the population is A with root R_i. If the protocol ever incorrectly outputs $f(R_j) \neq f(R_i)$, it will recognize this because the protocol would count enough symbols in $MORE_{i,j}$ to make $HAS\text{-}MORE_{i,j}$ a bitstring of all 1's. If the protocol outputs $f(R_i)$ and begins in a reset state, then there will never be a bitstring entry with all 1's. This protocol composition is a self-stabilizing protocol for f. See the full version of this paper for the proof [17].

Theorem (Self-Stabilizing Population Protocol Theorem).*] Let $f : \mathcal{X} \to Y$ be a function computable with population protocols on a complete interaction graph, where $\mathcal{X}$ is a set of multisets. Then f has a self-stabilizing protocol $\iff$ $(\forall A, B \in \mathcal{X}, A \subseteq B \implies f(A) = f(B))$.*

3   Conclusions and Generalizations

A function $f : \mathcal{X} \to Y$ in the basic computational model of population protocols is self-stabilizing if and only if for any multisets A and B in the domain, $A \subseteq B \implies f(A) = f(B)$. The principle insight yielding the forward implication is that we cannot have the scheduler isolate a subpopulation and have that subpopulation restabilize to another output. The converse holds because we only need to parse the root set of the domain and find a root that is a subset of the population to determine what the output is, as our protocol does.

The notion of a root set should be applicable to arbitrary interaction graphs as well. Angluin et al. [3] demonstrated that leader election in *non-simple* classes of graphs are impossible, which is effectively applying the idea that different subgraphs may converge on different answers. We can view input assignments as interaction graphs where the nodes are the input symbols; if there exists a subgraph of an input assignment that maps to a different output under f, then f could not admit a self-stabilizing protocol. We believe that the converse should hold as well. If we consider the class of all graphs of input assignments, we can use the subgraph relation $\subseteq$ as our partial order and find its corresponding minimal elements. Unfortunately there are infinite minimal elements in the class of rings, leading to an infinite root set. Perhaps taking a quotient on the domain of possible input assignments (e.g. calling all rings equivalent) may lead to a finite root set, though this is a subject for future research, as well as on characterizing self-stabilization in general population protocols and other distributed models.

Acknowledgments. We thank the anonymous reviewers for their helpful comments. This work is supported in part by DARPA under Cooperative Agreement No: HR0011-20-2-0025, NSF-BSF Grant 1619348, US-Israel BSF grant 2012366, Google Faculty Award, JP Morgan Faculty Award, IBM Faculty Research Award, Xerox Faculty Research Award, OKAWA Foundation Research Award, B. John Garrick Foundation Award, Teradata Research Award, and Lockheed-Martin Corporation Research Award. The views and conclusions contained herein are those of the authors and should not be interpreted as necessarily representing the official policies, either expressed or implied, of DARPA, the Department of Defense, or the U.S. Government. The U.S. Government is authorized to reproduce and distribute reprints for governmental purposes not withstanding any copyright annotation therein.

References

1. Angluin, D., Aspnes, J., Diamadi, Z., Fischer, M., Peralta, R.: Computation in networks of passively mobile finite-state sensors. Distrib. Comput. **18**, 235–253 (2006)
2. Angluin, D., Aspnes, J., Eisenstat, D., Ruppert, E.: The computational power of population protocols. Distrib. Comput. **20**(4), 279–304 (2007)
3. Angluin, D., Aspnes, J., Fischer, M., Jiang, H.: Self-stabilizing populationprotocols. ACM Trans. Auton. Adapt. Syst. **3** (2008)
4. Aspnes, J., Ruppert, E.: An introduction to population protocols. In: Garbinato, B., Miranda, H., Rodrigues, L. (eds.) Middleware for Network Eccentric and Mobile Applications. Springer, Heidelberg (2009). https://doi.org/10.1007/978-3-540-89707-1_5
5. Awerbuch, B., Ostrovsky, R.: Memory-efficient and self-stabilizing network RESET. In: PODC 1994, pp. 254–263 (1994)
6. Beauquier, J., Blanchard, P., Burman, J.: Self-stabilizing leader election in population protocols over arbitrary communication graphs. In: Baldoni, R., Nisse, N., van Steen, M. (eds.) OPODIS 2013. LNCS, vol. 8304, pp. 38–52. Springer, Cham (2013). https://doi.org/10.1007/978-3-319-03850-6_4
7. Beauquier, J., Burman, J., Clement, J., Kutten, S.: Brief announcement. In: PODC 2009. ACM Press (2009)
8. Beauquier, J., Burman, J., Kutten, S.: Making population protocols self-stabilizing. In: Guerraoui, R., Petit, F. (eds.) SSS 2009. LNCS, vol. 5873, pp. 90–104. Springer, Heidelberg (2009). https://doi.org/10.1007/978-3-642-05118-0_7
9. Burman, J., Doty, D., Nowak, T., Severson, E.E., Xu, C.: Efficient self-stabilizing leader election in population protocols (2020)
10. Cai, S., Izumi, T., Wada, K.: How to prove impossibility under global fairness: on space complexity of self-stabilizing leader election on a population protocol model. Theor. Comput. Syst. **50**, 433–445 (2012)
11. Cook, M., Soloveichik, D., Winfree, E., Bruck, J.: Programmability of chemical reaction networks. In: Condon, A., Harel, D., Kok, J., Salomaa, A., Winfree, E. (eds.) Algorithmic Bioprocesses, pp. 543–584. Springer, Heidelberg (2009). https://doi.org/10.1007/978-3-540-88869-7_27
12. Dickson, L.E.: Finiteness of the odd perfect and primitive abundant numbers with n distinct prime factors. Am. J. Math. **35**(4), 413–422 (1913)
13. Dijkstra, E.: Self-stabilizing systems in spite of distributed control. Commun. ACM, 643–644 (1974)
14. Fischer, M., Jiang, H.: Self-stabilizing leader election in networks of finite-state anonymous agents. In: Shvartsman, M.M.A.A. (ed.) OPODIS 2006. LNCS, vol. 4305, pp. 395–409. Springer, Heidelberg (2006). https://doi.org/10.1007/11945529_28
15. Goldwasser, S., Ostrovsky, R., Scafuro, A., Sealfon, A.: Population stability: regulating size in the presence of an adversary. In: PODC 2018, pp. 397–406 (2018)
16. Izumi, T., Kinpara, K., Izumi, T., Wada, K.: Space-efficient self-stabilizing counting population protocols on mobile sensor networks. Theoret. Comput. Sci. **552**, 99–108 (2014)
17. Mathur, S., Ostrovsky, R.: A combinatorial characterization of self-stabilizing population protocols. arXiv:2010.03869 (2020)

18. Mayer, A., Ofek, Y., Ostrovsky, R., Yung, M.: Self-stabilizing symmetry breaking in constant-space (extended abstract). In: Proceedings 24th ACM Symposium on Theory of Computing, pp. 667–678 (1992)
19. Sudo, Y., Ooshita, F., Kakugawa, H., Masuzawa, T., Datta, A.K., Larmore, L.L.: Loosely-Stabilizing Leader Election with Polylogarithmic Convergence Time. In: OPODIS 2018, vol. 125, pp. 30:1–30:16 (2018)

Smoothed Analysis of Leader Election in Distributed Networks

Anisur Rahaman Molla[1]([⊠])[iD] and Disha Shur[2][iD]

[1] Computer and Communication Sciences Division,
Indian Statistical Institute, Kolkata, Kolkata, India
molla@isical.ac.in
[2] R. C. Bose Centre for Cryptology and Security,
Indian Statistical Institute, Kolkata, Kolkata, India
disha.shur@gmail.com

Abstract. We study smoothed analysis of the leader election (LE) problem in distributed networks. Smoothed analysis is a hybrid between worst-case analysis and average-case analysis. It takes a worst-case instance for the algorithm and perturbs the input by adding some random noise and analyzes the algorithm on this perturbed input. We consider smoothed analysis, in which the topology of the input graph G is randomly perturbed by adding random edges to G. The complexity of the algorithm is parameterized by a smoothing parameter $0 \leq \epsilon(n) \leq 1$ which controls the amount of random edges to be added to the input graph G per round, where ϵ is a small function of n, e.g., n^{-4} (n is the number of nodes in the graph G). Informally, ϵ is the probability that a random edge can be added to a node per round.

We analyze the time and message complexity of leader election in the above smoothing model. We present the following three results in synchronous *CONGEST* distributed model:
(i) A simple randomized algorithm that elects a leader with high probability (With high probability (or w.h.p. in short) means with probability $\geq 1 - 1/n$.) in $O((\log n)/\epsilon)$ rounds and uses $O(\sqrt{n}\log^{2.5} n)$ messages. Note that both the time and the message bounds are optimal (up to a polylog n factor).
(ii) A time-improved randomized algorithm that elects a leader with high probability in $O\left(\frac{\log n}{\sqrt{\epsilon}}\right)$ rounds, but uses $O(m + n\log n)$ messages, where m is the number of edges in the input graph G.
(iii) A deterministic algorithm (except the randomized smoothing part) which solves leader election in $O\left(\frac{\log^2 n}{\sqrt{\epsilon}}\right)$ rounds and incurs $O(m + n\sqrt{\epsilon}\log^2 n)$ messages.

Our work extends the study of smoothed analysis of distributed problems one step further, an open direction raised by [7].

Keywords: Distributed algorithms · Smoothed analysis · Random model · Leader election

The work is supported, in part, by a project on IoT Security, funded by Govt. of India at R. C. Bose Centre for Cryptology and Security, Kolkata, India.

© Springer Nature Switzerland AG 2020
S. Devismes and N. Mittal (Eds.): SSS 2020, LNCS 12514, pp. 183–198, 2020.
https://doi.org/10.1007/978-3-030-64348-5_14

1   Introduction

Motivated by the work of Dinitz et al. [8], the smoothed analysis of distributed algorithms is first formally modeled by Chatterjee et al. [7] and studied for the minimum spanning tree (MST) problem. In this paper, we extend the study of smoothed analysis of distributed problems, an open direction raised by [7], by considering the smoothed analysis of leader election problem. Leader election is one of the fundamental and well studied problem in the field of distributed computing. It outlines the problem of electing a particular node in a network as the leader. A version of this problem requires only the leader node to be aware of its status. All the nodes other than the leader are simply aware that they are not the leader. They need not be aware the leader's identity. This version is called the *implicit* leader election. The *explicit* version of the leader election problem requires all the nodes in the network to be aware of the identity of the leader. The widespread application of the leader election can be found in many domains, e.g., sensor networks [29], IoT networks [26], grid computing [2], peer-to-peer networks [21,27] and cloud computing [32]. In IoT networks, a leader node performs crucial tasks such as gathering information, coordinate tasks among the nodes, generating encryption-decryption keys etc. [14,26].

We consider the same smoothing model as defined in the paper [7] to analyze distributed algorithms. In particular, we consider smoothed analysis, in which the topology of the input graph G is randomly perturbed by adding random edges to G. The perturbance is determined by a smoothing parameter $0 \leq \epsilon(n) \leq 1$ which controls the amount of random edges to be added to the input graph G per round. Typically ϵ is a small function of n, e.g., n^{-4}, where n is the number of nodes in the graph G. Smoothed edges are used for communication only. The study of the smoothing analysis investigates how these additional smoothed edges can be exploited to improve the time and message complexity of a distributed algorithm.

Kutten et al. [20] studied the (implicit) leader election problem in both complete networks and general networks. They presented an algorithm that takes in $O(1)$ time and uses only $O(\sqrt{n} \log^{3/2} n)$ messages to elect a leader in the complete graph. For the general graphs, they extend this algorithm which takes $O(\tau)$ time and $O(\tau \sqrt{n} \log^{3/2} n)$ messages, where τ is the mixing time of graph. The mixing time of a graph could be as large as $O(n^3)$ [23]. This algorithm requires to know the mixing time to be known as input. A major improvement was introduced in [12] where the same problem has been solved without any knowledge of the mixing time of the graph. All these works build on a technique of sampling smaller set of nodes (via random walks) and compute leader in the sampled set. This is a standard technique that is useful for reducing the message complexity. We also use the same idea in our randomized algorithms. The algorithms of [12,20] analyze the worst case time and message complexities. A smoothed analysis of the same problem is considered in this paper where the objective is to analyze both the time and message complexities of the algorithm. Smoothed analysis can be viewed as a hybrid between worst-case analysis and the average-case analysis. It takes a worst-case instance for the algorithm and perturbs the input by adding some random noise and analyzes the algorithm on this perturbed input.

The paper [7] analyzed only the time complexity of the MST problem. In this paper, we analyze both the time and the message complexity of the leader election problem. Similar to [7], we assume that the nodes of the input graph able to distinguish between the smoothed edges and the original graph edges. We present a simple algorithm that solves the problem with high probability in $O(\frac{\log n}{\epsilon})$ rounds using $O(\sqrt{n}\log^{5.2} n)$ messages. Then we present an improvement algorithm that takes $O(\frac{\log n}{\sqrt{\epsilon}})$ rounds and uses $O(m + n\log n)$ messages. We further present a deterministic algorithm that solves leader election in $O\left(\frac{\log^2 n}{\sqrt{\epsilon}}\right)$ rounds and incurs $O(m + n\sqrt{\epsilon}\log^2 n)$ messages. The algorithm is *deterministic* in the sense that except the randomized smoothing part, all other parts are deterministic. Note that one can directly solve leader election in general graphs in $O(D\log n)$ rounds and $O(m\log n)$ messages deterministically using algorithm in [19] or using a randomized algorithm in time $\tilde{O}(\tau)$ and $\tilde{O}(\tau\sqrt{n})$ messages [12,20], but the diameter D or mixing time τ of a graph could be large ($\tilde{O}$ hides a polylog n factor).

1.1 Model

Distributed Network Model. We model the communication network as an undirected, unweighted, connected graph $G = (V, E)$, where $|V| = n$, and $|E| = m$. Every node has limited initial knowledge. We assume anonymous network, i.e., nodes do not know their neighbors. We assume that nodes are associated with a distinct identity number (e.g., its IP address). If not, then each node can randomly pick a number in the range $[1, n^4]$ such that the numbers are distinct for all the nodes. The random number can be used as the ID of the nodes. The node may also accept some additional inputs as specified by the problem at hand. The nodes are allowed to communicate through the edges of the graph G. The communication occurs in synchronous rounds. In one round, nodes can send messages, receive messages (from neighbors) and perform some local computation. Our algorithms use only small-sized messages. In particular, in each round, each node sends a message of size $O(\log n)$ through its adjacent edges. This is a widely used standard model known as the *CONGEST* model of distributed computing [25], and captures the bandwidth constraints inherent in real-world computer networks.

Smoothing Model. There are many smoothing models exits in sequential settings, see [7,8,31]. We consider the smoothing model defined by Chatterjee et al. [7]. The smoothing model of [7] is appropriate to analyze distributed network algorithms.

Given an arbitrary graph G, smoothing allows to introduce some "perturbance" to the input graph. In particular, the smoothing model allows adding some random edges, determined by a smoothing parameter ϵ, to the input graph G. The smoothing parameter $0 \le \epsilon = \epsilon(n) \le 1$ is a function of n, which controls the amount of random edges to be added in the graph per round (typically ϵ is a small function of n, e.g., n^{-4}). Thereby the graph structure is altered. This model is called the ϵ-smoothing model [7]. More precisely, in every (smoothing) round, every node add an edge with probability ϵ to a randomly chosen node from V in the given graph G. The added edges are called smoothed edges. In case of multiple edges being present between two nodes, unless specified, only one smoothed edge is used for communication. The added edges persists in the network and can be used for communication in the later rounds. Let the induced graph formed by the random edges be $R(G) = R(V, S)$, where S is the set of only the smoothed edges. $R(G)$ is called smoothed graph.

As noted in [7], one can view the smoothing model as a generalization of the congested clique model. Suppose the graph G is embedded in a congested clique. A node, besides using its incident edges in E, can also choose to use a random edge in the clique (not in G) with probability ϵ in a round to communicate (once chosen, a random edge can be used subsequently till the end of computation). Thus, the smoothed edges are the clique edges.

Smoothed edges are used for faster communication. The study of the smoothing analysis investigates how these additional smoothed edges can be exploited to improve the time and message complexity of a distributed algorithm. In this paper, we consider only addition of edges to the input graph; while a smoothing model also allows deletion of edges from the graph.

A formal definition of the leader election problem.

Definition 1 (Leader Election). *Every node u maintains a variable $status_u$ that it can set to a value in $\{\perp, NON\text{-}ELECTED, ELECTED\}$; initially $status_u = \perp$ for all u. An algorithm A solves leader election in T rounds if, from round T on, exactly one node has its status set to ELECTED while all other nodes are in state NON-ELECTED. This is the requirement for standard (implicit) leader election.*

Note that the implicit leader election algorithm can be converted to an *explicit* leader election (where every node knows the ID of the leader) by simply the leader sends its ID to all the nodes.

Paper Organization. Section 2 discusses related works. Section 3 contains smoothed analysis of leader election algorithm. We first present a simple optimal time and message complexity leader election algorithm in Sect. 3.1. Then an improved time algorithm in Sect. 3.2 and a deterministic smoothed analysis in Sect. 3.3. Finally, we conclude in Sect. 4.

2   Related Work

Leader election is one of the fundamental problem in distributed networks. It has been extensively studied in various models and settings, starting from its introduction by Le Lann [22] in ring network and then by a seminal work of Gallager-Humblet-Spira [11] in general graphs. The problem is well studied in complete network itself [1,13,16–18,30], and reference there in, and also in general networks [12,15,19].

Some closely related leader election works in the classical congest model are [5,12,19,20]. Kutten et al. [20] studies the (implicit) leader election problem in both complete networks and general networks and showed efficient time and message bounds of leader election algorithms. In particular, they presented an algorithm that executes in $O(1)$ time and uses only $O(\sqrt{n}\log^{3/2} n)$ messages to elect a leader in a complete graph. They also showed an almost matching lower bound for randomized leader election. The standard technique of sampling smaller set of nodes and compute leader among themselves is further used in general graphs and achieve $\tilde{O}(\tau)$-time and $\tilde{O}(\tau\sqrt{n})$-message complexity leader election algorithm [12,20], where τ is the mixing time of a graph ($\tilde{O}$ hides a polylog n factor). We also use this sampling techniques in the randomized algorithms. Several tight results are shown in general graphs in [19], including a notable deterministic algorithm which solves leader election in $O(D\log n)$ rounds and $O(m\log n)$ messages. We adapt this deterministic algorithm in our deterministic smoothing analysis of leader election. The our paper is inspired mainly by the works of [12,19,20].

Smoothed algorithms have been explored in [28] as the next step in bridging the gap between the "theoretical predictions and empirical observation" in their performances. Since its introduction in [31], analyses have shown the practical running time of algorithms to be closer to their smoothed complexities than the worst-case ones. Smoothed analysis of some popular graph problems have been explored in [3,9,10] and [4]. The first smoothed analysis of distributed algorithms, to the best of our knowledge, has been conducted in [8] where in they study the robustness of their algorithm for dynamic networks. They analyse three problems, namely random walks, flooding and aggregation and their upper and lower bounds on dynamic networks. Their smoothing procedure consists of addition as well as deletion of edges from the evolving graph. Robustness is measured by monitoring the change in bounds with the magnitude of smoothing introduced in the model.

3   Leader Election in the Smoothing Model

Given an arbitrary graph $G = (V, E)$ and CONGEST model of communication, the goal is to compute a leader in the ϵ-smoothing model of G. We first present a simple algorithm that solves the leader election problem in $O(\log n/\epsilon)$ rounds using $O(\sqrt{n}\log^{2.5} n)$ messages in the ϵ-smoothing model. The message complexity is optimal (up to a polylog n factor) and the time complexity is also optimal (up to a polylog n factor) for any $\epsilon \geq 1/\text{polylog}\, n$. Then we present a time improved algorithm which solves leader election in $O(\log n/\sqrt{\epsilon})$ rounds, but uses $O(m + n\log n)$ messages. While these two algorithms are randomized, we present a deterministic algorithm that takes $O(\log^2 n/\sqrt{\epsilon})$ rounds and $O(m+n\sqrt{\epsilon}\log^2 n)$ messages to elect a leader. The deterministic algorithm builds on the similar idea of the improved algorithm. In all the algorithms, we assume that the nodes can distinguish between original edges in the given graph and the smoothed edges.

3.1 Simple and Efficient Algorithm

We present a simple, yet efficient algorithm for leader election in the ϵ-smoothing model. At a high-level, the algorithm consists of two parts: (I) Smoothing, i.e., adding random edges to the given graph in such a way that the smoothed graph becomes an expander, (II) Compute a leader in the smoothed graph. Note that the nodes are fixed; so the elected leader is a valid leader in the given graph. The smoothed edges are used for the communication only.

(I) Constructing Smoothed Graph (A Random Expander Graph). In the beginning, the algorithm adds $O(\log n)$ random edges per node according to the smoothing model. In particular, the algorithm runs the smoothing procedure for $\Theta(\log n/\epsilon)$ rounds. In every round, each node adds a smoothed edge with probability ϵ to one of the nodes selected uniformly at random. Thus, it is easy to show that with high probability a node will add $\Theta(\log n)$ random edges (smoothed edges). Let us call this graph as the smoothed graph $R(G) = (V, F)$ which is induced only by the smoothed edges F after $\Theta(\log n/\epsilon)$ rounds. It is intuitive that $R(G)$ is an Erdős-Rényi random graph (expander), which is formally shown in [7] and gives the following lemma.

Lemma 1 (Lemma 3.1, [7]). *The smoothed graph $R(G)$ has a constant conductance and $O(\log n)$ mixing time.*

(II) Computing a Leader. The second part of the algorithm computes leader among the n nodes. For this, the algorithm uses $R(G)$ as the communication graph. Our goal is to compute a leader using minimum number of messages and time. For this part, we adapt the leader election approach of [20] for general graphs. If the mixing time τ of a graph is known, then an algorithm in [20] computes a leader with high probability in $O(\tau)$ rounds and uses $O(\tau\sqrt{n}\log^{1.5} n)$ messages. In essence, we adapt this algorithm in the smoothed graph $R(G)$ since the mixing time $O(\log n)$ of $R(G)$ is known.

Let us discuss an outline of the algorithm. Recall that we consider anonymous network, i.e., nodes do not know each other's ID. We assume that nodes have unique IDs; otherwise each node can randomly pick a number (or rank) in the range $[1, n^4]$ such that the numbers are distinct for all the nodes. The rank can be used as the ID of the nodes. The idea of the algorithm is to select a smaller committee nodes (called *candidate nodes*) which are responsible for electing a leader node among themselves. In fact, the maximum ID node among the candidate nodes will be elected as the leader and all other nodes put themselves in the NON-ELECTED state. This is a standard technique to reduce the message complexity. A random set of candidate nodes of size $\Theta(\log n)$ is selected. For this, each node selects itself with probability $O(\log n/n)$ to become a candidate node. This selection is done locally at each node and hence the candidate nodes do not know each other initially. All the non-candidate nodes put themselves in the NON-ELECTED state. The candidate nodes communicate among themselves via some other nodes, called *referee nodes*. For this, each candidate node samples $2\sqrt{n\log n}$ random referee nodes in the network. This referee sampling is done via performing random walks on $R(G)$ by token forwarding. Each candidate node creates $2\sqrt{n\log n}$ random walk tokens and each token performs random walk of length $O(\log n)$ on the smoothed graph $R(G)$. Since the mixing time of $R(G)$ is $O(\log n)$, the tokens stop at random nodes after $O(\log n)$ steps. The ending nodes of the tokens after $O(\log n)$ steps act as the referee nodes. In this way each candidate node samples $2\sqrt{n\log n}$ random referee nodes. The reason behind sampling so many referee nodes is to make sure at least one common referee node between any pair of candidate nodes' referees. Each candidate node sends its ID with the random walk tokens. An intermediate node or referee node may receive IDs (or tokens) from multiple candidate nodes. Since our goal is to elect the maximum ID node to be the leader, the referee nodes send back a *winner* message $\langle$"WIN", *node_id*, *count*$\rangle$ to the maximum ID candidate node only, via back tracking the random walk paths. In the winner message, *node_id* carries the maximum ID of the candidate node and *count* variable stores the number of tokens having the maximum ID. During the token forwarding process, an intermediate node forwards only the tokens with maximum ID among all the tokens received in a round (and discards all other tokens with smaller ID). The intermediate node also stores this ID and the port numbers, through which it has received the maximum ID token, for back tracking. In subsequent rounds, if an intermediate node receives any token with higher ID than the previous one, then it updates its stored ID and the port number accordingly.

During back tracking, an intermediate node on receiving multiple winner messages $\langle$"WIN", *node_id*, *count*$\rangle$, adds up the *count* for the maximum *node_id* and forwards to the back track node (it discards the winner messages with lower *node_id*). When back track finishes, each candidate node sums up the *count* in the winner messages it has received. The candidate node which receives $2\sqrt{n\log n}$ winner messages enters the ELECTED state. All the other candidate nodes enter the NON-ELECTED state. It is easy to see that only the maximum ID candidate node receives $2\sqrt{n\log n}$ winner messages and elected as the leader.

One difficult part in the algorithm is the congestion over the edges when performing many random walks in parallel as we consider CONGEST model. The congestion is handled by sending only the count of random walk tokens that need to be sent by a particular candidate node, and not the tokens themselves.

To show the correctness of the algorithm, we first discuss the following two results. The first one says that there is at least one candidate node with high probability. Also the size of the candidate nodes is not too large– bounded by $O(\log n)$. The second one says that there is at least one common referee node between any pair of candidate nodes with high probability.

Lemma 2. *The size of the candidate nodes is $\Theta(\log n)$ with high probability.*

Proof. Each node selects itself with probability $O(\log n/n)$ to become a candidate node. Thus, in expectation, $O(\log n)$ candidate nodes are selected. Then one can show using a standard Chernoff bound that the number of the selected candidate nodes is $\Theta(\log n)$ with high probability.   □

Lemma 3 (Theorem 1, [20]). *With high probability, there is a common referee node between any pair of candidate nodes.*

This implies that the maximum ID candidate node has a common referee node with all other candidate nodes. Thus, those common referees generate the winner message for the maximum ID candidate node only and discards all other random walk tokens. Further, during the token forwarding procedure, the maximum ID candidate node's tokens dominate all other tokens. This means in subsequent rounds when the winner messages reach their respective candidate nodes, no candidate node, other than the one having maximum ID, would have received all $2\sqrt{n \log n}$ winner messages with high probability. Therefore, only the candidate node with the maximum ID enters the ELECTED state with a high probability.

Thus we get the following result.

Theorem 1. *Given an anonymous graph $G = (V, E)$ in the ϵ-smoothing model, there exists a randomized distributed algorithm that computes a leader in G with high probability in $O(\frac{\log n}{\epsilon})$ rounds and incurs $O(\sqrt{n} \log^{2.5} n)$ messages (assuming that the smoothed edges are added without sending any messages).*

Proof. We already discussed that the algorithm correctly elects a leader with high probability.

The time complexity of the algorithm is determined by two procedures: (I) Constructing smoothed graph– which takes $O(\log n/\epsilon)$ rounds. (II) Computing leader in the smoothed graph, which requires to perform random walks of length $O(\log n)$ and back tracking (in parallel). Further, there is no congestion due to performing multiple random walks in parallel as we are sending the token counts and not the tokens themselves. All other computations are done locally. This procedure takes $O(\log n)$ rounds. Thus, the time complexity of the algorithm is $O(\log n/\epsilon + \log n) = O(\log n/\epsilon)$ rounds.

The message complexity of the algorithm is determined by the leader election procedure in the smoothed graph $R(G)$. The number of the candidate nodes is $\Theta(\log n)$ with high probability. Each candidate node performs $2\sqrt{n \log n}$ random walks for $O(\log n)$ steps. Thus a total of $O(\sqrt{n} \log^{5/2} n)$ messages are required for this step. The back tracking of the winner messages may take at most the same number of messages. Therefore message complexity of the algorithm is $O(\sqrt{n} \log^{2.5} n)$.   □

Remark 1. In the above theorem, we assume that the smoothed edges are added by the system without sending any messages. If we consider one message is used per edge addition, then the message complexity of the smoothing process would be $O(n \log n)$, as each node adds $O(\log n)$ random edges. Then the message complexity of the algorithm would be $O(n \log n)$.

3.2   An Improved Algorithm

Now we present an improved algorithm which is a variant of the previous algorithm and has a lower time complexity. This algorithm solves the leader election in $O(\log n/\sqrt{\epsilon})$ rounds. It crucially applies the previous algorithm over a super-graph induced by minimum-spanning-tree (MST) fragments as the super nodes. Broadly, this algorithm consists of three parts: (I) Compute MST fragments. (II) Apply smoothing to add an expander over the super-graph induced by the MST fragments. (III) Compute a leader using the previous random walk based sampling algorithm on the smoothed super-graph.

Let us describe the algorithm and simultaneously its analysis.

(I) Computing MST Fragments. We use controlled Gallagher-Humblet-Spira (GHS) algorithm to construct MST fragments, see Sect. 7.4 in [24]. The main difference compared to the standard GHS algorithm is that the growth (size, diameter) of fragments are controlled during merging of fragments.

The controlled GHS algorithm runs in phases [24]. The algorithm starts with each individual node as a fragment and merges fragments in each phase. In each phase, the algorithm maintains the following invariant: Each MST fragment has a leader (which is the root of the tree) and all nodes know their respective parents and children. Initially, each node (a singleton fragment) is a leader node; subsequently each fragment will have one leader (root) node. Each fragment is identified by the identifier of its root (called the fragment ID) and each node in the fragment knows its fragment ID. Each fragment's operation is coordinated by the respective fragment's leader. Each phase consists of two major operations: (1) Finding minimum-weight-outgoing-edge (MOE) of all fragments and (2) Merging fragments via their MOEs. Note that, while the controlled GHS finds an MST in a weighted graph (where edges have weight), it also works on an unweighted graph where one can consider the edge-weights are 1. The details on finding MOE and merging fragments cane be found in Sect. 7.3.1 of [24].

The algorithm starts by running a controlled GHS algorithm for $\log(\frac{1}{\sqrt{\epsilon}})$ phases (ϵ is the smoothing parameter). The size and diameter of each MST fragment are $\Omega(1/\sqrt{\epsilon})$ and $O(1/\sqrt{\epsilon})$ respectively, and there will be $O(n\sqrt{\epsilon})$ such fragments ([24], Sect. 7.4). Each fragment will be treated as a super-node. Each node, that is not a root node, maintains two IDs: 1. Its fragment ID, denoted by FID, and 2. Its own ID in the given graph which we denote by GID in this section. For a root node, its FID would be same as its GID.

The following lemma shows that this MST fragments construction takes $O(\frac{\log^* n}{\sqrt{\epsilon}})$ rounds and uses $O(m)$ messages.

Lemma 4. *The number of MSTs formed after* $\log(\frac{1}{\sqrt{\epsilon}})$ *phases is* $O(n\sqrt{\epsilon})$. *The MST fragments construction takes* $O(\frac{\log^* n}{\sqrt{\epsilon}})$ *rounds and* $O(m)$ *messages, where* m *is the number of edges in the given graph.*

Proof. It is shown in Corollary 7.1 of [24] that the number of MST fragments at the beginning of phase i is at most $\frac{n}{2^i}$. So after $\log(\frac{1}{\sqrt{\epsilon}})$ phases, the number of

MSTs is at most $n/2^{\log(\frac{1}{\sqrt{\epsilon}})} = O(n\sqrt{\epsilon})$. It also follows from Lemma 7.3 of [24] that the diameter of each MST at the beginning of phase i is bounded by 2^i. Hence the diameter of the MST fragments after $\log(\frac{1}{\sqrt{\epsilon}})$ phases is $O(2^{\log(\frac{1}{\sqrt{\epsilon}})}) = O(1/\sqrt{\epsilon})$.

The time and message complexities follows from the 3 sub-procedures.

1. Finding MOE from each node in parallel takes $O(2^i)$ rounds for the ith phase. So for the $\log(\frac{1}{\sqrt{\epsilon}})$ phases, it takes: $O\left(\sum_{i=1}^{\log(\frac{1}{\sqrt{\epsilon}})} 2^i\right) = O\left(\frac{2}{\sqrt{\epsilon}}(1-\sqrt{\epsilon})\right) = O\left(\frac{1}{\sqrt{\epsilon}}\right)$ rounds.

   The number of messages required is: $O\left(\sum_{v\in V} 2 \cdot d(v) + \sum_{i=1}^{\log(\frac{1}{\sqrt{\epsilon}})} \sum_{v\in V} O(1)\right)$ $= O\left(m + n\log(\frac{1}{\sqrt{\epsilon}})\right)$.

2. Selecting MOE for merging fragments takes $O(2^i \log^* n)$ rounds and $O(n\log^* n)$ messages per phase. Thus in total it takes, $O\left(\sum_{i=1}^{\log(\frac{1}{\sqrt{\epsilon}})} (2^i \log^* n)\right) = O\left(\frac{\log^* n}{\sqrt{\epsilon}}\right)$ rounds and $O\left(n\log^* n\log\left(\frac{1}{\sqrt{\epsilon}}\right)\right)$ messages.

3. Merging fragments takes $O(2^i)$ rounds and $O(n)$ messages per phase. So in total it takes, $O\left(\frac{1}{\sqrt{\epsilon}}\right)$ rounds and $O\left(n\log(\frac{1}{\sqrt{\epsilon}})\right)$ messages.

   Thus, MST fragments construction takes $O(\frac{\log^* n}{\sqrt{\epsilon}})$ rounds and $O(m)$ messages. □

(II) Constructing a Smoothed Graph. After constructing the MST forest, perform $O(\frac{\log n}{\sqrt{\epsilon}})$ rounds of smoothing. Let S denotes the set of smoothed edges added in the graph. The probability that a smoothed edge added between two nodes in G is $\Theta(\frac{\sqrt{\epsilon}\log n}{n})$. Consider each MST fragment as a super-node. Let the set of super-node be V'; then $|V'| = O(n\sqrt{\epsilon})$ as there are so many MST fragments. Let $S' \subseteq S$ be the set of inter-super-node smoothed edges. Consider the super-graph $R'(V', S')$; we call it as smoothed super-graph. Then it is easy to show that $R'(V', S')$ is an Erdős-Rényi random graph [7]. Thus, the mixing time of $R'(V', S')$ is $O(\log(n\sqrt{\epsilon})) = O(\log n)$.

(III) Computing a Leader. Now we apply the similar random walk based approach on the smoothed super-graph $R'(V', S')$ to elect a leader in G. In particular, we run the "Computing a leader" procedure of the previous algorithm over this super-graph $R'(V', S')$, where the root node in each super-node simulates and coordinates the tasks of a node there (recall that each MST fragment has a specified root node). First, $O(\log n)$ random set of candidate super-nodes are selected. For this, each super-node selects itself with probability $\Theta(\frac{\log n}{n\sqrt{\epsilon}})$. Note that when we say a "super-node does something", it means the root node inside the super-node does this and communicate to all the nodes in the fragment. To implement this in a super-node, the root node coordinates the tasks with the fragment nodes. Thus, an extra term of $O(1/\sqrt{\epsilon})$ rounds may be incurred due to the communication within a super-node. This is because, the diameter of each MST fragment is $O(1/\sqrt{\epsilon}))$, so the communication within a fragment takes $O(1/\sqrt{\epsilon})$ rounds. Further recall that each super-node has an ID, the fragment ID which is essentially the ID of the root node in the fragment. Then in the similar way, each candidate super-node samples $O(\sqrt{n \log n})$ referee nodes (super-nodes)[1] by performing $O(\sqrt{n \log n})$ random walks of length $O(\log n)$ (since the mixing time of R' is $O(\log n)$). Then the referee nodes send back winner message to the maximum ID super-node. Finally, the super-node with maximum ID becomes the leader. Since the super-node carries the ID of its root node, the root node in the maximum ID super-node becomes the leader in G.

One crucial part remains to discuss is– how the super-nodes perform multiple random walks on $R'(V', S')$ in parallel. Let r be the root node in a super-node T. In the beginning of this procedure – "computing a leader" – every node in T sends the number of its inter-super-node smoothed edges (i.e., outgoing smoothed edges) to the root r. The root node computes the total number of outgoing smoothed edges from T and also stores the ID of the fragment nodes and their outgoing edge number. Suppose a super-node T has k random walk tokens to forward in the next round. For each token, r (locally) selects one outgoing smoothed edge randomly among all the outgoing edges of T. Then r sends a count of the number of tokens to the corresponding fragment nodes. A count ℓ of a fragment node v indicates that ℓ number of random walk tokens are selected to move over the outgoing edges of the node v. The root sends this random walk count information to all the selected nodes in T in parallel. Then the fragments nodes forward their tokens to the outgoing neighbors selected uniformly at random (among the outgoing edges). The node forwards the tokens together with the count through its corresponding outgoing smoothed edges to avoid any congestion. When a node receives random walk tokens from a different super-node, it sends the count (of the tokens) to the root node of its super-node. The root node sums up the count to know the total number of received tokens from the same candidate node. Then it forwards the tokens, in the same way as described above, for the next step.

Theorem 2. *Given an anonymous graph $G = (V, E)$ in the ϵ-smoothing model, there exists a randomized distributed algorithm that computes a leader in G with high probability in $O(\frac{\log n}{\sqrt{\epsilon}})$ rounds and incurs $O(m + n \log n)$ messages.*

Proof. The algorithm correctly elects a leader as the previous algorithm does.

Time Complexity. The first procedure ('computing MST fragments') takes $O(\frac{\log^* n}{\sqrt{\epsilon}})$ rounds (follows from Lemma 4). The second procedure ('constructing smoothed graph') takes $O(\frac{\log n}{\sqrt{\epsilon}})$ rounds as we apply smoothing for so many rounds. The third procedure ('computing a leader') takes $O(\frac{\log n}{\sqrt{\epsilon}})$ rounds. This is because, the random walks are performed over the super-nodes in parallel for $O(\log n)$ rounds and one step of the random walk may take extra $\frac{1}{\sqrt{\epsilon}}$

---

[1] Since $|V'| = O(n\sqrt{\epsilon})$, it would suffice to sample $O(\sqrt{n\sqrt{\epsilon} \log n})$ referee nodes to ensure a common referee node between any pair of candidate super-nodes.

rounds for communication inside a super-node. In the calculation, we implicitly assume $O(\log(n\sqrt{\epsilon})) = O(\log n)$. Therefore, the time complexity the algorithm is: $O(\frac{\log^* n}{\sqrt{\epsilon}} + \frac{\log n}{\sqrt{\epsilon}} + \frac{\log n}{\sqrt{\epsilon}}) = O(\frac{\log n}{\sqrt{\epsilon}})$ rounds.

Message Complexity. The first procedure uses $O(m)$ messages (follows from Lemma 4). The second procedure uses no messages if we assume that the smoothing process (addition of random edges) is done by the system without incurring any messages. It may use $O(n \log n)$ messages if we assume one message cost per one edge addition. The third procedure uses $O(\sqrt{n} \log^{2.5} n + n \log n)$ messages. The term $O(\sqrt{n} \log^{2.5} n)$ comes from performing $O(\sqrt{n \log n})$ random walks for $O(\log n)$ steps from $O(\log n)$ candidate nodes. The term $O(n \log n)$ comes from the communication inside super-nodes or MST fragments via the spanning tree edges for $O(\log n)$ rounds. Thus, the message complexity of the algorithm is $O(m + n \log n)$.   □

3.3   Deterministic Algorithm

In this section, we describe a deterministic algorithm for leader election in the smoothing model. Note that the smoothing part uses random bits, which we cannot avoid in this smoothing model. The rest of the algorithm is deterministic. That's why we are calling the algorithm *deterministic*. The algorithm builds on the similar ideas of the previous improved algorithm (cf. Sect. 3.2). Analogously, it has three procedures: (I) Compute MST fragments, (II) Construct smoothed graph, and (III) Compute a leader. The first two procedures (compute MST fragments, construct smoothed graph) follow the same procedures as in the previous algorithm. Note that the controlled GHS algorithm (which is used to compute MST fragments) is deterministic. Now for the third procedure, we use a deterministic approach instead of applying random walks. In fact, we use a deterministic algorithm from [19] which had solved the leader election problem in $O(D \log n)$ rounds and $O(m \log n)$ messages.

After running the first two procedures, we obtain the smoothed graph $R'(V', S')$ where V' is the set of super-nodes (MST fragments), $|V'| = O(n\sqrt{\epsilon})$ and S' is the set of inter-super-node smoothed edges. Since $R'(V', S')$ is a random expander graph, its diameter is $O(\log(n\sqrt{\epsilon})) = O(\log n)$, e.g., see Theorem 8.13 of [6].

Now for the third procedure (computing a leader), we use the deterministic algorithm from [19] and describe how to run it on R' to elect a leader. This algorithm runs in phases, and each phase consists of 4 stages. First all the nodes become candidate nodes. In each phase i, every candidate node is required to develop a BFS tree of depth 2^{i-1} (for $i = 1, 2, \ldots \log(n\sqrt{\epsilon})$ and then carry out operations in 4 stages. In the first stage, every candidate node v (which translates as the root node of the BFS tree) sends a token **ELECT(phase, ID, counter)** on its BFS tree, where phase refers to the phase i that this candidate node is in, ID is candidate node's ID and counter refers to the depth of BFS tree in any phase i that this candidate node needs to develop. As the tree develops, this counter is decremented by 1. At the beginning, that is, at the candidate node, it is set to 2^{i-1}. In the 2nd stage, it receives an ACK token from all its BFS children of the form: **ACK(ID, max, phase_status)**, where ID is the ID of the candidate node that generated this token, max is the maximum ID encountered by that candidate node, and phase_status is about whether its phase status is same as or lower than this candidate node's phase. Next, the candidate node v updates its field **v.max** with the highest ID it has received, that was contained in the max field among the ACK tokens and sends a **CONFIRM(v.max)** token along its BFS tree in the 3rd stage. In the 4th stage, v receives a **VICTOR(phase, ID)** token from all its neighbors which contains the **v.max** – the highest ID node – encountered by them. If this **v.max** is the same as the ID of the candidate node, v, then v continues to the next phase; otherwise assumes a NON-ELECTED state. In every phase, all the above 4 stages are repeated. After $\log(n\sqrt{\epsilon})$ phases, only the maximum ID node will be ELECTED as a leader and all the other nodes have NON-ELECTED status.

We now explain how to adapt this procedure on the smoothed graph $R'(V', S')$. The super-nodes are MST fragments which contain a root node. Since the root node coordinates the tasks inside a super-node, all the root nodes mark themselves as the candidate nodes in the beginning of the algorithm (essentially, all the super-nodes become candidate nodes). Each super-node acts as a single node and implement all the phases on the smoothed graph $R'(V', S')$ only, e.g., the BFS trees are constructed on $R'(V', S')$. Inside a super-node, the root controls and coordinates the tasks and all the communication done through MST edges inside the supernode. Thus, an extra factor of $O(1/\sqrt{\epsilon})$ rounds may be incurred for each step of the above algorithm due to the communication within a super-node. This is because, the diameter of each MST fragment is $O(1/\sqrt{\epsilon})$), so the communication within a fragment takes $O(1/\sqrt{\epsilon})$ rounds. In the end, one super-node will be elected as the leader. The root node of this super-node will be the leader in the graph.

We now discuss the time and message complexity of the entire deterministic algorithm.

Theorem 3. *Given an anonymous graph $G = (V, E)$ in the ϵ-smoothing model, there exists a deterministic distributed algorithm that solves the leader election problem in $O\left(\frac{\log^2 n}{\sqrt{\epsilon}}\right)$ rounds using $O(m + n\sqrt{\epsilon}\log^2 n)$ messages.*

Proof. It follows from Theorem 2 that the first two procedures takes $O\left(\frac{\log n}{\sqrt{\epsilon}}\right)$ rounds and $O\left(m + n\log^* n\log(\frac{1}{\sqrt{\epsilon}})\right)$ messages.

It follows Lemma 4.8 in [19] that third procedure can have $O(\log(n\sqrt{\epsilon}))$ phases. Further, each of the 4 stages can take at most the diameter time of the smoothed graph R'. Since the diameter of R' is $O(\log(n\sqrt{\epsilon}))$, and communication inside a super-node takes $O(\frac{1}{\sqrt{\epsilon}})$ rounds, the total time for the third procedure is:
$$O((\tfrac{1}{\sqrt{\epsilon}}) \cdot \log(n\sqrt{\epsilon}) \cdot \log(n\sqrt{\epsilon})) = O\left(\frac{\log^2(n\sqrt{\epsilon})}{\sqrt{\epsilon}}\right) = O\left(\frac{\log^2 n}{\sqrt{\epsilon}}\right) \text{ rounds.}$$
Thus, the time complexity of the algorithm is $O\left(\frac{\log^2 n}{\sqrt{\epsilon}}\right)$ rounds.

Let us calculate the message complexity of the third procedure. Consider a phase i of the algorithm. The message exchanges within a fragment happen over the MST edges. Thus, it uses $O(n\sqrt{\epsilon}) \times O(\frac{1}{\sqrt{\epsilon}}) = O(n)$ messages inside all the MST fragments (super-nodes). Message exchanges between the MST fragments happen over the inter-super-node smoothed edges, which is $O(n\sqrt{\epsilon}\log n)$. Therefore, total number of messages uses in $O(\log(n\sqrt{\epsilon}))$ phases is: $O(\log(n\sqrt{\epsilon})) \times O(n + n\sqrt{\epsilon}\log n) = O(n\sqrt{\epsilon}\log^2 n)$.

Thus, the message complexity of the algorithm is: $O\left(m + n\log^* n\log(\frac{1}{\sqrt{\epsilon}})\right) + O(n\sqrt{\epsilon}\log^2 n) = O(m + n\sqrt{\epsilon}\log^2 n)$. □

4   Conclusion

We studied smoothed analysis of leader election, one of the fundamental problem in distributed networks. We consider the same smoothing model as introduced by Chaterjee et al. [7] in distributed networks. We present two randomized algorithms and a deterministic algorithm and discuss their smoothed complexity of time and messages. The time and message complexity of our first algorithm are optimal, up to a polylog n factor. For the second algorithm there is a trade off as it solves the problem in less number of rounds, but incurs more messages. We present a third algorithm which is deterministic but takes slightly more time to solve the leader election problem.

We believe this work extends the study of smoothed analysis of distributed problems. An obvious next step is to investigate how tight these complexities are by analyzing the lower time and message bound of these algorithms. Another line of work could probe the behavior of these algorithms in a different smoothing model and when the nodes can not differentiate between the input edges and the smoothed edges.

References

1. Afek, Y., Gafni, E.: Time and message bounds for election in synchronous and asynchronous complete networks. SIAM J. Comput. **20**(2), 376–394 (1991)
2. Anderson, D.P., Kubiatowicz, J.: Introduction to distributed algorithms. The worldwide computer. Sci. Am. **286**(3), 28–35 (2002)
3. Angel, O., Bubeck, S., Peres, Y., Wei, F.: Local max-cut in smoothed polynomial time. In: STOC (2017)
4. Arthur, D., Manthey, B., Röglin, H.: Smoothed analysis of the k-means method. J. ACM **58**(5), 1–31 (2011)
5. Augustine, J., Molla, A.R., Pandurangan, G.: Sublinear message bounds for randomized agreement. In: PODC, pp. 315–324. ACM (2018)
6. Blum, A., Hopcroft, J., Kannan, R.: Foundations of Data Science. Cambridge University Press (2020). https://doi.org/10.1017/9781108755528
7. Chatterjee, S., Pandurangan, G., Pham, N.D.: Distributed MST: a smoothed analysis. In: ICDCN, pp. 15:1–15:10 (2020)

8. Dinitz, M., Fineman, J., Gilbert, S., Newport, C.: Smoothed analysis of dynamic networks. In: Moses, Y. (ed.) DISC 2015. LNCS, vol. 9363, pp. 513–527. Springer, Heidelberg (2015). https://doi.org/10.1007/978-3-662-48653-5_34

9. Elsässer, R., Tscheuschner, T.: Settling the complexity of local max-cut (almost) completely. In: Aceto, L., Henzinger, M., Sgall, J. (eds.) ICALP 2011. LNCS, vol. 6755, pp. 171–182. Springer, Heidelberg (2011). https://doi.org/10.1007/978-3-642-22006-7_15

10. Etscheid, M., Röglin, H.: Smoothed analysis of local search for the maximum-cut problem. ACM Trans. Algorithms **13**(2), 1–12 (2017)

11. Gallager, R.G., Humblet, P.A., Spira, P.M.: A distributed algorithm for minimum-weight spanning trees. ACM Trans. Program. Lang. Syst. **5**(1), 66–77 (1983)

12. Gilbert, S., Robinson, P., Sourav, S.: Leader election in well-connected graphs. In: Proceedings of the ACM Symposium on Principles of Distributed Computing (PODC), pp. 227–236 (2018)

13. Humblet, P.: Electing a leader in a clique in o(n log n) messages. Intern. Memo., Laboratory for Information and Decision Systems. M.I.T., Cambridge, MA (1984)

14. Kadjouh, N., et al.: A dominating tree based leader election algorithm for smart cities IoT infrastructure. Mob. Netw. Appl., 1–14 (2020). https://doi.org/10.1007/s11036-020-01599-z

15. Khan, M., Kuhn, F., Malkhi, D., Pandurangan, G., Talwar, K.: Efficient distributed approximation algorithms via probabilistic tree embeddings. Distrib. Comput. **25**(3), 189–205 (2012). https://doi.org/10.1007/s00446-012-0157-9

16. Korach, E., Kutten, S., Moran, S.: A modular technique for the design of efficient distributed leader finding algorithms. ACM Trans. Program. Lang. Syst. **12**(1), 84–101 (1990)

17. Korach, E., Moran, S., Zaks, S.: The optimality of distributive constructions of minimum weight and degree restricted spanning trees in a complete network of processors. SIAM J. Comput. **16**(2), 231–236 (1987)

18. Korach, E., Moran, S., Zaks, S.: Optimal lower bounds for some distributed algorithms for a complete network of processors. Theor. Comput. Sci. **64**(1), 125–132 (1989)

19. Kutten, S., Pandurangan, G., Peleg, D., Robinson, P., Trehan, A.: On the complexity of universal leader election. J. ACM **62**(1), 7:1–7:27 (2015)

20. Kutten, S., Pandurangan, G., Peleg, D., Robinson, P., Trehan, A.: Sublinear bounds for randomized leader election. Theor. Comput. Sci. **561**, 134–143 (2015)

21. Kutten, S., Zinenko, D.: Low communication self-stabilization through randomization. In: Lynch, N.A., Shvartsman, A.A. (eds.) DISC 2010. LNCS, vol. 6343, pp. 465–479. Springer, Heidelberg (2010). https://doi.org/10.1007/978-3-642-15763-9_45

22. Lann, G.L.: Distributed systems - towards a formal approach. In: Information Processing, Proceedings of the 7th IFIP Congress 1977, pp. 155–160 (1977)

23. Levin, D.A., Peres, Y., Wilmer, E.L.: Markov Chains and Mixing Times. American Mathematical Society, Providence (2006)

24. Pandurangan, G.: Distributed network algorithms (2018). http://sites.google.com/site/gopalpandurangan/dnabook.pdf

25. Peleg, D.: Distributed Computing: A Locality-Sensitive Approach. SIAM, Philadelphia (2000)

26. Rahman, M.U.: Leader election in the Internet of Things: challenges and opportunities. CoRR abs/1911.00759 (2019). http://arxiv.org/abs/1911.00759

27. Ratnasamy, S., Francis, P., Handley, M., Karp, R.M., Shenker, S.: A scalable content-addressable network. In: SIGCOMM, pp. 161–172. ACM (2001)

28. Roughgarden, T.: Beyond worst-case analysis. Commun. ACM **62**(3), 88–96 (2019)
29. Shi, E., Perrig, A.: Designing secure sensor networks. IEEE Wirel. Commun. **11**(6), 38–43 (2004)
30. Singh, G.: Efficient distributed algorithms for leader election in complete networks. In: ICDCS, pp. 472–479. IEEE Computer Society (1991)
31. Spielman, D.A., Teng, S.H.: Smoothed analysis of algorithms: why the simplex algorithm usually takes polynomial time. J. ACM **51**(3), 385–463 (2004)
32. Wright, A.: Contemporary approaches to fault tolerance. Commun. ACM **52**(7), 13–15 (2009)

Brief Announcement: Byzantine Geoconsensus

Joseph Oglio, Kendric Hood, Gokarna Sharma$^{(\boxtimes)}$, and Mikhail Nesterenko

Department of Computer Science, Kent State University, Kent, OH 44242, USA
{joglio,khood}@kent.edu, {sharma,mikhail}@cs.kent.edu

Abstract. We define and investigate the consensus problem for a set of N processes embedded on the d-dimensional plane, $d \geq 2$, which we call the *geoconsensus* problem. The processes have unique coordinates and can communicate with each other through oral messages. In contrast to the literature where processes are individually considered Byzantine, it is considered that all processes covered by a finite-size convex fault area F are Byzantine and there may be one or more processes in a fault area. Similarly as in the literature where correct processes do not know which processes are Byzantine, it is assumed that the fault area location is not known to the correct processes.

In this paper, we first prove that the geoconsensus is impossible if all processes may be covered by at most three areas where one is a fault area. We then prove the following results on the constructive side considering the 2-dimensional embedding. For $M \geq 1$ fault areas F of arbitrary shape with diameter D, we present a consensus algorithm that tolerates $f \leq N-(2M+1)$ Byzantine processes provided that there are $9M+3$ processes with pairwise distance between them greater than D. For square F with side ℓ, we provide a consensus algorithm that lifts this pairwise distance requirement and tolerates $f \leq N - 15M$ Byzantine processes given that all processes are covered by at least $22M$ axis aligned squares of the same size as F. For a circular F of diameter ℓ, this algorithm tolerates $f \leq N - 57M$ Byzantine processes if all processes are covered by at least $85M$ circles. Finally, we extend these results to various size combinations of fault and non-fault areas as well as d-dimensional process embeddings, $d \geq 3$.

1 Introduction

The problem of *Byzantine consensus* [10,14] has been attracting extensive attention from researchers and engineers in distributed systems. It has applications in distributed storage [1,2,4,5,9], secure communication [6], safety-critical systems [16], blockchain [12,17,19], and Internet of Things (IoT) [11].

Consider a set of N processes with unique IDs that can communicate with each other. Assume that f processes out of these N processes are Byzantine. Assume also that which process is Byzantine is not known to correct processes,

© Springer Nature Switzerland AG 2020
S. Devismes and N. Mittal (Eds.): SSS 2020, LNCS 12514, pp. 199–204, 2020.
https://doi.org/10.1007/978-3-030-64348-5_15

except possibly the size f of Byzantine processes. The Byzantine consensus problem here requires the $N - f$ correct processes to reach to an agreement tolerating arbitrary behaviors of the f Byzantine processes.

Pease *et al.* [14] showed that the maximum possible number of faults f that can be tolerated depends on the way how the (correct) processes communicate: through oral messages or through unforgable written messages (also called signatures). An *oral* message is completely under the control of the sender, therefore, if the sender is Byzantine, then it can transmit any possible message. This is not true for a signed, written message. Pease *et al.* [14] showed that the consensus is solvable only if $f < N/3$ when communication between processes is through oral messages. For signed, written messages, they showed that the consensus is possible tolerating any number of faulty processes $f \leq N$.

The Byzantine consensus problem discussed above assumes nothing about the locations of the processes, except that they have unique IDs. Since each process can communicate with each other, it can be assumed that the N processes work under a complete graph (i.e., clique) topology consisting of N vertices and $N(N-1)/2$ edges. Byzantine consensus has also been studied in arbitrary graphs [14,18] and in wireless networks [13], relaxing the complete graph topology requirement so that a process may not be able to communicate with all other $N - 1$ processes. The goal in these studies is to establish necessary and sufficient conditions for consensus to be solvable. For example, Pease *et al.* [14] showed that the consensus is solvable through oral messages tolerating f Byzantine processes if the communication topology is $3f$-regular. Furthermore, there is a number of studies on a related problem of *Byzantine broadcast* when the communication topology is not a complete graph topology, see for example [8,15]. Byzantine broadcast becomes fairly simple for a complete graph topology.

Recently, motivated by IoT-blockchain applications, Lao *et al.* [11] proposed a consensus protocol, which they call Geographic-PBFT or simply G-PBFT, that extends the well-known PBFT consensus protocol by Castro and Liskov [4] to the geographic setting. The authors considered the case of fixed IoT devices embedded on geographical locations for data collection and processing. The location data can be obtained through recording location information at the installation time or can also be obtained using low-cost GPS receivers or location estimation algorithms [3,7]. They argued that the fixed IoT devices have more computational power than other mobile IoT devices (e.g.., mobile phones and sensors) and are less likely to become malicious nodes. They then exploited (geographical) location information of fixed IoT devices to reach consensus. They argued that G-PBFT avoids Sybil attacks, reduces the overhead for validating and recording transactions, and achieves high consensus efficiency and low traffic intensity. However, G-PBFT is validated only experimentally and no formal analysis is given.

In this paper, we formally define and study the Byzantine consensus problem when processes are embedded on the geographical locations in fixed unique coordinates, which we call the *Byzantine geoconsensus* problem. If fault locations are not constrained, the geoconsensus problem differs little from the Byzantine consensus. This is because the unique locations serve as IDs of the processes and same set of results can be established depending on whether communication between processes is through oral messages or unforgable written messages. Therefore, we relate the fault locations to the geometry of the problem, assuming that the faults are limited to a *fault area* F (going beyond the limitation of mapping Byzantine behavior to individual processes). In other words, the fault area lifts the restriction of mapping Byzantine behavior to individual processes in the classic setting and now maps the Byzantine behavior to all the processors within a certain area in the geographical setting. Applying the classic approaches of Byzantine consensus may not exploit the collective Byzantine behavior of the processes in the fault area and hence they may not provide benefits in the geographical setting. Furthermore, we are not aware of prior work in Byzantine consensus where processes are embedded in a geometric plane while faulty processes are located in a fixed area.

In light of the recent development on location-based consensus protocols, such as G-PBFT [11], discussed above, we believe that our setting deserves a formal study. In this paper we consider the Byzantine geoconsensus problem in case the processes are embedded in a d-dimensional plane, $d \geq 2$. Formally, we define the problem as follows. Consider the binary consensus where every correct process is input a value $v \in \{0, 1\}$ and must output an irrevocable decision with the following three properties.

Agreement – no two correct processes decide differently;
Validity – if all the correct processes input the same value v, then every correct process decides v;
Termination – every correct process eventually decides.

Definition 1 (Byzantine Geoconsensus). *An algorithm solves the Byzantine geoconsensus Problem (or geoconsensus for short) for fault area set $\mathcal{F}$, if every computation produced by this algorithm satisfies the three consensus properties.*

We study the possibility and bounds for a solution to geoconsensus. We demonstrate that geoconsensus allows quite robust solutions: all but a fixed number of processes may be Byzantine. We discuss in detail our contributions below.

Contributions. Let N denotes the number of processes, M denotes the number of fault areas F, D denotes the diameter of F, and f denotes the number of faulty processes. Assume that each process can communicate with all other $N - 1$ processes and the communication is through oral messages. Assume that all the processes covered by a faulty area F are Byzantine. The correct processes know the size of each faulty area (such as its diameter, number of edges, area, etc.) and the total number M of them but do not know their exact location.

In this paper, we made the following five contributions:

(i) An impossibility result that geoconsensus is not solvable if all N processes may be covered by 3 equal size areas F and one of them may be fault area.

This extends to the case of N processes being covered by $3M$ areas F with M areas being faulty.

(ii) The algorithm *BASIC* that solves geoconsensus tolerating $f \leq N - (2M+1)$ Byzantine processes, provided that there are $9M + 3$ processes with pairwise distance between them greater than D.

(iii) The algorithm *GENERIC* that solves geoconsensus tolerating $f \leq N - 15M$ Byzantine processes, provided that all N processes are covered by $22M$ axis-aligned squares of the same size as the fault area F, removing the pairwise distance assumption in the algorithm *BASIC*.

(iv) An extension of the *GENERIC* algorithm to circular F tolerating $f \leq N - 57M$ Byzantine processes if all N processes are covered by $85M$ circles of same size as F.

(v) Extensions of the results (iii) and (iv) to various size combinations of fault and non-fault areas as well as to d-dimensional process embeddings, $d \geq 3$.

Our results are interesting as they provide trade-offs among N, M, and f, which is in contrast to the trade-off provided only between N and f in the Byzantine consensus literature. For example, the results in Byzantine consensus show that only $f < N/3$ Byzantine processes can be tolerated, whereas our results show that as many as $f \leq N - \alpha M$, Byzantine processes can be tolerated provided that the processes are placed on the geographical locations so that at least βM areas (same size as F) are needed to cover them. Here α and β are both integers with $\beta \geq c \cdot \alpha$ for some constant c.

Furthermore, our geoconsensus algorithms reduce the message and space complexity in solving consensus. In the Byzantine consensus literature, every process sends communication with every other process in each round. Therefore, in one round there are $O(N^2)$ messages exchanged in total. As the consensus algorithm runs for $O(f)$ rounds, in total $O(f \cdot N^2)$ messages are exchanged in the worst-case. In our algorithms, let N processes are covered by X areas of size the same as fault area F. Then in a round only $O(X^2)$ messages are exchanged. Since the algorithm runs for $O(M)$ rounds to reach geoconsensus, in total $O(M \cdot X^2)$ messages are exchanged in the worst-case. Therefore, our geoconsensus algorithms are message (equivalently communication) efficient. The improvement on space complexity can also be argued analogously.

Finally, Pease *et al.* [14] showed that it is impossible to solve consensus through oral messages when $N = 3f$ but there is a solution when $N \geq 3f + 1$. That is, there is no gap on the impossibility result and a solution. We can only show that it is impossible to solve consensus when all N processes are covered by $3M$ areas that are the same size as F but there is a solution when all N processes are covered by at least $22M$ areas (for the axis-aligned squares case). Therefore, there is a general gap between the condition for impossibility and the condition for a solution.

Techniques. Our first contribution is established extending the impossibility proof technique of Pease *et al.* [14] for Byzantine consensus to the geoconsensus setting. The algorithm *BASIC* is established first through a leader selection to compute a set of leaders so that they are pairwise more than distance D away from each other and then running carefully the Byzantine consensus algorithm of Pease *et al.* [14] on those leaders.

For the algorithm *GENERIC*, we start by covering processes by axis-aligned squares and studying how these squares may intersect with fault areas of various shapes and sizes. Determining optimal axis-aligned square coverage is NP-hard. We provide constant-ratio approximation algorithms. We also discuss how to cover processes by circular areas. Then, we use these ideas to construct algorithm *GENERIC* for fault areas that are either square or circular, which does not need the pairwise distance requirement of *BASIC* but requires the bound on the number of areas in the cover area set. Finally, we extend these ideas to develop covering techniques for higher dimensions. These covering techniques then provide tolerance bounds for Byzantine consensus in higher dimensions.

Future Work. Our results show the dependency of the tolerance guarantees on the shapes and sizes of the fault areas. Therefore, for future work, it would also be interesting to consider fault area F shapes beyond circles and squares that we studied; to investigate process coverage by non-identical squares, circles or other shapes to see whether better bounds on the set $\mathcal{A}$ and fault-tolerance guarantee f can be obtained. It would also be interesting to close or reduce the gap between the condition for impossibility and a solution (as discussed in Contributions).

References

1. Abd-El-Malek, M., Ganger, G.R., Goodson, G.R., Reiter, M.K., Wylie, J.J.: Fault-scalable byzantine fault-tolerant services. ACM SIGOPS Operating Syst. Rev. **39**(5), 59–74 (2005)
2. Adya, A., et al.: FARSITE: federated, available, and reliable storage for an incompletely trusted environment. ACM SIGOPS Operating Syst. Rev. **36**(SI), 1–14 (2002)
3. Bulusu, N., Heidemann, J., Estrin, D., Tran, T.: Self-configuring localization systems: design and experimental evaluation. ACM Trans. Embed. Comput. Syst. (TECS) **3**(1), 24–60 (2004)
4. Castro, M., Liskov, B.: Practical byzantine fault tolerance and proactive recovery. ACM Trans. Comput. Syst. (TOCS) **20**(4), 398–461 (2002)
5. Castro, M., Rodrigues, R., Liskov, B.: BASE: using abstraction to improve fault tolerance. ACM Trans. Comput. Syst. (TOCS) **21**(3), 236–269 (2003)
6. Cramer, R., Gennaro, R., Schoenmakers, B.: A secure and optimally efficient multi-authority election scheme. Eur. Trans. Telecommun. **8**(5), 481–490 (1997)
7. Hightower, J., Borriello, G.: Location systems for ubiquitous computing. Computer **34**(8), 57–66 (2001)
8. Koo, C.Y.: Broadcast in radio networks tolerating byzantine adversarial behavior. In: PODC, pp. 275–282 (2004)
9. Kubiatowicz, J., et al.: OceanStore: an architecture for global-scale persistent storage. ACM SIGOPS Operating Syst. Rev. **34**(5), 190–201 (2000)
10. Lamport, L., Shostak, R., Pease, M.: The byzantine generals problem. ACM Trans. Programm. Lang. Syst. **4**(3), 382–401 (1982)
11. Lao, L., Dai, X., Xiao, B., Guo, S.: G-PBFT: a location-based and scalable consensus protocol for IOT-Blockchain applications. In: IPDPS, pp. 664–673 (2020)

12. Miller, A., Xia, Y., Croman, K., Shi, E., Song, D.: The honey badger of BFT protocols. In: Proceedings of the 2016 ACM SIGSAC Conference on Computer and Communications Security, pp. 31–42 (2016)
13. Moniz, H., Neves, N.F., Correia, M.: Byzantine fault-tolerant consensus in wireless Ad Hoc networks. IEEE Trans. Mobile Comput. **12**(12), 2441–2454 (2012)
14. Pease, M., Shostak, R., Lamport, L.: Reaching agreement in the presence of faults. J. ACM **27**(2), 228–234 (1980), https://doi.org/10.1145/322186.322188
15. Pelc, A., Peleg, D.: Broadcasting with locally bounded byzantine faults. Inf. Process. Lett. **93**(3), 109–115 (2005)
16. Rushby, J.: Bus architectures for safety-critical embedded systems. In: Henzinger, T.A., Kirsch, C.M. (eds.) EMSOFT 2001. LNCS, vol. 2211, pp. 306–323. Springer, Heidelberg (2001). https://doi.org/10.1007/3-540-45449-7_22
17. Sousa, J., Bessani, A., Vukolic, M.: A byzantine fault-tolerant ordering service for the hyperledger fabric blockchain platform. In: DSN, pp. 51–58. IEEE (2018)
18. Vaidya, N.H., Tseng, L., Liang, G.: Iterative approximate byzantine consensus in arbitrary directed graphs. In: PODC, pp. 365–374 (2012)
19. Zamani, M., Movahedi, M., Raykova, M.: RapidChain: scaling blockchain via full sharding. In: CCS, pp. 931–948 (2018)

Brief Announcements: Verifiable Data Sharing in Distributed Computing

Kun Peng[(⊠)]

Huawei Technology, Shenzhen, China
dr.kun.peng@gmail.com

Abstract. When being applied to share data in distributed computing in practice, even the VSS scheme with the highest efficiency and strongest information-theoretic security cannot be totally assumption free and is vulnerable to attack if its secure communication channel is publicly implemented and verified.

1 Introduction

The VSS by Peng [12] claims to be free of computational assumption or costly exponentiation computation. However, when deploying it in practice, we discover that it still needs some computational assumption and costly exponentiation computation although they can be confined to a very limited extend. In addition, a key technique in [12] is not sufficiently implemented and may cause disputes between the computing nodes, which need PVSS (publicly verifiable secret sharing) to settle. Unfortunately, extension of Peng's VSS to PVSS [13] is not always sound.

2 Background

Among the existing VSS schemes [1,3–5,8–10,14] and PVSS schemes [2,7,11, 15,16], Peng's design [12,13] have unique properties in assumption-freeness and high efficiency. Peng's VSS [12] works as follows for a sharer A to ensure share holders $P_1, P_2, \ldots, P_n$ that their shares are fragments of and can reconstruct the same secret where N is a large prime.

1. Sharing
   A builds a polynomial $F(x) = \sum_{j=0}^{t-1} a_j x^j$ where $a_0 = s$ and a_j for $j = 1, 2, \ldots, t-1$ are random integers chosen from Z_N. Each P_i's share is $s_i = F(i) \bmod N$ for $i = 1, 2, \ldots, n$.
2. Verification
   (a) A builds a polynomial $G(x) = \sum_{j=0}^{t-1} b_j x^j$ where b_j for $j = 0, 1, \ldots, t-1$ are random integers chosen from Z_N. Each P_i's verification commitment is $k_i = G(i) \bmod N$.
   (b) The share holders choose a random integer r from Z_N as a challenge to A.

© Springer Nature Switzerland AG 2020
S. Devismes and N. Mittal (Eds.): SSS 2020, LNCS 12514, pp. 205–210, 2020.
https://doi.org/10.1007/978-3-030-64348-5_16

(c) A publishes $\gamma_j = b_j + ra_j \bmod N$ for $j = 0, 1, \ldots, t - 1$.

(d) Each P_i can verify (1) and accepts his share only if (1) is satisfied.

$$k_i + rs_i = \sum_{j=0}^{t-1} \gamma_j i^j \bmod N \tag{1}$$

8 years after its publication Peng's design is still the simplest and most efficient VSS protocol. In Sect. 2 of [13], when the share for every P_i does not overflow the message space of the encryption algorithm used to securely deliver the share to him, his VSS is extended to PVSS as follows to settle any dispute between the participants.

– Paillier encryption is employed to encrypt each P_i's share such that

$$c_i = E_i(s_i) = g_i^{s_i} r_i^{N_i} \bmod N_i^2 \tag{2}$$

$$c_i' = E_i(k_i) = g_i^{k_i} r_i'^{N_i} \bmod N_i^2 \tag{3}$$

where N_i is the product of two secret primes and larger than N, g_i is generator of a cyclic subgroup defined in Paillier encryption and r_i is a random integer in Z_{N_i}.

– Facing the random challenge r and returning responses γ_j for $j = 0, 1, \ldots, t - 1$, to convince P_i the sharer has to publish $R_i \in Z_{N_i}$ and prove knowledge of z_i to satisfy

$$c_i^r c_i' = g_i^{S_i} R_i^{N_i} (g_i^N)^{z_i} \bmod N_i^2 \tag{4}$$

using ZK proof of knowledge of logarithm as detailed in [13] where $S_i = \sum_{j=0}^{t-1} \gamma_j i^j \bmod N$.

3   Drawback of VSS

When applying Peng's VSS in practice, we find that it needs a secure communication channel for the sharer to securely distribute the shares to the share holders. The abstract concept of reliable and confidential communication, whose implementation is taken for granted in Peng's design, usually depends on encryption in the communication channel in practice. No matter which encryption algorithm is employed, it must rely on some computational assumption for its security. Therefore, when being applied in practice, it is actually impossible to deploy VSS in an environment free of any computational assumption, although we do agree that by employing a symmetric cipher to implement the needed secure communication channel, application of Peng's simple and efficient VSS can be very efficient and does not need any exponentiation of large integers (except a couple of exponentiations in session key exchange).

Although Peng's VSS proof and verification in [12] is very efficient and exponentiation-freed, its realisation depends on randomness of a challenge chosen by the share holders. Our concern in implementation is that it is not specified in detail how the share holders can cooperate to generate the random challenge all of them are satisfied with. The explanation in [12] about this question is not directive or helpful to actual implementation work. We have no clue how to generate the random integer in practice, no to mention practical methods to generate the random challenge accepted by all the share holders probably need some computational assumption to compromise its claimed perfect information-theoretic security. We notice that a random integer to satisfy any single share holder can be chosen by the share holder himself for verification of his own share. Namely, each P_i can choose his own challenge r_i and then A sends a response $\gamma_{i,j} = b_j + r_i a_j \bmod N$ for $j = 0, 1, \ldots, t - 1$ to P_i such that it can verify $k_i + r_i s_i = \sum_{j=0}^{t-1} \gamma_{i,j} i^j \bmod N$. Unfortunately, after this modification, Theorem 2 in [12] cannot guarantee that all the share holders' shares are generated by a unique polynomial and soundness of VSS may fail. Moreover, if any two share holders P_u and P_v collude and put their received responses together, they will have

$$\gamma_{u,j} = b_j + r_u a_j \bmod N \text{ for } j = 0, 1, \ldots, t - 1$$
$$\gamma_{v,j} = b_j + r_v a_j \bmod N \text{ for } j = 0, 1, \ldots, t - 1.$$

Then they can use the $2t$ equations to easily obtain the $2t$ secret coefficients $a_0, a_1, \ldots, a_{t-1}$ and $b_0, b_1, \ldots, b_{t-1}$ and thus $s = a_0$.

If all the n share holders must cooperatively generate the random challenge, each P_i must contribute a seed r_i to r. As none of them wants to reveal his own seed first, they have to commit to their seeds first before any commitment is open to reveal the committed seed. However, the commitment function will depend on some additional computational assumption.

4   Upgrade to PVSS—Resolution and More Challenge

To handle possible dispute between the sharer and share holders and practically implement gneration of the random challenge, Peng's VSS in is upgraded to PVSS in [13]. When the shares are guaranteed to be smaller than N, we can employ the simpler and more efficient PVSS in Sect. 2 of [13] instead of the more complicated version in Sect. 3 of [13]. Another benefit of PVSS is that the random challenge can be publicly calculated as $r = H(c_1, c_2, \ldots, c_n, c'_1, c'_2, \ldots, c'_n)$ using Fiat-Shamir heuristic [6] where $H()$ is a one-way and collision-resistent hash function and the share holders do not need to generate it any longer. However, as implied in Sect. 3 of [13] the PVSS in Sect. 2 of [13] is vulnerable to the following attack such that it cannot fairly solve the disputes caused by a dishonest sharer who maliciously distributes incorrect and inconsistent shares.

1. For $i = 1, 2, \ldots, n$, a malicious sharer does not build any share-generating polynomial but chooses any s_i in Z_N and any k_i as he likes, and then he calculates and publishes c_i and c'_i following (2) and (3) as the encrypted shares.
2. The sharer chooses $\gamma_j \in Z_N$ for $j = 0, 1, \ldots, t - 1$ in any way he likes and publishes them.

3. Although the shares are not generated by the same share-generating polynomial and have nothing to do with any γ_j, the sharer can prove correctness of each P_i's share as follows:

   (a) The sharer can employ the Extended Euclidean Algorithm with an overwhelmingly large probability to calculate two integers μ and ν to satisfy

   $$rs_i + k_i - S_i = \mu N + \nu N_i.$$

   where $S_i = \sum_{j=0}^{t-1} \gamma_j i^j \bmod N$ since $GCD(N, N_i) = 1$ except for a negligible probability.

   (b) By setting $R_i = r_i^r r_i' g_i^\nu \bmod N_i^2$ and $z_i = \mu$ the sharer can pass the verification (4) as

   $$c_i^r c_i' = (g_i^{s_i} r_i^{N_i})^r g_i^{k_i} r'_i^{N_i} = g_i^{s_i r + k_i} (r_i^r r_i')^{N_i} = g_i^{S_i + \mu N + \nu N_i} (r_i^r r_i')^{N_i}$$
   $$= g_i^{S_i} g_i^{N\mu} (g_i^\nu)^{N_i} (r_i^r r_i')^{N_i} = g_i^{S_i} g_i^{N\mu} (g_i^\nu r_i^r r_i')^{N_i} = g_i^{S_i} R_i^{N_i} (g_i^N)^{z_i} \bmod N_i^2$$

Success of the attack means that Peng's PVSS protocol in Sect. 2 of [13] is still not compeletely sound.

5   Further Optimisation

The PVSS to covercome the problem is as follows.

1. Public commitment
   The dealer publicly publishes a commitment of his secret.
   (a) p and q are large primes such that $p = 2q + 1$. G is the cyclic subgroup of Z_p^* with order q. g and h are two generators of G such that $\log_g h$ is secret.
   (b) A has a secret s in Z_q to share. He randomly chooses an additional secret k in Z_q.
   (c) A publishes $u = g^s h^x \bmod p$ and $v = g^k h^y \bmod p$ where x and y are randomly chosen from Z_q.

2. Public share distribution
   The dealer encrypts the shares and publicly sends them to the share holders. Unlike the existing PVSS schemes, which employ costly asymmetric encryption algorithms, our new solution can employ a much more efficient symmetric encryption algorithm to encrypt the shares. However, for fairness in comparison with the existing PVSS schemes, we include distribution of the symmetric keys through asymmetric cipher in our share distribution procedure. The dealer publicly distributes the shares of secret s as follows.
   (a) A builds a polynomial $G(x) = \sum_{j=0}^{t-1} b_j x^j$ where $b_0 = k$ and b_j for $j = 1, 2, \ldots, t-1$ are random integers chosen from Z_q.
   (b) A calculates $k_i = G(i) \bmod q$ and publishes $d_i = E_i(k_i)$ for each P_i where $E_i()$ denotes encryption using P_i's public key and an asymmetric encryption algorithm (e.g. RSA or ElGamal).

(c) A builds a polynomial $F(x) = \sum_{j=0}^{t-1} a_j x^j$ where $a_0 = s$ and a_j for $j = 1, 2, \ldots, t-1$ are random integers chosen from Z_q.

(d) A calculates $s_i = F(i) \bmod q$ and publishes $e_i = E_{k_i}(s_i)$ for each P_i where $E_{k_i}()$ denotes symmetric encryption using session key k_i and an symmetric encryption algorithm.

3. Reconstruction

If at least t share holders submit their shares s_i, the secret s is reconstructed:

$$s = \sum_{i \in V} s_i w_i \bmod q \text{ where } w_i = \prod_{j \in V, j \neq i} \frac{j}{j-i} \bmod q$$

and V is the set containing the indexes of the t shares.

6 Conclusion

Peng's VSS and PVSS proposal can work better in practical applications of distributed computing after being upgraded and optimised.

References

1. Benaloh, J.C.: Secret sharing homomorphisms: keeping shares of a secret secret (extended abstract). In: Odlyzko, A.M. (ed.) CRYPTO 1986. LNCS, vol. 263, pp. 251–260. Springer, Heidelberg (1987). https://doi.org/10.1007/3-540-47721-7_19

2. Boudot, F., Traoré, J.: Efficient publicly verifiable secret sharing schemes with fast or delayed recovery. In: Varadharajan, V., Mu, Y. (eds.) ICICS 1999. LNCS, vol. 1726, pp. 87–102. Springer, Heidelberg (1999). https://doi.org/10.1007/978-3-540-47942-0_8

3. Cachin, C., Kursawe, K., Lysyanskaya, A., Strobl, R.: Asynchronous verifiable secret sharing and proactive cryptosystems. In: ACM CCS 2002, pp. 88–97 (2002)

4. Chaum, D., Crepeau, C., Damgård, I.: Multiparty unconditionally secure protocols (extended abstract). In Proceedings of the Twentieth Annual ACM Symposium on Theory of Computing, STOC 1988, pp. 11–19 (1988)

5. Feldman, P.: A practical scheme for non-interactive verifiable secret sharing. In: FOCS 1987, pp. 427–437 (1987)

6. Fiat, A., Shamir, A.: How to prove yourself: practical solutions to identification and signature problems. In: Odlyzko, A.M. (ed.) CRYPTO 1986. LNCS, vol. 263, pp. 186–194. Springer, Heidelberg (1987). https://doi.org/10.1007/3-540-47721-7_12

7. Fujisaki, E., Okamoto, T.: A practical and provably secure scheme for publicly verifiable secret sharing and its applications. In: Nyberg, K. (ed.) EUROCRYPT 1998. LNCS, vol. 1403, pp. 32–46. Springer, Heidelberg (1998). https://doi.org/10.1007/BFb0054115

8. Gennaro, R., Micali, S.: Verifiable secret sharing as secure computation. In: Guillou, L.C., Quisquater, J.-J. (eds.) EUROCRYPT 1995. LNCS, vol. 921, pp. 168–182. Springer, Heidelberg (1995). https://doi.org/10.1007/3-540-49264-X_14

9. Pedersen, T.P.: Distributed provers with applications to undeniable signatures. In: Davies, D.W. (ed.) EUROCRYPT 1991. LNCS, vol. 547, pp. 221–242. Springer, Heidelberg (1991). https://doi.org/10.1007/3-540-46416-6_20

10. Pedersen, T.P.: Non-interactive and information-theoretic secure verifiable secret sharing. In: Feigenbaum, J. (ed.) CRYPTO 1991. LNCS, vol. 576, pp. 129–140. Springer, Heidelberg (1992). https://doi.org/10.1007/3-540-46766-1_9

11. Peng, K., Bao, F.: Efficient publicly verifiable secret sharing with correctness, soundness and ZK privacy. In: Youm, H.Y., Yung, M. (eds.) WISA 2009. LNCS, vol. 5932, pp. 118–132. Springer, Heidelberg (2009). https://doi.org/10.1007/978-3-642-10838-9_10

12. Peng, K.: Efficient vss free of computational assumption. J. Parallel Distrib. Comput. **71**(12), 23–31 (2011)

13. Peng, K.: Threshold distributed access control with public verification – a practical application of PVSS. Int. J. Inf. Secur. **11**(1), 1592–1597 (2012)

14. Rabin, T., Ben-Or, M.: Verifiable secret sharing and multiparty protocols with honest majority. In: ACM STOC 1989, pp. 73–85 (1989)

15. Schoenmakers, B.: A simple publicly verifiable secret sharing scheme and its application to electronic voting. In: Wiener, M. (ed.) CRYPTO 1999. LNCS, vol. 1666, pp. 148–164. Springer, Heidelberg (1999). https://doi.org/10.1007/3-540-48405-1_10

16. Stadler, M.: Publicly verifiable secret sharing. In: Maurer, U. (ed.) EUROCRYPT 1996. LNCS, vol. 1070, pp. 190–199. Springer, Heidelberg (1996). https://doi.org/10.1007/3-540-68339-9_17

Fast Uniform Scattering on a Grid for Asynchronous Oblivious Robots

Pavan Poudel and Gokarna Sharma$^{(\boxtimes)}$

Department of Computer Science, Kent State University, Kent, OH 44242, USA
{ppoudel,sharma}@cs.kent.edu

Abstract. We consider $K = (k+1) \times (k+1)$ autonomous mobile robots operating on an anonymous $N = (n + 1) \times (n + 1)$-node grid, $n = k \cdot d, d \geq 2, k \geq 2$, following *Look-Compute-Move* cycles under the *classic oblivious robots* model. Starting from any initial configuration of robots positioned on distinct grid nodes, we consider the *uniform scattering* problem of repositioning them on the grid nodes so that each robot reaches to a static configuration in which they cover uniformly the grid. In this paper, we provide the first $O(n)$ time, collision-free algorithm for this problem in the asynchronous setting, given that the robots have common orientation, knowledge of n and k, $O(1)$-bits of memory, and visibility range of $2 \cdot \max\{n/k, k\}$. The best previously known algorithm for this problem on a grid has runtime $O(n^2/d)$ (or $O(nk)$) with the same robot capabilities in the asynchronous setting except the visibility range $2 \cdot n/k$. The proposed algorithm is asymptotically time-optimal since there is a time lower bound of $\Omega(n)$.

1 Introduction

The well-studied model in distributed computing by a team of autonomous mobile robots is the *classic oblivious robots* (COR) model [8] where the robots in the team are *points* (do not occupy any space), *autonomous* (no external control), *anonymous* (no unique identifiers), *indistinguishable* (no external identifiers), *disoriented* (no agreement on coordinate systems and units of distance measures), *oblivious* (no memory of past computation), and *silent* (no direct communication and actions are coordinated via only vision and mobility). The robots operate on a plane and execute the same algorithm. The robots perform their computation in *Look-Compute-Move* (LCM) cycles: an active robot first gets a snapshot of its surroundings (*Look*), computes a destination point based on the snapshot (*Compute*), and finally moves to the destination point (*Move*).

In this paper, we assume that robots operate on a grid G. We assume that nodes and edges of G are *unlabeled*, i.e., robots cannot differentiate one node (edge) from another. The robots reside at nodes of G and they can move from one node to another following the edges of G. The non-neighbor nodes in G are visited following the intermediate nodes of G. We assume that, if a robot at a

© Springer Nature Switzerland AG 2020
S. Devismes and N. Mittal (Eds.): SSS 2020, LNCS 12514, pp. 211–228, 2020.
https://doi.org/10.1007/978-3-030-64348-5_17

Table 1. Results for UNIFORM SCATTERING on a grid.

Algorithm	Model	Visibility range	Runtime	Setting
Barriere *et al.* [2]	Classic oblivious	$2 \cdot n/k$	$O(nk)$	Asynchronous
Poudel and Sharma [18]	Robots with lights	$2 \cdot n/k$	$\Theta(n)$	Fully Synchronous
Theorem 1	**Classic oblivious**	$2 \cdot \max\{n/k, k\}$	$\Theta(n)$	Asynchronous

node computes a destination point (to move) in one LCM cycle, then that destination point is the neighboring node of the node where that robot is currently positioned.

We study the fundamental UNIFORM SCATTERING problem on an anonymous (square) grid G of $N = (n+1) \times (n+1)$ nodes for a set of $K = (k+1) \times (k+1)$ robots, which is defined as follows: Given any initial configuration of K robots positioned on distinct nodes of G, the robots reposition to reach a configuration in which each robot is on a distinct node of G and they uniformly cover G (see Fig. 1).

This problem has practical applications when a team of randomly deployed robots in a region have to cover the region uniformly to maximize the coverage for different purposes, such as intruder detection. An essential requirement is clearly that the robots will reach a state of *static equilibrium* and that scattering is completed as fast as possible. It is assumed that $n = k \cdot d$, $d \geq 2$, $k \geq 2$ to guarantee a final UNIFORM SCATTERING configuration.

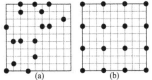

Fig. 1. (a) Initial configuration; (b) UNIFORM SCATTERING.

Barriere *et al.* [2] studied UNIFORM SCATTERING for the first time in the COR model, providing a deterministic algorithm in the asynchronous setting given that the robots have the following capabilities:

– *common orientation* – each robot has consistent notion of North-South and West-East, e.g., as provided by a compass,
– *knowledge* of parameters n and k,
– a *visibility range* of $2 \cdot \lfloor n/k \rfloor$ (i.e., a robot can see robots within distance $2 \cdot \lfloor n/k \rfloor$),
– $O(1)$-*bits of memory* in each robot to store the different states of the system.

Barriere *et al.* [2] did not formally analyze the runtime; however, it is easy to show that their algorithm has runtime $O(n^2/d)$ (or $O(nk)$). Recently, Poudel and Sharma [18] proposed a $\Theta(n)$-time algorithm in the fully synchronous setting for UNIFORM SCATTERING in the *robots with lights* (RWL) model [4], where robots have an externally visible light that can assume a distinct color at a time from a given constant sized set. In this paper, our goal is to design a faster algorithm in the asynchronous setting for UNIFORM SCATTERING in the COR model (see Table 1 for the comparison).

Contributions. We consider the robots and problem setting on a grid as in Barriere *et al.* [2], except the visibility range $2 \cdot \max\{\lfloor n/k \rfloor, k\}$. *Unobstructed visibility* is considered where a robot sees all other robots within its visibility range. *Asynchronous* setting is considered where the robots perform their LCM cycles at arbitrary times. Two robots cannot move to the same node in G. This would constitute a *collision*. We prove:

Theorem 1. *For any initial configuration of $K = (k + 1) \times (k + 1)$ robots positioned on distinct nodes of an anonymous square grid G of $N = (n + 1) \times (n+1)$ nodes with each robot having the visibility range $2 \cdot \max\{n/k, k\}$,* UNIFORM SCATTERING *can be solved in $\Theta(n)$ time in the asynchronous setting avoiding collisions, when robots have common orientation, knowledge of n and k, and $O(1)$ bits of internal memory.*

Theorem 1 improves significantly on the $O(nk)$ (or $O(n^2/d)$) runtime of Barriere *et al.* [2] for the COR model under the same capabilities, except that our algorithm has visibility range $2 \cdot \max\{n/k, k\}$ whereas Barriere *et al.* [2] has $2 \cdot n/k$. Interestingly when $k \le \sqrt{n}$, our visibility range matches with the visibility range of Barriere *et al.*

Techniques. The time lower bound can be established by showing the minimum number of times some robot has to move to reach a UNIFORM SCATTERING configuration. The time lower bound established in Poudel and Sharma [18] immediately proves this lower bound. For the time upper bound, we provide a deterministic algorithm that works in three phases, Phase 1 to Phase 3, executed sequentially.

- **Phase 1 (Gather):** In this phase, all K robots are repositioned on the distinct nodes in the top-right part of G forming a square sub-grid G' (which we call the *gathering configuration* C_{gather}; formal definition in Sect. 2). Essentially what happens in C_{gather} is that robots occupy the $(k + 1) \times (k + 1)$ sub-grid G', one robot on each node of G'. C_{gather} is obtained through two kinds of moves: (i) Northeast moves and (ii) Balancing moves. A robot performs Northeast moves to reach G'. The robot moves either vertically North or horizontally East during Northeast moves. After reaching G', the robot switches to Balancing moves. In the Balancing moves, based on the configuration of other robots, the robot may move North, East, South or West inside G', facilitating new incoming robots to be accommodated inside G' fast. We show that this process results C_{gather} in $O(n)$ time.
- **Phase 2 (Pre-scatter):** The robots in C_{gather} move horizontally West to occupy $(k + 1)$ columns of G with distance between two subsequent columns exactly $d = n/k$. In each such column, the $k + 1$ robots occupy $k + 1$ consecutive positions from the top boundary line of G, which we call the pre-scatter configuration $C_{pre-scatter}$. In this phase, when a robot sees at least one robot on the same horizontal line in the East at distance less than d and the neighboring node in the West is empty, it moves to the West. We show that this phase finishes in $O(n)$ time.

- **Phase 3 (Scatter):** The $k + 1$ robots in each of the $k + 1$ columns move vertically South maintaining a fixed distance $d = n/k$ between consecutive robots. There are $k + 1$ final positions on each line, and hence a UNIFORM SCATTERING configuration is achieved when robots move to the final positions on those lines. The algorithm then terminates. We will show that this phase also finishes in $O(n)$ time.

Therefore, the overall runtime of the algorithm becomes $O(n)$, which is asymptotically optimal given the time lower bound of $\Omega(n)$. Executing Phases 1–3 becomes relatively straightforward in the fully synchronous setting. The challenge is on how to execute Phases 1–3 correctly in the asynchronous setting. For simplicity in understanding, we present the synchronous case first and extend it later to the asynchronous case.

Related Work. UNIFORM SCATTERING is the subject of extensive research in several fields. The research literature is vast and we only discuss in brief the aspects related to our work. In cooperative mobile swarm robotics, this question has been studied in terms of *scattering*, *coverage*, and a special case of *formation* [1,3,5,9,10,13,15,22]. UNIFORM SCATTERING has also been studied in terms of *self-deployment* in mobile sensor networks and in networks of robotic sensors [11,12,16,17,19,21].

The existing works differ on whether robots (sensors) operate on plane or on graphs. They also differ on various parameters, e.g., (i) synchronization settings (fully synchronous, semi-synchronous, or asynchronous), (ii) the robots are oblivious or have persistent memory, (iii) unlimited or limited visibility range, (iv) exact or approximate covering, (v) termination guarantees, (vi) knowledge of the number of robots in the system, (vii) obstructed/unobstructed visibility, (viii) knowledge of the global coordinate system/common orientation/chirality/one-axis agreement, etc.

Our work is on graphs, particularly grids. The grid setting was heavily used in self-deployment and covering problems. We build upon the only previous work of Barriere et al. [2] in the COR model. Furthermore, scattering is considered in [14] in the Euclidean plane under limited visibility and non-trivial time bounds (lower and upper) were reported. Scattering on a ring is considered in [6,7,20].

Paper Organization. We discuss model and some preliminaries in Sect. 2. The algorithm in the fully synchronous setting is presented in Sect. 3, and the extension to the asynchronous setting is provided in Sect. 4. Finally, we conclude in Sect. 5. Some proofs and pseudocodes are omitted due to space constraints.

2   Model and Preliminaries

Graph. Let $G = (V, E)$ be a grid of $N = (n + 1) \times (n + 1)$ nodes, where $V = \{v_1, v_2, \ldots\}$ denotes the node sets and $E \subseteq V \times V$ denotes the edge sets. Each node v_i represents a point location and each edge $(v_i, v_j), i \neq j$, represents a line connecting any two nodes v_i and v_j of V. We assume that the grid is anonymous, i.e., nodes and edges of G are unlabeled. We assume that each edge of G is of unit distance length.

Robots. Let $\mathcal{R} = \{r_1, r_2, \ldots, r_K\}$ be a set of $K = (k+1) \times (k+1)$ robots residing on the nodes of grid G. No robot can reside on the edges of G at any time (except in motion). Moreover, no two robots can occupy the same node of G. In the initial configuration, we assume that robots in $\mathcal{R}$ are at distinct nodes of G and maintain this property throughout the execution of the algorithm. In the algorithm description, we denote by r_i the robot r_i and v_i the node on which r_i resides. The robots have the visibility range $2 \cdot \max\{n/k, k\}$. We assume unobstructed visibility, i.e., a robot sees all other robots within distance $2 \cdot \max\{n/k, k\}$ (even if the robots are collinear). Following Barriere et al. [2], we assume that robots can detect the boundary lines of G when they are $\leq 2 \cdot \max\{n/k, k\}$ distance away from the boundary of G.

Common Orientation. The common orientation means that each robot has a consistent notion of "North-South" and "West-East", e.g., as provided by a compass [2]. For the common orientation, no access to any global localization system is required, i.e., the robots do not need to know their own position on the grid G. We assume that the edges of G are consistently labeled *North* (top), *South* (bottom), *West* (left), and *East* (right), and edge labels are visible to robots.

Look-Compute-Move. At any time, a robot $r_i \in \mathcal{R}$ could be active or inactive. When a robot r_i becomes active, it performs the "Look-Compute-Move" cycle as follows.

- *Look:* For each robot r_j that is visible to it, r_i can observe the position of r_j on G. Robot r_i can also know its own position.
- *Compute:* In any LCM cycle, r_i may perform an arbitrary computation using only the positions observed during the "look" portion of that cycle. This includes determination of a (possibly) new position (which is a node of G) and internal memory storage for r_i for the start of the next cycle. Robot r_i maintains this new memory information from that cycle to the next.
- *Move:* At the end of the LCM cycle, r_i changes its memory to the new information and moves to its new position.

Robot Activation and Time. In the fully synchronous setting ($\mathcal{F}SYNC$), every robot is active in every LCM cycle. In the semi-synchronous setting ($\mathcal{S}SYNC$), at least one robot is active, and over an infinite number of LCM cycles, every robot is active infinitely often. In the asynchronous setting ($\mathcal{A}SYNC$), there is no common notion of time and no assumption is made on the number and frequency of LCM cycles in which a robot can be active; nevertheless, each robot is active infinitely often. For the $\mathcal{F}SYNC$, time is measured in *rounds*. For the $\mathcal{S}SYNC$ and $\mathcal{A}SYNC$, time is measured in epoch. An *epoch* is the smallest interval of time within which each robot is guaranteed to be active at least once.

Configuration. A *configuration* $C_t = \{(r_1^t, mem_1^t), \ldots, (r_K^t, mem_K^t)\}$ defines the positions of the robots in $\mathcal{R}$ on the nodes of G and their internal memory for any time $t \geq 0$. A configuration for a robot $r_i \in \mathcal{R}$, $C_t(r_i)$, defines the positions of the robots in $\mathcal{R}$ that are visible to r_i (including r_i) and their memory, i.e., $C_t(r_i) \subseteq C_t$, at time t. Since each robot has visibility range $2 \cdot \max\{n/k, k\}$, $C_t(r_i)$ has the robots that are within distance $2 \cdot \max\{n/k, k\}$ from r_i. For simplicity and clarity, we sometime write $C, C(r_i)$ to denote $C_t, C_t(r_i)$, respectively. The configuration C_t at $t = 0$ is called the *initial configuration* C_{init}, in which K robots are on K distinct nodes of G.

Uniform Scattering. *Given an anonymous grid $G = (V, E)$ of $N = (n + 1) \times (n+1)$ nodes and a team of $K = (k+1) \times (k+1)$ robots with $n = k \cdot d, k \geq 2, d \geq 2$, positioned initially arbitrarily on the distinct nodes of G, reposition the robots autonomously to reach an equilibrium such that the nodes $(i \cdot d, j \cdot d)$ of G with $i, j \in [0, k]$ hosting exactly one robot each. We say nodes $(i \cdot d, j \cdot d)$ with $i, j \in [0, k]$ the final positions. We say a node (x, y) of G occupied (or non-empty), if there is a robot positioned on it.*

Gathering Configuration. Let $\mathcal{R}$ be a set of K robots positioned on the distinct nodes of G. Let L_N, L_S, L_W, L_E be the North, South, East and West boundary lines of G, respectively. Let L_W' and L_S' be the vertical and horizontal lines parallel to L_E and L_N and passing through k hops West and South of L_E and L_N, respectively. Let G' be the sub-grid of G enclosed by lines L_E, L_N, L_W', and L_S' (including the nodes of G on L_E, L_N, L_W', L_S') in G. We say that a robot $r_i \in \mathcal{R}$ is in a gathering configuration C_{gather} if r_i lies on G' and r_i sees all the nodes in G' are occupied (Fig. 2). We say that

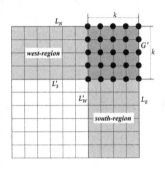

Fig. 2. C_{gather}.

the robots in the set $\mathcal{R}$ are in C_{gather}, if each robot in $\mathcal{R}$ is in C_{gather}. Therefore, in C_{gather}, the $(k + 1) \times (k + 1)$ sub-grid on the topright part of G is occupied with robots. Moreover, we define two regions w.r.t. G'. The grid area of G in the West of G' between L_N and L_S' is denoted as *west-region* of G'. The grid area of G in the South of G' between L_E and L_W' is denoted as *south-region* of G'.

3   Uniform Scattering **Algorithm in** $\mathcal{FSYNC}$

We now describe our collision-free, time-optimal $O(n)$-round Uniform Scattering algorithm in the $\mathcal{FSYNC}$ setting. The pseudocode is given in Algorithm 1. The robots have the common orientation, knowledge of parameters n and k, visibility range of $2 \cdot \max\{\lfloor n/k \rfloor, k\}$, and $O(1)$-bits of memory internal to each robot. We describe the algorithm with respect to a single robot $r_i \in \mathcal{R}$. Figure 3 depicts what intuitively Phases 1–3 do to solve Uniform Scattering starting from any arbitrary C_{init}.

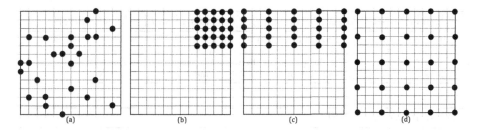

Fig. 3. (a) Initial configuration C_{init}; (b) Gathering configuration C_{gather} (Phase 1); (c) Pre-scatter Configuration (Phase 2); (d) Uniform Scattering Configuration (Phase 3).

Algorithm 1: UNIFORM_SCATTER(r_i, n, k, G)

1 $C(r_i) \leftarrow$ configuration C for robot r_i (including r_i);
2 $L_E, L_W, L_N, L_S \leftarrow$ East, West, North and South boundary lines of G, respectively;
3 $L'_W \leftarrow$ vertical line parallel to L_E at distance k west from L_E;
4 $L'_S \leftarrow$ horizontal line parallel to L_N at distance k south from L_N;
5 $G' \leftarrow$ subgraph of G enclosed by lines L_E, L_N, L'_W and L'_S;
6 $H(r_i), V(r_i) \leftarrow$ horizontal and vertical lines on G passing through r_i, respectively;
7 $r_i \cdot state \leftarrow 0$ (initial state of r_i); $d \leftarrow n/k$;
8 **if** $r_i \cdot state = 0$ **then** $GATHER(r_i, H(r_i), V(r_i), L_E, L_N, L'_S, L'_W, G', C(r_i))$;
9 **else if** $r_i \cdot state = 1$ **then** $PRE\_SCATTER(r_i, d, H(r_i), V(r_i), L_E, L_W, L'_S, C(r_i))$;
10 **else if** $r_i \cdot state = 2$ **then** $SCATTER(r_i, d, V(r_i), L_N, L_S, C(r_i))$;

Algorithm 2: GATHER($r_i, H(r_i), V(r_i), L_E, L_N, L'_S, L'_W, G', C(r_i)$)

1 $(x_i, y_i) \leftarrow$ current position of r_i in G;
2 **if** $(x_i, y_i) \in G'$ **then**
3   **if** r_i sees all the nodes of G' occupied **then** $r_i \cdot state \leftarrow 1$;
4   **else** $BALANCE(r_i, H(r_i), V(r_i), L_E, L_N, L'_W, L'_S, G', C(r_i))$;
5 **else if** $(x_i, y_i + 1)$ is empty $\land$ $((x_i, y_i + 1) \notin G' \lor ((x_i, y_i + 1) \in G' \land (x_i + 1, y_i + 1)$ is empty $\land$ r_i sees no robot in the West of G' between L_N and L'_S)) **then**
6   r_i moves to $(x_i, y_i + 1)$;
7 **else if** $(x_i + 1, y_i)$ is empty $\land$ $(((x_i + 1, y_i) \in G' \land (x_i + 1, y_i + 1)$ is empty$)$ $\lor((x_i + 1, y_i) \notin G' \land (x_i + 1, y_i - 1)$ is empty$))$ **then** r_i moves to $(x_i + 1, y_i)$;

Phase 1 (Gather). The purpose of Phase 1 is to reach a gathering configuration C_{gather} starting from C_{init} (Fig. 3(a)–(b)). The pseudocode is given in Algorithm 2. Phase 1 has two sub-phases, Phase 1.1 (Northeast moves) and Phase 1.2 (Balancing moves), which execute sequentially one after another. In Phase 1.1, robots move towards the North-East of G until they reach G'. After a robot reaches G', it switches to Phase 1.2 doing balancing moves to reposition itself inside G'. We guarantee that after a robot enters G', it never moves out of G' during Phase 1. By the end of Phase 1, all K robots are positioned on the distinct nodes of G' achieving C_{gather}. We will prove that Phases 1.1 and 1.2 each run for $O(n)$ rounds. We describe Phase 1.1 and 1.2 in detail below.

Phase 1.1 (Northeast Moves). Let (x_i, y_i) be the current position of robot r_i in G. In Phase 1.1, r_i does the following in each LCM cycle.

Algorithm 3: BALANCE($r_i, H(r_i), V(r_i), , L_E, L_N, L'_W, L'_S, G', C(r_i)$)

---
1 **if** r_i sees no robot in the South of G' between L_E and $L'_W \lor r_i$ sees at least a robot at distance $\leq (2k - 2)$ in the West of G' between L_N and L'_S **then** MoveSE();
2 **else if** r_i sees no robot in the West of G' between L_N and $L'_S \land r_i$ sees at least a robot at distance $\leq (2k - 2)$ in the South of G' between L_E and L'_W **then** MoveWN();

Algorithm 4: MoveSE()

---
1 $r_{south} \leftarrow$ southmost robot seen by r_i in the West of G';
2 $L_{ref} \leftarrow$ horizontal reference line passing through r_{south};
3 $d_{ref} \leftarrow$ distance between L_N and L_{ref};
4 $dx, dy \leftarrow$ distance from r_i to L'_W and L_{ref}, respectively;
5 **if** $(x_i, y_i - 1)$ is empty $\land$ $dx \geq dy$ $\land$ there exist less than $(k - d_{ref})$ robots on $V(r_i)$ in the South of r_i **then** r_i moves to $(x_i, y_i - 1)$;
6 **else if** $(x_i + 1, y_i)$ and $(x_i + 1, y_i + 1)$ are empty **then** r_i moves to $(x_i + 1, y_i)$;

- Move to $(x_i, y_i + 1)$, if that position (i.e., grid node) is empty and either:
    i. $(x_i, y_i + 1)$ does not lie on G', or
    ii. $(x_i, y_i + 1)$ lies on G', $(x_i + 1, y_i + 1)$ is empty and r_i sees no robot in the *west-region* of G'. (Note: This condition prevents possible collision of r_i with another robot r_j inside G' due to the balancing move of r_j.)
- Otherwise, move to $(x_i + 1, y_i)$, if $(x_i + 1, y_i)$ is empty, and either:
    i. $(x_i + 1, y_i)$ lies on G' and there is no robot on $(x_i + 1, y_i + 1)$, or
    ii. $(x_i + 1, y_i)$ does not lie on G' and there is no robot on $(x_i + 1, y_i - 1)$.
- Switch to Phase 1.2, if it lies on G' and G' is not fully occupied.
- Switch to Phase 2, if G' is fully occupied.

Phase 1.2 (Balancing Moves). The pseudocode for Phase 1.2 is in Algorithm 3. When a robot r_i reaches G', it performs balancing moves as follows. (Note that the robot r_i never moves outside of G' during Phase 1.2.).

Case 1 $- r_i$ **sees no robot in the *south-region* of G' OR r_i sees at least a robot at distance $\leq (2k - 2)$ in the *west-region* of G':** r_i moves either South or East. r_i first checks for possible move towards South, and then towards East. Let (x_i, y_i) be the current position of r_i in G and $H(r_i), V(r_i)$ be the horizontal and vertical lines passing through r_i, respectively. Let r_{south} be the southmost robot seen by r_i in the *west-region* of G' and L_{ref} be the horizontal reference line passing through r_{south}. Let L'_W be the westmost vertical line of G'. Let d_{ref} be the distance between L_N and L_{ref}, dx be the distance from r_i to L'_W and dy be the distance from r_i to L_{ref}. Then, r_i moves South to $(x_i, y_i - 1)$, if the following conditions are satisfied: (i) $(x_i, y_i - 1)$ is empty, (ii) $dx \geq dy$, and (iii) r_i sees less than $(k - d_{ref})$ robots on $V(r_i)$ in the South of r_i.

Else if $(x_i + 1, y_i)$ and $(x_i + 1, y_i + 1)$ are empty, then r_i moves East to $(x_i + 1, y_i)$.

Case 2 $- r_i$ **sees no robot in the *west-region* of G' but sees at least a robot at distance $\leq (2k - 2)$ in the *south-region* of G':** r_i moves either West or North. r_i first checks for possible move towards West, and then towards

Algorithm 5: MoveWN()

1 $r_{west} \leftarrow$ westmost robot seen by r_i in the South of G';
2 $L'_{ref} \leftarrow$ vertical reference line passing through r_{west};
3 $d'_{ref} \leftarrow$ distance between L_E and L'_{ref};
4 $dx', dy' \leftarrow$ distance from r_i to L'_{ref} and L'_S, respectively;
5 **if** $(x_i - 1, y_i)$ is empty $\wedge\ dy' \geq dx'\ \wedge$ there exist less than $(k - d'_{ref})$ robots on $H(r_i)$ in the West of r_i **then** r_i moves to $(x_i - 1, y_i)$;
6 **else if** $(x_i, y_i + 1)$ and $(x_i + 1, y_i + 1)$ are empty **then** r_i moves to $(x_i, y_i + 1)$;

North. Let r_{west} be the westmost robot seen by r_i in the *south-region* of G' and L'_{ref} be the vertical line passing through r_{west}. Let L'_S be the southmost horizontal line of G'. Let d'_{ref} be the distance between L_E and L'_{ref}. Let dx' and dy' be the distances from r_i to L'_{ref} and L'_S, respectively. Then, r_i moves one unit West, if the following conditions are satisfied: (i) $(x_i - 1, y_i)$ is empty, (ii) $dy' \geq dx'$, and (iii) r_i sees less than $(k - d'_{ref})$ robots on $H(r_i)$ in the West of r_i. Else if $(x_i, y_i + 1)$ and $(x_i + 1, y_i + 1)$ are empty, r_i moves North to $(x_i, y_i + 1)$.

Recall that a robot reaches Phase 1.2 after Phase 1.1; however, two different sets of robots execute Phase 1.1 and Phase 1.2 in parallel. While the robots inside G' are performing *Balancing* moves, the robots outside G' are performing *Northeast* moves. Phase 1.2 starts after at least a robot reaches G'. Phase 1.1 ends when all the robots reach Phase 1.2. When all the robots reach Phase 1.2, gathering configuration C_{gather} is achieved and Phase 1.2 also ends. That means, Phase 1.1 and 1.2 both end together.

Lemma 1. *Phase 1.2 starts in at most $O(n)$ rounds after Phase 1.1.*

Proof. If a robot r_j lies inside G' in the initial configuration C_{init}, then r_j directly reaches Phase 1.2. In this case, both Phase 1.1 and Phase 1.2 start at the same time. Let us analyze the case where no robot lies inside G' in C_{init}. Let r be the topmost and rightmost robot in the initial configuration C_{init} of K robots in G. Let Phase 1.1 starts and r executes Algorithm 2. Since, r is the topmost and rightmost robot in G, it moves North until it reaches either G', or north boundary line L_N of G. If r reaches G', it has taken less than n rounds and Phase 1.2 starts. Otherwise, r takes at most n rounds to reach L_N, and it moves East on L_N until it reaches G' in less than next n rounds. Since, r is the topmost and rightmost robot, there is no other robot that blocks the movement of r. Hence, in less than $2n$ rounds, r reaches G' and Phase 1.2 starts.   □

Lemma 2. *Phase 1.1 is collision-free.*

Lemma 3. *Phase 1.2 is collision-, deadlock-, and livelock-free.*

Lemma 4. *Phase 1 finishes in $O(n)$ rounds.*

Proof. We have two sub-phases of Phase 1 (Phase 1.1 and Phase 1.2). Phase 1.2 starts after at least a robot reaches G'. Thus, the total runtime of Phase 1 can be divided into two parts: (i) time elapsed in Phase 1.1 and (ii) runtime of Phase

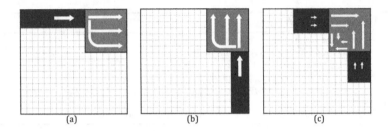

Fig. 4. Illustration of movement of robots during Phase 1; (a) if all robots reach *west-region* of G' in Phase 1.1, they move South/East inside G' in Phase 1.2; (b) if all robots reach *south-region* of G' in Phase 1.1, they move North/West inside G' in Phase 1.2; (c) if robots reach in both *west-region* and *south-region* of G' in Phase 1.1, they may perform all four types of moves (East, West, North or South) inside G' in Phase 1.2.

1.2. From Lemma 1, the time elapsed in Phase 1.1 before the start of Phase 1.2 is $O(n)$ rounds.

Now, let us analyze the runtime of Phase 1.2 with three different cases.

Case I: All robots reach the *west-region* of G' during Phase 1.1 (Fig. 4(a)). Let $R_i = \{r_i^0, r_i^1, \ldots, r_i^{p-1}\}, i = 1, 2, \ldots, n-k$, be the set of $p \leq k+1$ robots on each column at i distance west of L'_W where r_i^0 represents the robot at the northmost horizontal line, r_i^1 represents the robot on the next horizontal line below it and so on. Note that each horizontal line contains $\leq n-k$ robots and each column in the West of L'_W contains $\leq p$ robots on it. When robots in set R_1 move East, they reach G' (i.e. L'_W) and Phase 1.2 starts. In round 1 of Phase 1.2, the robots on L'_W (initially in set R_1) of G' execute Algorithm 3 to perform balancing moves. If $p = k + 1$, all the robots on L'_W move East. This process repeats for all other sets of robots and it is easy to see that all the robots reach G' in $2(k+1)$ rounds. Let us analyze the scenario of $p < k+1$. In this case, the southmost robot (r_1^{p-1}) on L'_W moves South and the remaining ones move East leaving behind the top p positions on L'_W empty. During this round, the next set of robots (R_2) move East and occupy the previous positions of R_1 in the West of L'_W. In round 2 of Phase 1.2, the robots in R_2 reach L'_W, the robots which are already in G' (i.e. R_1), move further East or South (the southmost, r_1^{p-2}, moves South and others move East). This provides empty nodes for the robots currently on L'_W (i.e. R_2) to move East or South in the next round. Also, in round 2, the robots in set R_3 reach to the initial positions of R_2. In round 3, robots in R_4 reach to the initial positions of R_3, robots in R_3 reach to the initial positions of R_1 and the robots in R_2 (currently on L'_W) move East or South in G'. The robots in R_1 move further East or South by one unit. When the south-most robot r_1^{p-1} of R_1 reaches the South boundary line L'_S of G', it moves East in the next round where it meets r_1^{p-2} on its North neighboring node. In the next round, these both robots move East and meet r_1^{p-3}. Following this process, all the p robots of set R_1 ultimately reach the consecutive nodes on L_E in the South part of G'. That means, the southmost p rows of G' will be occupied by the first $k + 1$ sets of robots (i.e. R_1 to R_{k+1}). Similarly, next p rows of G' will be occupied by the next $k + 1$ sets of robots (i.e. R_{k+2} to R_{2k+2}). Recall that, in this case, a robot always search for a possible East or South move inside G', thus creating an empty node for each incoming robot from next column in the West. That means, in every two rounds, one column of p robots enter G'. Thus, in $2(n - k) \leq 2n$ rounds, all the robots reach G'. This achieves the gathering configuration C_{gather} and Phase 1.2 terminates.

Case II: All robots reach the *south-region* of G' during Phase 1.1 (Fig. 4(b)). This case is analogous to Case I. Here, each vertical line contains $\leq n - k$ robots on it. In this case, robots reach G' performing North move from each of the eastmost $q \leq k + 1$ vertical lines. Once a robot reaches G', it performs West or North move inside G'. Following the arguments of Case I analogously, in every 2 rounds, one robot each from the q vertical lines reaches G'. That means, in $\leq 2n$ rounds, all the robots reach G' having the gathering configuration C_{gather} and Phase 1.2 terminates.

Case III: There are robots in both sides (*west-region* and *south-region*) of G' during Phase 1.1 (Fig. 4(c)). This case is the combination of Case I and Case II. In Phase 1.1, the robots in the *south-region* of G' do not move to G' until they see robots in the *west-region* of G'. That means, first all the robots in the *west-region* of G' move to G' and then the robots in the *south-region* of G' move to G'. The robots in the *west-region* follow case I and the robots in the *south-region* follow case II to reach and move inside G'. However, as soon as all the robots in the *west-region* reach G', the robots in the *south-region* may not be able to move immediately to G' as there might not be empty positions. Because, in case I, the robots inside G' move South/East to occupy South/East part of G'. But, when there are no robots in the *west-region* of G', the robots inside G' also satisfy case II and start moving North/West. This may take at most $2k$ time to have empty nodes in the southmost horizontal line L'_N of G'. As soon as there are empty nodes on L'_N, the robots in the *south-region* start moving to G' following case II. Case I and II execute for $\leq 2n$ rounds each. Thus, all the robots reach gathering configuration in $\leq 4n + 2k$ rounds and Phase 1.2 terminates.

Hence, Phase 1 finishes in total at most $O(n) + 4n + 2k = O(n)$ rounds.   □

Phase 2 (Pre-Scatter). The pseudocode of the algorithm for Phase 2 is given in Algorithm 6. The purpose of Phase 2 is to distribute the robots on $k + 1$ vertical lines separated at d distance apart, such that each vertical line contains $k + 1$ robots achieving the pre-scatter configuration $C_{pre-scatter}$ (Fig. 3(c)). In this phase, robots move horizontally West in G. Let $H(r_i)$ and $V(r_i)$ be the horizontal and vertical line passing through r_i in G, respectively. Let L_N be the north boundary line of G and L'_S be the horizontal line parallel to L_N and passing through k distance South of L_N. In each LCM cycle, r_i moves one unit West if the node is empty and it sees a robot on $H(r_i)$ in the East at distance less than d. When a robot reaches to the West boundary line L_W, it changes its state to Phase 3. r_i also changes its state to Phase 3, if it sees a robot in the South of L'_S at horizontal distance $d \cdot x$ from $V(r_i)$ where $x = 0, 1, 2, \ldots$. We prove the following lemma.

Algorithm 6: PRE_SCATTER$(r_i, d, H(r_i), V(r_i), L_E, L_W, L'_S, C(r_i))$

---

1 **if** $(x_i, y_i) \in L_W$ **then** $r_i \cdot state \leftarrow 2$;
2 **else if** r_i sees a robot in the south of L'_S at distance $d \cdot x$ from $V(r_i)$ (on, left or right of $V(r_i)$) where $x = 0, 1, 2, \ldots$ **then** $r_i \cdot state \leftarrow 2$;
3 **else if** $(x_i - 1, y_i)$ is empty $\wedge$ r_i sees a robot on $H(r_i)$ at distance less than d in the East **then** r_i moves to $(x_i - 1, y_i)$;

Algorithm 7: SCATTER$(r_i, d, V(r_i), L_N, L_S, C(r_i))$

---

1 $L_V^d \leftarrow$ vertical line parallel to $V(r_i)$ at distance d east of $V(r_i)$;
2 **if** $(x_i, y_i) \in L_N \vee (x_i, y_i) \in L_S$ **then** r_i terminates;
3 **else if** r_i sees robots on $V(r_i)$ exactly at distance $d \cdot x$ from r_i where $x = 1, 2, 3, \ldots$ **then**
4   r_i terminates;
5 **else if** r_i sees a robot on $V(r_i)$ in the North at distance less than $d \wedge r_i$ sees no robot between $V(r_i)$ and $L_V^d \wedge (x_i, y_i - 1)$ is empty **then** r_i moves to $(x_i, y_i - 1)$;

Lemma 5. *Phase 2 finishes in $O(n)$ rounds avoiding robot collisions.*

Phase 3 (Scatter). Phase 3 executes after Phase 2 and the pseudocode is given in Algorithm 7. The purpose of Phase 3 is to uniformly scatter the robots in G achieving the UNIFORM SCATTERING configuration as depicted in Fig. 3(d). In this phase, robots move vertically towards South in G. Let L_N, L_S be the North and South boundary lines of G, respectively and $H(r_i), V(r_i)$ be the horizontal and vertical lines passing through r_i, respectively. Let L_V^d be the vertical line parallel to $V(r_i)$ and passing through d distance East of $V(r_i)$. All the robots on L_N terminate without moving as they are already at the final positions. When a robot r_i at (x_i, y_i) sees another robot on $V(r_i)$ in the North at distance less than d, it moves one unit South to $(x_i, y_i - 1)$ if the node is empty and r_i sees no robot between $V(r_i)$ and L_V^d. When r_i reaches to L_S, it terminates. r_i also terminates when it sees all the robots on $V(r_i)$ (up to the visibility range) are exactly at d distance apart.

Lemma 6. *Phase 3 finishes in $O(n)$ rounds avoiding robot collisions.*

Proof of Theorem 1. The analysis above proves Theorem 1 for the $\mathcal{F}SYNC$. $\qquad\square$

4   UNIFORM SCATTERING **Algorithm in $\mathcal{A}SYNC$**

In this section, we extend the algorithm for the $\mathcal{F}SYNC$ to the $\mathcal{A}SYNC$ setting. We describe a collision-free, time-optimal $\mathcal{A}SYNC$ $O(n)$-epoch algorithm. The algorithm has four phases: Phase 0 (Pre-Gather), Phase 1 (Gather), Phase 2 (Pre-Scatter) and Phase 3 (Scatter). Unlike $\mathcal{F}SYNC$, the $\mathcal{A}SYNC$ algorithm has one more phase called Phase 0 (Pre-Gather) before Phase 1. Phases 1, 2 and 3 of the $\mathcal{A}SYNC$ are equivalent to the $\mathcal{F}SYNC$ but each phase is modified appropriately. In the $\mathcal{F}SYNC$ algorithm, all the robots switch to Phase 2 from Phase 1 synchronously when C_{gather} is achieved. But in the $\mathcal{A}SYNC$ algorithm,

---

Phase 0: *Pre-Gather*
  - All the robots on L_N of G move South to L'_N and reach Phase 1 avoiding collision.

Phase 1: *Gather*
  - All the robots perform *Northeast move* (avoiding collisions due to the movement of robots at L_N to L'_N in Phase 0) to reach sub-grid G' in the North-East part of G below L_N.
  - When a robot reaches G', it performs *Balancing move* inside G' to achieve C_{gather}.
  - All the robots reach Phase 2 after achieving gathering configuration C_{gather}.

Phase 2: *Pre-Scatter*
  Phase 2.1:
  - Robot at the North-East corner v'_{ne} of G' moves North to the North-East corner v_{ne} of G.
  - Robot at the South-West corner v'_{sw} of G' moves West after the robot at v'_{ne} moved to v_{ne}.
  - Then, the remaining robots on the westmost boundary line L'_W of G' move West.
  - After all the robots on L'_W moved one unit west of L'_W, the southmost robot among them moves further West; the remaining others in the column also follow the West move after it.

  Phase 2.2:
  - When a robot inside G' sees next robot in the West on its horizontal line at distance 3 and all the nodes in the East are occupied, it moves one unit West.
  - The robot also moves West when it sees the next robot in the East on its horizontal line has already started moving West.

  Phase 2.3:
  - A robot moves West when it sees another robot in the East on its horizontal line at distance less than d.
  - When the $k + 1$ robots reach the westmost boundary line L_W of G, the northmost robot among them moves North to L_N.
  - Among every next column of $k + 1$ robots at d distance apart, the northmost robot moves North to L_N after seeing the robot from the previous column at distance d moved to L_N.
  - The robots on L_N reach Phase 3. The remaining robots on L_E move one unit West and reach Phase 3. The other remaining robots move one unit East and reach Phase 3.

Phase 3: *Scatter*
  - Each robot moves South avoiding collision and maintaining a gap of at most distance d to the next robot in the North on its vertical line.
  - When a robot reaches the south boundary line L_S of G, if it lies one unit West of L_E, it moves one unit East and terminates; Otherwise, it moves one unit West and terminates.
  - Any other robot when sees no robot in South on its vertical line but a robot at d distance South on the next vertical line in the West (East), it moves one unit West (East) and terminates.

Fig. 5. Algorithm for UNIFORM SCATTERING in the $\mathcal{ASYNC}$ setting.

robots may become active asynchronously and some robots may never see C_{gather} when they become active. So, we need a different mechanism to switch from Phase 1 to Phase 2 in $\mathcal{ASYNC}$. To handle this situation, we introduce Phase 0 before Phase 1 which makes the northmost boundary line of G empty. Later in Phase 1, when C_{gather} is achieved, one robot is moved to the North-East corner of G which becomes a reference for other robots to switch from Phase 1 to Phase 2. The detail mechanism is explained later in the description of each Phase. Each robot passes through Phases 0–3 sequentially. Figure 5 outlines the algorithm in high level. Figure 6 illustrates the configuration of robots at different stages of each phase.

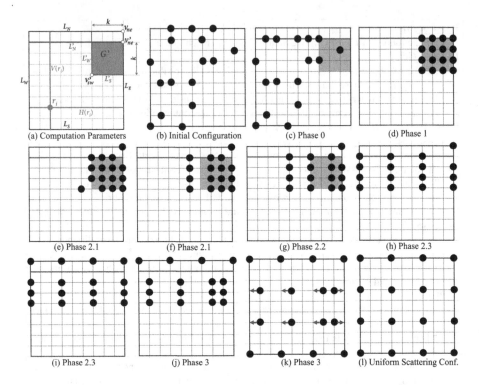

Fig. 6. Configuration of robots at different phases executing algorithm for $\mathcal{ASYNC}$.

Phase 0 (Pre-Gather). The purpose of this phase is to make the North boundary line L_N of G empty. If any robot r_i is located on L_N in C_{init}, the robot is moved South during Phase 0. For this, first, r_i checks if the South neighboring node on the next horizontal line below L_N (i.e. L'_N) is empty or not. If the South node is empty, r_i moves to it. Otherwise, if the West neighboring node is empty, r_i moves one unit West on L_N. The movement of r_i on L_N towards West helps it to find the empty node on L'_N fast because the robots on L'_N move East. Once a robot moves South of L_N, it never moves again to L_N. Phase 0 ends when no robot is positioned on L_N (e.g. Fig. 6(c)).

Lemma 7. *Phase 0 ends in $O(n)$ epochs avoiding robot collisions.*

Phase 1 (Gather). Similar to the Phase 1 of $\mathcal{FSYNC}$, the purpose of this phase is to reach a gathering configuration C_{gather} on the North-East part of G; the only difference is that the sub-grid G' for C_{gather} in $\mathcal{ASYNC}$ lies one unit South of G' in $\mathcal{FSYNC}$. Let L'_N be the next horizontal line below the North boundary line L_N of G. Then the sub-grid G' is bounded by the East boundary line L_E, L'_N, the vertical line parallel to L_E at k distance West of L_E (say L'_W) and the horizontal line parallel to L'_N at k distance South of L'_N (say L'_S). C_{gather} is said to be achieved if all the robots reach in G' at the distinct nodes. Phase 1 in $\mathcal{ASYNC}$ is also divided into two sub-phases, Phase 1.1 and 1.2 that execute sequentially. We describe each sub-phase below.

Phase 1.1 (Northeast moves). This phase is analogous to the Phase 1.1 of $\mathcal{FSYNC}$ after removing the North boundary line L_N of G. A robot r_i never moves North towards L_N from L'_N. If r_i is already on L_N in C_{init}, it moves South to L'_N during Phase 0. Any robot below L_N (except the robot at L'_N) first searches empty position on its North neighboring node for possible North move. If the North move is not possible, it searches an empty node in the East neighboring node for possible East move. A robot below L'_N does not move North to L'_N if it sees a robot on L_N in the same vertical line. Similarly, a robot at L'_N does not move East if it sees a robot on L_N in its North-East neighboring node. This handles the possible collision due to movement of robot at L_N to L'_N.

Phase 1.2 (Balancing moves). Phase 1.2 of $\mathcal{ASYNC}$ directly follows Phase 1.2 of $\mathcal{FSYNC}$ to reach the gathering configuration C_{gather}. Since robots perform their LCM cycles asynchronously, how they change their states to reach Phase 2 in $\mathcal{ASYNC}$ is slightly different than in $\mathcal{FSYNC}$. If a robot r_i sees C_{gather} configuration, it changes its state to Phase 2. Otherwise, r_i changes its state to Phase 2 after seeing the robot in the North-East corner v'_{ne} of G' moved North to L_N. In the mean time, r_i ensures that there is no robot in the South of G', the remaining k nodes on L_E of G' are occupied and the nodes in the East of r_i on $H(r_i)$ (except v'_{ne}) are also occupied.

Complying with Lemma 2–4, we have following lemma for Phase 1 in $\mathcal{ASYNC}$:

Lemma 8. *Phase 1 finishes in $O(n)$ epochs. Phase 1 is collision-free and deadlock-free.*

Phase 2 (Pre-Scatter). In this phase, robots move West to reach the pre-scatter configuration $C_{pre-scatter}$. Unlike $\mathcal{FSYNC}$, in $C_{pre-scatter}$ of $\mathcal{ASYNC}$, the horizontal line below L_N (i.e. L'_N) is empty, instead the horizontal line at $k+1$ distance South of L_N (i.e. L'_S) contains the $k+1$ robots separated at d distance apart. Figure 6(e–i) illustrate the movements of robots during Phase 2 to reach $C_{pre-scatter}$ from C_{gather}.

Phase 2 is divided into three sub-phases, Phase 2.1–2.3. In Phase 2.1, only the robots on L'_W and the robot at the North-East corner of G' (say v'_{ne}) are moved. The robot at v'_{ne} moves North to the North-East corner of G (say v_{ne}) and reaches Phase 2.3. Then, the robot at the South-West corner of G' (say v'_{sw}) moves one unit West to the neighboring node (say v_{ref}). After that, the remaining robots on L'_W also move one unit West. Now, the robot at v_{ref} sees all the nodes on its vertical line towards North occupied and L'_W empty, then it moves one unit West and reaches Phase 2.2. The remaining robots in the North of v_{ref} also move one unit West after it and reach Phase 2.2.

In Phase 2.2, when a robot r_i is inside G' and sees next robot in the West on the same horizontal line $H(r_i)$ at distance 3, it moves one unit West and waits for next robot in the East on $H(r_i)$ to move one unit West. When r_i sees the next robot in the East on $H(r_i)$ moved one unit West (i.e. r_i sees a robot at distance 2 in the East on $H(r_i)$), it also moves one unit West and reaches Phase 2.3.

In Phase 2.3, robot r_i moves West when it sees another robot in the East at distance less than d on $H(r_i)$. Since v'_{ne} is empty, as a special case, the eastmost robot on L'_N can move up to d distance West of L_E without seeing a robot in the East on L'_N. When the westmost $k+1$ robots reach L_W of G, the northmost robot among them moves North to L_N. Among every next column of $k+1$ robots at d distance apart, the northmost robot moves to L_N. Then, the pre-scatter configuration $(C_{pre-scatter})$ is achieved. Since, in every 2 epochs, at least one column of robots move one unit West, it is immediate that the westmost $k+1$ robots reach L_W in $2(n-k)$ epochs. In next $2k$ epochs, the k robots on L'_N move to L_N. Thus, $C_{pre-scatter}$ is achieved in at most $2n$ epochs.

All the robots change their states to Phase 3 after achieving $C_{pre-scatter}$. Additionally, if $d > 2$, any robot r_i south of L_N moves one unit East as well (except the robots on L_E which move one unit West) (Fig. 6(j)). Thus, Phase 2 also finishes in $O(n)$ epochs in $\mathcal{ASYNC}$. Since no robot moves South in Phase 2 and no robot reaches Phase 3 before $C_{pre-scatter}$, the movements of robots in Phase 2 of $\mathcal{ASYNC}$ are collision-free.

Lemma 9. *Phase 2 finishes in $O(n)$ epochs in $\mathcal{ASYNC}$ avoiding robot collisions.*

Phase 3 (Scatter). In this phase, robots move South to achieve UNIFORM SCATTERING configuration and terminate. Figure 6(j–l) provide an illustration. If $d = 2$, robot r_i has the visibility range of n and hence can see all the final positions on it's vertical line. Then, r_i moves South by directly following the Phase 3 of $\mathcal{FSYNC}$ to reach the final position and terminates. If $d > 2$, the algorithm works as follow: Let $H(r_i), V(r_i)$ be the horizontal and vertical lines passing through r_i, respectively. Let $V'(r_i)$ be the vertical line parallel to $V(r_i)$ and passing through one unit West of r_i (for the robots at one unit West of L_E, consider L_E as $V'(r_i)$). When a robot r_i at (x_i, y_i) sees another robot on $V(r_i)$ in North at distance less than d, then r_i moves to $(x_i, y_i - 1)$ if $(x_i, y_i - 1)$ and $(x_i - 1, y_i - 1)$ are both empty. If r_i it on one unit West of L_E, it ensures that $(x_i + 1, y_i - 1)$ is empty instead of $(x_i - 1, y_i - 1)$ to move South. When r_i reaches the south boundary line L_S, it moves West (except the eastmost robot on L_S which moves East to L_E) to reach the final position and terminates. When r_i sees another robot r_j on $V'(r_i)$ at exactly d distance South of $H(r_i)$, r_i moves horizontally to $V'(r_i)$ to occupy the final position and terminates. The southmost robots on the $k + 1$ vertical lines take at most $2(n - k)$ epochs to reach L_S. By that time all the robots on each of those $k+1$ vertical lines are at d distance apart. In at most next $2k$ epochs, all those robots reach to the final positions and terminate. Thus, in at most $2n$ epochs, Phase 3 terminates.

Lemma 10. *Phase 3 finishes in $O(n)$ epochs in the $\mathcal{ASYNC}$ setting.*

Proof of Theorem 1. Combine results of Lemmas 7, 8, 9 and 10.   □

5   Concluding Remarks

We have provided the first optimal $O(n)$ time algorithm to the UNIFORM SCATTERING problem in a square grid graph of $N = (n+1) \times (n+1)$ nodes in the COR model under the $\mathcal{ASYNC}$ setting. This is the $O(n/d) = O(k)$ improvement compared to the best previously known algorithm with runtime $O(N/d) \equiv O(n^2/d)$ in the COR model. In the future work, it will be interesting to extend our algorithm to consider faults.

References

1. Barrameda, E.M., Das, S., Santoro, N.: Uniform dispersal of asynchronous finite-state mobile robots in presence of holes. In: ALGOSENSORS, pp. 228–243 (2013)
2. Barriere, L., Flocchini, P., Mesa-Barrameda, E., Santoro, N.: Uniform scattering of autonomous mobile robots in a grid. In: IPDPS, pp. 1–8 (2009)
3. Cohen, R., Peleg, D.: Local spreading algorithms for autonomous robot systems. Theor. Comput. Sci. **399**(1–2), 71–82 (2008)
4. Das, S., Flocchini, P., Prencipe, G., Santoro, N., Yamashita, M.: Autonomous mobile robots with lights. Theor. Comput. Sci. **609**, 171–184 (2016)
5. Défago, X., Souissi, S.: Non-uniform circle formation algorithm for oblivious mobile robots with convergence toward uniformity. Theor. Comput. Sci. **396**(1–3), 97–112 (2008)
6. Elor, Y., Bruckstein, A.M.: Uniform multi-agent deployment on a ring. Theor. Comput. Sci. **412**(8–10), 783–795 (2011)
7. Flocchini, P., Prencipe, G., Santoro, N.: Self-deployment of mobile sensors on a ring. Theor. Comput. Sci. **402**(1), 67–80 (2008)
8. Flocchini, P., Prencipe, G., Santoro, N.: Distributed Computing by Oblivious Mobile Robots. Synthesis Lectures on Distributed Computing Theory, vol. 3, no. 2, pp. 1–185 (2012)
9. Flocchini, P., Prencipe, G., Santoro, N., Viglietta, G.: Distributed computing by mobile robots: uniform circle formation. Distrib. Comput. **30**(6), 413–457 (2016). https://doi.org/10.1007/s00446-016-0291-x
10. Flocchini, P., Prencipe, G., Santoro, N., Widmayer, P.: Arbitrary pattern formation by asynchronous, anonymous, oblivious robots. Theor. Comput. Sci. **407**(1–3), 412–447 (2008)
11. Heo, N., Varshney, P.K.: Energy-efficient deployment of intelligent mobile sensor networks. Trans. Sys. Man Cyber. Part A **35**(1), 78–92 (2005)
12. Howard, A., Matarić, M.J., Sukhatme, G.S.: An incremental self-deployment algorithm for mobile sensor networks. Auton. Rob. **13**(2), 113–126 (2002)
13. Hsiang, T.-R., Arkin, E.M., Bender, M.A., Fekete, S.P., Mitchell, J.S.B.: Algorithms for rapidly dispersing robot swarms in unknown environments. In: Boissonnat, J.-D., Burdick, J., Goldberg, K., Hutchinson, S. (eds.) Algorithmic Foundations of Robotics V. STAR, vol. 7, pp. 77–93. Springer, Heidelberg (2004). https://doi.org/10.1007/978-3-540-45058-0_6
14. Izumi, T., Kaino, D., Potop-Butucaru, M.G., Tixeuil, S.: On time complexity for connectivity-preserving scattering of mobile robots. Theor. Comput. Sci. **738**, 42–52 (2018)

15. Katreniak, B.: Biangular circle formation by asynchronous mobile robots. In: Pelc, A., Raynal, M. (eds.) SIROCCO 2005. LNCS, vol. 3499, pp. 185–199. Springer, Heidelberg (2005). https://doi.org/10.1007/11429647_16

16. Lin, Z., Zhang, S., Yan, G.: An incremental deployment algorithm for wireless sensor networks using one or multiple autonomous agents. Ad Hoc Netw. **11**(1), 355–367 (2013)

17. Poduri, S., Sukhatme, G.S.: Constrained coverage for mobile sensor networks. In: ICRA, pp. 165–171 (2004)

18. Poudel, P., Sharma, G.: Time-optimal uniform scattering in a grid. In: ICDCN, pp. 228–237 (2019)

19. Sharma, G., Krishnan, H.: Tight bounds on localized sensor self-deployment for focused coverage. In: ICCCN, pp. 1–7. IEEE (2015)

20. Shibata, M., Mega, T., Ooshita, F., Kakugawa, H., Masuzawa, T.: Uniform deployment of mobile agents in asynchronous rings. In: PODC, pp. 415–424 (2016)

21. Sinan Hanay, Y., Gazi, V.: Distributed sensor deployment using potential fields. Ad Hoc Netw. **67**(C), 77–86 (2017)

22. Suzuki, I., Yamashita, M.: Distributed anonymous mobile robots: formation of geometric patterns. SIAM J. Comput. **28**(4), 1347–1363 (1999)

Brief Announcement: Leader Election in the ADD Communication Model

Sergio Rajsbaum[1], Michel Raynal[2,3], and Karla Vargas[1(✉)]

[1] Instituto de Matemáticas, UNAM, Mexico City, Mexico
karla.vargas@ciencias.unam.mx
[2] Univ Rennes IRISA, 35042 Rennes, France
[3] Department of Computing, Polytechnic University,
Hong Kong, People's Republic of China

Abstract. A channel from a process p to a process q satisfies the *ADD property* if there are two constants K and D, unknown to the processes, such that in any sequence of K consecutive messages sent by a process p to a process q, at least one of them is delivered to q at most D time units after it has been sent. This paper studies implementations of an eventual leader, namely an Ω failure detector, in a (not necessarily complete) connected network of eventual ADD channels, where processes may fail by crashing. It presents an algorithm that assumes that the processes initially know n, the total number of processes, sending messages of size $O(\log n)$.

Keywords: ADD channel · Arbitrarily connected networks · Distributed algorithm · Eventual leader · Fault-tolerance · Process crash · Synchrony · System model · Weak channel

1 Introduction

Leader Election. This is a classical problem encountered in distributed computing. Each process p_i has a local variable $leader_i$, and it is required that all the local variables $leader_i$ forever contain the same identity, which is the identity of one of the processes. If processes may crash, the system is fully asynchronous, and the elected leader must be a process that does not crash, leader election cannot be solved [9]. Not only the system must no longer be fully asynchronous, but the leader election problem must be weakened to the *eventual leader election problem*. This problem is denoted Ω in the failure detector parlance [3,4]. Notice that the algorithm must elect a new leader each time the previously elected leader crashes.

The ADD Distributed Computing Model. ADD channels were introduced by S. Sastry and S. M. Pike in [10] as a realistic model for partially synchronous systems in which channels can lose and reorder messages. Each channel guarantees that some subset of the messages sent on it will be delivered in a timely

© Springer Nature Switzerland AG 2020
S. Devismes and N. Mittal (Eds.): SSS 2020, LNCS 12514, pp. 229–234, 2020.
https://doi.org/10.1007/978-3-030-64348-5_18

manner and such messages are not too sparsely distributed in time. More precisely, for each channel there exist two constants K and D, not known to the processes and not necessarily the same for all channels, such that for every K consecutive messages sent in one direction, at least one is delivered within D time units after it has been sent.

Even though ADD channels seem so weak, S. Kumar and J.L. Welch showed in [6] that it is possible to implement an *eventually perfect failure detector* in an arbitrarily connected network of ADD channels. An efficient implementation of $\Diamond P$ in such a system, using messages of size $O(n \log n)$, is presented [11].

Contribution. Considering the election of an eventual leader, called Ω in the failure detector parlance [3], this paper shows that it is possible to implement such a leader in an arbitrarily connected network where asynchronous processes may fail by crashing in a weaker ADD model than the one used in [11]. More precisely, it presents an implementation of Ω where the size of the messages is $O(\log n)$, reducing significantly the message size with respect to [11]. The proposed algorithm works under very weak assumptions, requiring only that a directed spanning tree from the leader exists, composed of channels that eventually satisfy the ADD property.

We put particular attention to the size of the messages but also, we make sure our solution is efficient in terms of the time it takes for the processes to agree on the same leader. When designing leader election ADD-based algorithms using messages whose size is bounded, a difficult challenge comes from the uncertainty created by the fact that, while the constants K and D do exist, a process knows neither them nor the time at which the channels satisfy them. This is the type of difficulty encountered in the design of leader election algorithms under weak eventual synchrony assumptions, e.g., [1,5,9,11]. Also, the hopbound technique we used is reminiscent of the one used in self-stabilizing algorithms [2] and in [11].

2   Model of Computation

System Model. The system consists of a finite set of processes $\Pi = \{p_1, p_2, ..., p_n\}$. Any number of processes may fail by crashing. A process is *correct* if it does not crash, otherwise, it is *faulty*. The communication network is represented by a directed graph $G = (\Pi, E)$, where an edge $(p_i, p_j) \in E$ means that there is a unidirectional channel that allows the process p_i to send messages to p_j. Is required the existence of a spanning tree containing all correct processes whose root will be the leader (the correct process with the smallest identity).

A directed channel (p_i, p_j) satisfies the $\Diamond$ *ADD property* if there is a finite time (unknown to the processes) after which there are two constants K and D (unknown to the processes) such that for every K consecutive messages sent by p_i to p_j, at least one is delivered to p_j within D time units after it has been sent. The other messages from p_i to p_j can be lost or experience arbitrary delays.

Eventual Leader Election. Assuming a read-only local variable $leader_i$ at each process p_i, the leader failure detector Ω satisfies the follwing properies [3,8]:

- *Validity:* Each read of $leader_i$ returns a process name to p_i.
- *Eventual leadership:* There is a finite (but unknown) time after which the local variables $leader_i$ of all the correct processes contain forever the same process name, which is the name of one of them.

3   Eventual Leader Election in the $\Diamond$ADD Model

This section presents an algorithm that implements Ω, assuming that each process knows n, the number of processes. The parameter T denotes an arbitrary duration. Its value can affect the efficiency of the algorithm, but not its correctness. (If T is too big, the failure detection of a process currently considered as a leader can be delayed. On the contrary, a too small value of T can entail false suspicions of the current eventual leader p_j until the corresponding timer $timer_i[j]$ has been increased to an appropriate timeout value.)

3.1   General Principle of the Algorithm

The algorithm uses a single type of message denoted ALIVE. Such a message carries two values: a process identity and an integer $x \in \{2, \ldots, n-1\}$. In the following, "*" stands for any process identity. A message ALIVE($*, n-1$) is called *generating* message, while a message ALIVE($*, n-k$) such that $1 < k < n-1$, is called *forwarding* message. Moreover, the value $n-k$ is called *hopbound value*.

When process p_i starts the algorithm, it proposes itself as candidate to be leader. It sends a generating ALIVE($i, n-1$) message to its neighbors every T time units.

When a process p_i receives an ALIVE($j, n-k$) message such that $1 < k < n-1$, it learns that (a) p_j is candidate to be leader, and (b) there is a path with k hops from p_j to itself. If $j < i$, p_i adopts p_j as current leader, and forwards messages ALIVE($j, n-(k+1)$) to its neighbors. It follows that a generating message ALIVE($j, n-1$) (which can be issued only by p_j) can give rise to a finite number of forwarding messages ALIVE($j, n-2$), ALIVE($j, n-3$),..., possibly up to ALIVE($j, 2$).

As it is possible that there are several paths from p_j to p_i of different lengths, p_i can receive different hopbound values of forwarding messages ALIVE($j, n-k$) with leader j. As it does not know which of those paths will satisfy the $\Diamond$ADD property, p_i manages a timer for each value of $n-k$ for each potential leader. In this way, p_i can associate increasing penalties with each hopbound value, namely, every time a hopbound value does not arrive on time, its penalty is increased. Assuming p_j will be the elected leader, p_i selects a hopbound value associated with p_j with the smallest penalty, which allows p_i to identify a path satisfying the $\Diamond$ADD property from an ALIVE($j, n-k$) message.

Let us observe that, whatever the delays of the messages is, the number of messages sent in the whole execution by faulty processes is finite. It follows that whatever their transfer delays are, there is time after which eventually there are no messages in transit sent by faulty processes

3.2 Local Variables at Each Process p_i

Each process p_i manages the following local variables.

- $in\_neighbors_i$ (resp., $out\_neighbors_i$) is a (constant) set containing the identities of the processes p_j such that there is channel from p_j to p_i (resp., there is channel from p_i to p_j).
- $leader_i$: when stable, will contain the identity of the elected leader.
- $timeout_i[1..n, 1..n]$ is a matrix of timeout values and $timer_i[1..n, 1..n]$ is a matrix of timers, such that the pair $\langle timer_i[j, n - k], timeout_i[j, n - k]\rangle$ is used by p_i to monitor the cycle-free paths from p_j to p_i whose length is k.
- $hopbound_i[1..n]$ is an array of non-negative integers; $hopbound_i[i]$ is initialized to n, while each other entry $hopbound_i[j]$ is initialized to 0. Then, when $j \neq i$, $hopbound_i[j] = n - k \neq 0$ means that, if p_j is currently considered as leader by p_i, the information carried by the last message ALIVE$(j, n - 1)$ sent by p_j to its out-neighbors (which forwarded ALIVE$(j, n - 2)$ to their out-neighbors, etc.) went through a path of k different processes before being received by p_i. The identifier $hopbound$ stands for "upper bound on the number of forwarding" that – due to the last message ALIVE$(j, -)$ received by p_i– the message ALIVE$(j, -)$ sent by p_i has to undergo to be received by all processes. It is similar to a $time$-to-$live$ value.
- $penalty_i[1..n, 1..n]$ is a matrix of integers such that p_i increases $penalty_i[j, n - k]$ each time the $timer_i[j, n-k]$ expires. It is a penalization counter monitored by p_i with respect to the elementary paths of length k starting at p_j and ending at p_i.
- $not\_expired_i$ is an auxiliary local variable.

3.3 Underlying Behavioral Assumption and Proof Sketc.h

The proof assumes there is a time τ after which there is a directed spanning tree (i) that includes all the correct processes and only them, (ii) its root is the correct process with the smallest identity, and (iii) its channels satisfy the $\Diamond$ ADD property.

The structure of the correctness proof is the following. First, let us observe that the local variables $leader_i$ of all the processes always contain a process identity.

Then, the proof shows that, for the *Diamond* ADD channels, there is a constant Δ indicating what is the maximum delay between two consecutive correct delivery of messages on a channel. Then, the proof shows that eventually there are no *alive* messages in the network from a crashed process. And finally, it shows that there is a finite time after which the variables $leader_i$ of all the correct processes contain the smallest identity from the set of correct processes.

Page limitation prevents us from giving a detailed description of the algorithm and its correctness proof. The interested reader will find them in [7].

```
initialization                                                    —-Code for p_i—-
(1)    leader_i ← i; hopbound_i[i] ← n; set timer_i[i,n] to +∞;
(2)    for each j ∈ {1,···,n} \ {i} and each x ∈ {1,···,n} do
(3)        timeout_i[j,x] ← 1; set timer_i[j,x] to timeout_i[j,x];
(4)        set penalty_i[j,x] to −1; hopbound_i[j] ← 0

(5)    every T time units of clock_i() do
(6)        if (hopbound_i[leader_i] > 1) then
(7)            for each j ∈ out_neighbors_i do
                   send ALIVE(leader_i, hopbound_i[leader_i] − 1) to p_j

(8)    when ALIVE(ℓ, hb) such that ℓ ≠ i is received   % from a process in in_neighbors_i
(9)        if (ℓ ≤ leader_i) then
(10)           leader_i ← ℓ;
(11)           if ([timer_i[leader_i, hb] expired) then
                   timeout_i[leader_i, hb] ← timeout_i[leader_i, hb] × 2
(12)           set timer_i[leader_i, hb] to timeout_i[leader_i, hb];
(13)           not_expired_i ← {x | timer_i[leader_i, x] not expired };
(14)           hopbound_i[leader_i] ←
                   max{x ∈ not_expired with smallest non-negative penalty_i[leader_i, x]}

(15)   when timer_i[leader_i, hb] expires and (leader_i ≠ i) do
(16)       penalty_i[leader_i, hb] ← penalty_i[leader_i, hb] + 1;
(17)       if ( ∧_{1≤x≤n} ([timer_i[leader_i, x] expired)) then
(18)           leader_i ← i
(19)       else   % same as lines 13-14
(20)               not_expired_i ← {x | timer_i[leader_i, x] not expired };
(21)               hopbound_i[leader_i] ←
                       max{x ∈ not_expired with smallest non-negative penalty_i[leader_i, x]}
```

Algorithm 1: Eventual leader election in the $\Diamond$ADD model

Acknowledgments. We thank the reviewers for the constructive comments. This work was partially supported by UNAM-PAPIIT grant IN106520.

References

1. Aguilera, M., Delporte-Gallet, C., Fauconnier, H., Toueg, S.: Communication-efficient leader election and consensus with limited link synchrony. In: 23th ACM Symposium on Principles of Distributed Computing (PODC 2004), pp. 328–337. ACM press (1996)
2. Altisen, K., Devismes, S., Dubois, S., Petit, F.: Introduction to Distributed Self-Stabilizing Algorithms. Morgan and Claypool series on Distributed Computing Theory, p. 148 (2019)
3. Chandra, T.D., Hadzilacos, V., Toueg, S.: The weakest failure detector for solving consensus. J. ACM **43**(4), 685–722 (1996)
4. Chandra, T.D., Toueg, S.: Unreliable failure detectors for reliable distributed systems. J. ACM **43**(2), 225–267 (1996)
5. Fernández, A., Jimenez, E., Raynal, M., Trédan, G.: A timing assumption and two t-resilient protocols for implementing an eventual leader service in asynchronous shared-memory systems. Algorithmica **56**(4), 550–576 (2010)
6. Kumar, S., Welch, J.L.: Implementing $\Diamond P$ with bounded messages on a network of ADD channels. Parallel Process. Lett. **29**(1), 1950002 (2019)

7. Rajsbaum, S., Raynal, M., and Vargas K., Leader election in arbitrarily connected networks with process crashes and weak channel reliability. Technical report, p. 21 (2020)
8. Raynal, M.: A short introduction to failure detectors for asynchronous distributed systems. ACM SIGACT News **36**(1), 53–70 (2005)
9. Raynal M.: Fault-tolerant message-passing distributed systems: an algorithmic approach, p. 492. Springer (2018). https://doi.org/10.1007/978-3-319-94141-7, ISBN 978-3-319-94140-0
10. Sastry, S., Pike, S.M.: Eventually perfect failure detectors using ADD channels. In: Stojmenovic, I., Thulasiram, R.K., Yang, L.T., Jia, W., Guo, M., de Mello, R.F. (eds.) ISPA 2007. LNCS, vol. 4742, pp. 483–496. Springer, Heidelberg (2007). https://doi.org/10.1007/978-3-540-74742-0_44
11. Vargas, K., Rajsbaum, S., Raynal, M.: An eventually perfect failure detector for networks of arbitrary topology connected with ADD channels using time-to-live values. Parallel Process. Lett. **30**(2), 2050006 (2020). A preliminary version appeared in the 49th IEEE/IFIP International Conference on Dependable Systems and Networks (DSN 2019), pp. 264–275 (2019)

Physical Zero-Knowledge Proof
for Suguru Puzzle

Léo Robert[1], Daiki Miyahara[2,3(✉)], Pascal Lafourcade[1],
and Takaaki Mizuki[4]

[1] University Clermont Auvergne, LIMOS, CNRS UMR 6158, Aubière, France
{leo.robert,pascal.lafourcade}@uca.fr
[2] Graduate School of Information Sciences, Tohoku University, Sendai, Japan
daiki.miyahara.q4@dc.tohoku.ac.jp
[3] National Institute of Advanced Industrial Science and Technology, Tokyo, Japan
[4] Cyberscience Center, Tohoku University, Sendai, Japan
mizuki+lncs@tohoku.ac.jp

Abstract. Suguru is a paper and pencil puzzle invented by Naoki Inaba. The goal of the game is to fulfil a grid with numbers between 1 and 5 and to respect three simple constraints. In this paper we design a physical Zero-Knowledge Proof (ZKP) protocol for Suguru. A ZKP protocol allows a prover (P) to prove that he knows a solution of a Suguru grid to a verifier (V) without leaking any information on the solution. For constructing such a physical ZKP protocol, we only rely on a small number of physical cards and an adapted encoding. For a grid of Suguru with n cells, we only use $5n + 5$ cards. Moreover, we prove the three classical security properties of a ZKP: completeness, extractability, and zero-knowledge.

Keywords: Physical zero-knowledge proof · Suguru · Security · Completeness · Extractability · Zero-knowledge

1 Introduction

Zero-Knowledge Proofs (ZKP) were introduced in 1985 by Goldwasser et al. [8]. Two parties are involved in such a ZKP protocol: a prover P and a verifier V. At the end of the protocol, the verifier V is convinced that P knows the solution s to the instance $\mathcal{I}$ of a problem $\mathcal{P}$, without revealing any information about s. A zero-knowledge proof prevents the verifier from gaining any knowledge about the solution other than its correctness. In fact, when both randomization and interaction are allowed, the proofs that can be verified in polynomial time are exactly those proofs that can be generated within polynomial space [19].

© Springer Nature Switzerland AG 2020
S. Devismes and N. Mittal (Eds.): SSS 2020, LNCS 12514, pp. 235–247, 2020.
https://doi.org/10.1007/978-3-030-64348-5_19

Formally, for a solution s to any instance $\mathcal{I}$ of a problem P, a convincing interactive zero-knowledge protocol between P and V must then satisfy the three following properties[1]:

Completeness: If P knows s, then he is able to convince V.

Extractability[2]: If P does not know s, then he is not able to convince V except with some *small* probability. More precisely, we want a negligible probability, *i.e.*, the probability should be a function f of a security parameter λ (for example the number of repetitions of the protocol) such that f is negligible, that is for every polynomial Q, there exists $n_0 > 0$ such that:

$$\forall\, x > n_0, f(x) < \frac{1}{Q(x)}.$$

Zero-Knowledge: V learns *nothing* about s except $\mathcal{I}$, *i.e.* there exists a probabilistic polynomial time algorithm $\texttt{Sim}(\mathcal{I})$ (called the simulator) such that outputs of the real protocol and outputs of $\texttt{Sim}(\mathcal{I})$ follow the same probability distribution.

There exist two kinds of ZKP: *interactive* and *non-interactive*. In an interactive ZKP the prover can exchange messages with the verifier in order to convince him, while in the non-interactive case the prover can just create the proof in order to convince the verifier.

ZKPs are usually executed by computers. They are often used in electronic voting to prove that some parties correctly mix some ballots without cheating, or in multi-party computation [3,4,16]. Moreover, there exist generic cryptographic zero-knowledge proofs for all problems in NP [6], via a reduction to an NP-complete problem with a known zero-knowledge proof.

In [15], the authors explained simply this concept to some children using a circular cave. This was the first proposition of a physical ZKP. Later, Gradwohl et al. [9] proposed a ZKP for the famous Nikoli's puzzle called Sudoku[3]. They just used some physical cards to construct a ZKP protocol. It was one of the first interactive physical ZKP protocols for such puzzles. Our aim is to design a ZKP protocol for Suguru puzzles in the same spirit as the one done for Sudoku.

Suguru: It was designed by Naoki Inaba, the original name of the game was *"Nanba Burokku'* but it is also known as *Tectonics* or *Number Blocks*. Suguru is a paper and pencil puzzle in which a grid is divided into outlined blocks called *region*. Each region containing up to five cells. Every cell of the grid must contain

---

[1] Moreover, if $\mathcal{P}$ is NP-complete, then the ZKP should be run in a polynomial time [7]. Otherwise it might be easier to find a solution than proving that a solution is a correct solution, making the proof pointless.

[2] This implies the standard soundness property, which ensures that if there exists no solution of the puzzle, then the prover is not able to convince the verifier regardless of the prover's behavior.

[3] https://www.nikoli.co.jp/en/puzzles/sudoku.html.

a number from 1 to 5 (according to the number of cells in the region). Each cell should be filled such that no two identical numbers touch—not even diagonally.

Suguru's Rule: This puzzle is formed by a rectangular grid where blocks divide the overall area. Those blocks called *region* contain up to five cells. The goal is to fill all the cells with integers under the following constraints:

- **Number region rule:** A region composed of k cells must be filled with integers $1, \ldots, k$.
- **Neighbour rule:** For every cell, all of its eight neighbours must have different values from the cell's value.

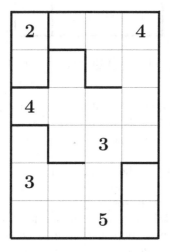

Fig. 1. Initial Suguru grid

In Fig. 1, we give an example of an initial Suguru grid and in Fig. 2 we give its unique solution.

Contributions: We propose a simple ZKP protocol for Suguru using a small number of cards. Our construction is simple and can be used as a pedagogical example to explain the role of ZKP protocols. We propose an encoding of the number using simple cards. Using this encoding, the prover places some cards on the grid according to its solution. We use these cards to prove that the two rules of Suguru are satisfied. We start with the first number of cell in a region, after having verified the validity of all regions and replacing the cards placed by the prover, we reuse them to prove the second rule of Suguru about the eight neighbours of each cell of the grid. Here is the difficulty of Suguru, since we need to prove that all values of the eight neighbours of each cell are different without revealing any information to the verifier. Here, we use a trick of our encoding of the values of the cards in order not to leak any information. Our encoding

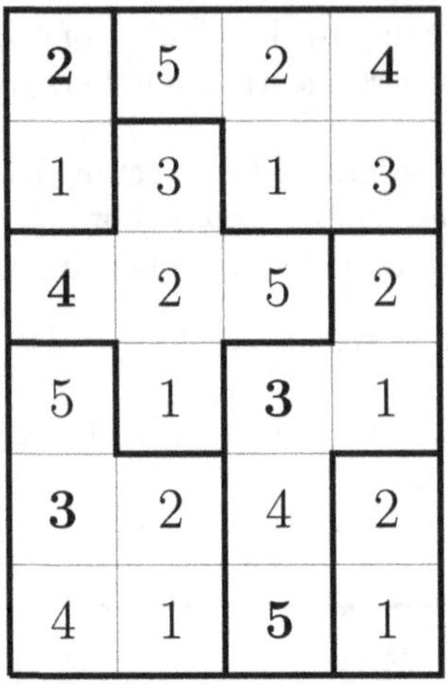

2	5	2	4
1	3	1	3
4	2	5	2
5	1	3	1
3	2	4	2
4	1	5	1

Fig. 2. Solution of the Suguru grid of Fig. 1

requires five cards per cell; therefore if a Suguru grid has n cells to guess, our protocol only requires $5n+5$ cards. Finally, we prove the three security properties of our construction *i.e.*, completeness, extractability, and zero-knowledge.

Related Work: In [18] the authors proposed an improved ZKP protocol for Sudoku that follows the pioneer work of [9]. In [5], the authors proposed a ZKP protocol for the Nikoli's puzzle Norinori.

In [12], a method to take into account one feature of several puzzles that consists to construct a single loop, has been invented. This technique used a topological approach with successive interactive transformations.

Recently several ZKP proofs have been proposed for different Nikoli's puzzles. In [13], card-based ZKP Protocols for Takuzu and Juosan have been proposed. In [17] a physical ZKP proof for Numberlink has been designed. In [14], a card-based physical ZKP for Kakuro have been given that improves the first version proposed by Bultel et al. in [1] with the ZKP protocols for three other Nikoli's games: Akari, Takuzu, and Kenken.

All these works clearly demonstrate that designing physical ZKP is clearly an interesting topic of research. Each game has its own particular rules and requires an adapted construction.

Although the existence of all those previous works, one cannot reuse or adapt directly them for the Suguru game. Indeed, the main reason is the "strong" neighbour rule where no cell can have its eight neighbours with the same value. Other puzzles have a similar rule but with relaxed restrictions. For instance, Makaro has a neighbour rule but only for adjacent cells (and not in diagonal). Thus a naive adaptation would imply a loss in terms of efficiency and of zero-knowledge (no information about the solution can be leaked). Furthermore, it is worth noting that the encoding for the proof of NP-completeness of Makaro cannot be applied for Suguru. Thus ZKP for Suguru cannot be directly adapted from the ZKP of Makaro. Further discussion is given in Sect. 5.

Outline: In Sect. 2, we present our notations, and all subprotocols needed to construct our ZKP. In Sect. 3, we design our ZKP protocol for Suguru. In Sect. 4, we prove the security of our protocol. In Sect. 5, we discuss a complexity of Suguru and of Makaro. In the last section, we conclude the paper.

2   Preliminaries

We introduce some notations of cards and shuffles used in our construction.

2.1   Notations

Card: A deck of cards used in our protocol consists of blacks ♣ and reds ♡ whose back sides are identical ?. Each integer $i \in \{1, \ldots, 5\}$ is encoded as:

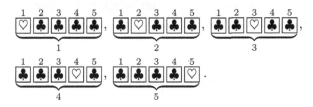

We call such face-down five cards ? ? ? ? ? corresponding to an integer according to the above encoding rule a *commitment* to the respective integer.

We also use *numbered* cards such as 1 2 3 4 5 whose backs are identical ?.

Neighbour Cell: Consider a target cell denoted c_t on a grid. A cell is a *neighbour* of c_t if it is next to c_t. It can be on the left, the right, the top, or the bottom of c_t, and also on its diagonal:

Thus, a cell can have at most eight neighbours.

Pile-Scramble Shuffle: A shuffle used in our protocol is a *pile-scramble shuffle*, which was first used by Ishikawa et al. [10] and was used in other physical ZKP protocols for puzzles (e.g., Sudoku [18]). Consider that we have a sequence of ℓ piles of cards, each of which consists of the same number of face-down cards, denoted by $(p_1, p_2, \ldots, p_\ell)$ for some positive integer ℓ. Applying a pile-scramble shuffle to the sequence results in $(p_{r^{-1}(1)}, p_{r^{-1}(2)}, \ldots, p_{r^{-1}(\ell)})$ where permutation r is uniformly and randomly chosen from the symmetric group of degree ℓ. That is, it randomly permutes a sequence of piles and nobody knows the order of the resulting sequence.

One can easily implement a pile-scramble shuffle by using physical tools that can fix each pile of cards such as rubber bands and envelopes; a player (or players) randomly shuffle them until nobody traces the order of the piles.

3   ZKP Protocol for Suguru

We propose a ZKP protocol for Suguru composed of two phases, the setup phase and the verification phase.

3.1   Setup Phase

The verifier V and the prover P place commitments corresponding to the integers on the initial grid of a Suguru puzzle. In addition, when a region of k cells is already filled with $k - 1$ cells, then P and V agreed on the last cell to complete and place the commitment accordingly[4].

Then, P continues to place commitments on all the remaining cells by himself according to the solution of the puzzle.

3.2   Verification Phase

There are two verifications to ensure the number region rule and the neighbour rule.

Number Region Rule: V wants to check that a region of k cells contains all the consecutive integers from 1 to k.

1. For every i, $1 \leq i \leq k$, V picks all cards of the i-th cell (in any ordering) to form a pile p_i. Then, V attaches a numbered card $\boxed{i}$ to p_i. Thus, there are $p_1, \ldots, p_k$ piles, each of which consists of six cards.
2. Apply the pile-scramble shuffle [10].
3. V reveals the cards of each pile except for the numbered card. The revealed output is of the form (up to a permutation in the rows), i.e., all the k (opened) commitments corresponding to 1 through k should appear. For example, if $k = 4$, the revealed output should be of the following form (up to a permutation in the rows):

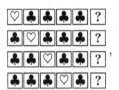

where the face-down cards on the right side are the numbered cards. If the revealed output is not of this form, V aborts.

---

[4] For example, in Fig. 1 the upper left region can be directly completed with a 1.

4. Turn over the face-up cards and apply the pile-scramble shuffle again to the piles.
5. Reveal only the numbered cards of all piles. Because these revealed cards indicate the initial positions for each pile, V rearranges each pile back to their initial place. The revealed numbered cards can be reused for the remaining verifications.

Neighbour Rule: V wants to check if a given cell has no neighbour having the same integer as the cell.

1. V picks the first card of the target commitment and then picks each first card of the commitments on its neighbour cells (in any ordering) to form the pile p_1. The following is an example when there are eight neighbours:

2. V repeats the same operation until the pile p_5 is formed.
3. V attaches a numbered card $\boxed{i}$ to p_i. (If the target cell is the last one, V does not perform this step.)
4. If an integer is written on the target cell, then go to the next step. Otherwise, apply the pile-scramble shuffle to the piles.
5. V reveals the first card of each pile, which corresponds to the target commitment. Let p_t denotes the pile where a red card appears.
6. V reveals all the cards in the pile p_t except for the numbered cards. If there are two red cards in the pile then V aborts; otherwise, V goes to the next step.
7. As Steps 4. and 5. in the previous verification, V rearranges all the cards in the piles back to their initial places. (If the target cell is the last one, V does not perform this step.)

3.3 Evaluation

If the size of the grid is represented by n then the number of cards used in this protocol is equal to $5n + 5$. Indeed, each cell must be encoded with 5 cards (4 blacks and 1 red)[5] and 5 numbered cards are used for all target cell.

---

[5] We could have encoded each cell with a total of ℓ cards where ℓ is the number of cells in the region (thus, a region with two cells has its cell encoded with only two cards, a red and a black). Yet, this would lead to inconstancy in the encoding rule which is required in the neighbour verification.

4 Security Proofs

We give the theorems along with their proofs to provide that our protocol respects the security properties. A ZKP protocol is secure if the following three properties are satisfied:

Correctness: If the prover P commits its cards according to the actual solution, then all verifications will not abort. Hence, if P knows a solution, then it can always convince the verifier V. Correctness is proven in Theorem 1.

Extractability: If P's input is invalid, the protocol will point errors out to V. Therefore, if P does not know the solution, it cannot convince V. Extractability is proven in Theorem 2.

Zero-knowledge: V learns nothing about P's solution. Zero-knowledge is proven in Theorem 3.

Theorem 1 (Completeness). *If P knows a solution of a Suguru grid, then it can convince V.*

Proof. Suppose that the prover P knows the solution of the Suguru grid. It runs with the verifier V the Setup phase (Sect. 3.1). We show that P can perform both verification phases without aborting.

Number Region Verification: In this phase, the goal of P is to show that each region of size k contains consecutive integers from 1 to k (note that the lower bound of k is 1 and its upper bound 5). Since P places the cards accordingly with the solution, each region of size k contains the numbers 1 to k. Without loss of generality, suppose that the pile p_i corresponds to the number i with $i = 1 \ldots k$. The pile p_1 is composed of the sequence (in this order):

The pile p_2 is composed of the sequence (in this order):

More generally, the pile p_i is a sequence of black cards where the red card is placed on position i.

Since the pile-scramble shuffle applied on Step 2. does not modify the order of the sequence, the red card of pile p_i is on position i. As $i = 1 \ldots k$, all numbers from 1 to k are represented. Thus, V is convinced that the number region rule is verified by revealing the piles in Step 3..

Neighbour Verification: The goal of P is to convince V that no cell has the same number of its neighbours (there are eight neighbours as defined in Sect. 2). Let c_t be the target cell placed on the center of the 3×3 square. Since P placed the commitments according to the solution, there is no cell with the same value of c_t in this square. Let i be the position of the red card of c_t (with $i = 1 \ldots 5$). Since no neighbour cell has the same value of c_t, there is no other red card with position i. Since each pile is composed of card with the same indices, the pile p_i (before the shuffle) contains exactly one red card. Hence, V is convinced that the neighbour rule is verified.

Finally, since all verifications are checked, we proved that if P has the solution then the verifications will always succeed.   □

Theorem 2 (Extractability). *If P does not provide a solution of the Suguru puzzle, it is not able to convince V.*

Proof. Suppose that P does not know a solution for the puzzle. We want to show that V will always detect it.

Since P cannot provide the solution, at least one of the two rules is not verified (if both can be verified, this is the solution). We can distinguish two cases corresponding to each verification:

- The number region rule is not respected. That is, suppose w.l.o.g. that a region of size k with $k > 1$ does not contain the number 1. Hence, the sequence corresponding to this number is missing, meaning that V cannot reveal at Step 3. the sequence:

$$\boxed{\heartsuit}\,\boxed{\clubsuit}\,\boxed{\clubsuit}\,\boxed{\clubsuit}\,\boxed{\clubsuit}\,.$$

Thus, V will abort the protocol and detect that P cannot provide the solution.
- The neighbour rule is not respected. Suppose that we have the following configuration:

with blank cards of unimportant value but different from 2.
We encode number 2 as:

$$\boxed{\clubsuit}\,\boxed{\heartsuit}\,\boxed{\clubsuit}\,\boxed{\clubsuit}\,\boxed{\clubsuit}\,.$$

Thus, the pile p_2 corresponding to all the cards with indices 2 (before the shuffle) will contain exactly two red cards. Thus, V will abort the protocol.

We proved that if P does not have the solution then the verifications will abort in both cases meaning that P cannot convince V.   □

Theorem 3 (Zero-knowledge). *V learns nothing about P's solution of the given grid G.*

Proof. We use the same proof technique as in [9]: zero-knowledge is caused by a description of an efficient *simulator* which simulates interaction between a cheating verifier and a real prover. However, the simulator does not have a solution but it can swap cards for different ones during shuffles. The simulator acts as follows:

– During the number region verification at Step 2., the simulator swaps the piles to replace them by the sequences (up to a permutation in the rows):

– During the neighbour verification, when revealing the cards at Step 6., the simulator swaps the pile with a pile containing $k - 1$ black cards and 1 red card.

The simulated proofs and the real proofs are indistinguishable; thus, V learns nothing about P's solution.   □

5   Discussion

Among related work introduced in Sect. 1, the closest work is [2], where a ZKP for Makaro has been proposed. This game is close to the Nikoli's game called Makaro[6] where regions are called rooms and some extra black cells indicate thanks to an arrow the position of the biggest number among the (up to) four cells around (over, under, left, right) the cell with the arrow (black) is in the cell the arrow points at.

In [11], the authors proved that the two games Herugolf and Makaro are NP-complete. Solving Makaro was shown to be NP-complete via a reduction from 3-SAT.

Makaro is a game close to Suguru, with different constraints as follows:

1. *Room condition*: Each room contains all the numbers from 1 up to the number of cells in the room.
2. *Neighbour condition*: A number cannot be next (adjacent) to the same number in another room.
3. *Arrow condition*: Every black arrow cell must point at the largest number among the numbers in the adjacent cells of the black cell (possibly the fours cells: right, left, above, and bottom).

In Fig. 3, we give a simple example of a Makaro game, where all black cells are arrow cells and all white cells are empty cells except for one filled cell with three. It is easy to verify that the three constraints are satisfied in the solution given in Fig. 4. We remark that in a solution all white cells are filled with numbers between 1 and k, where k is the maximum size of all the rooms of the grid.

The room condition is the same, however the neighbour condition is different in Suguru. In Suguru for each cell, we consider the eight neighbours while in Makaro we consider only the four neighbours. Moreover in Suguru the size of

---

[6] http://nikoli.co.jp/en/puzzles/makaro.html.

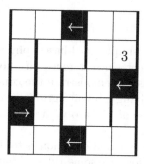

Fig. 3. Example of a Makaro grid

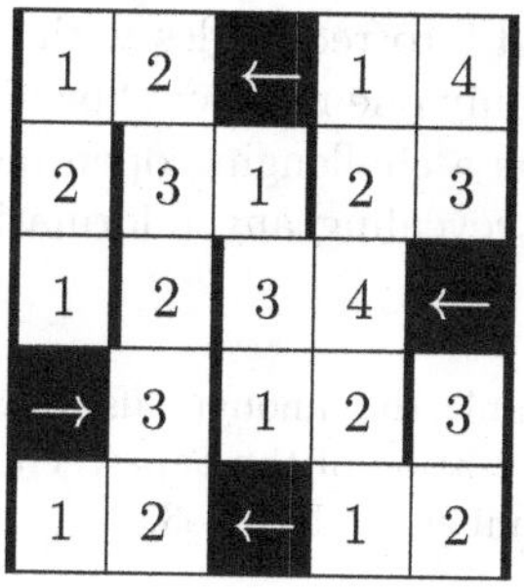

1	2	←	1	4
2	3	1	2	3
1	2	3	4	←
→	3	1	2	3
1	2	←	1	2

Fig. 4. Solution of a Makaro grid of Fig. 3

the regions is limited to five while in Makaro there is no limit on the size of the rooms.

These differences avoid us to reuse the 3-SAT encoding used in the proof of NP-completeness of Makaro [11]. If we remove the limit of the maximum of five cells in each region, we can use the same gadgets as the ones used for Makaro to prove in a similar way that Suguru is also NP-complete. But with the limit of five cells per regions and the change of four neighbours into eight neighbours, it is not clear if we can prove the NP-completeness of Suguru. These extra constraints seem to remove some difficulty in finding some solutions for Suguru. At the moment we are not able to prove NP-completeness and we conjecture that solving a Suguru grid should be in P.

Even if solving Suguru is in P, our physical ZKP is an interesting approach since it requires only $5n+5$ cards as we show when we present our ZKP protocol in Sect. 3. (In addition, it could always happen that one cannot solve a Suguru puzzle.) This is clearly a real ZKP protocol that can be used by Suguru players in practice.

6   Conclusion

In this paper we propose a simple card-based physical ZKP for Suguru. Our solution is simple and efficient since it relies on only $5n + 5$ cards. One open question left for the future is to demonstrate the conjecture states in the introduction *i.e.*, proving that solving a Suguru grid is in P. We clearly cannot adapt the proof of NP-completeness of Makaro with the rules of Suguru, it is why we conjecture that Suguru is in P.

Moreover, our long term research direction is to design physical ZKP protocols for all Nikoli's games. However some rules of some games like Shakashaka[7] that requires to draw rectangles which is not easy to model without leaking any information. Another example of challenging game is Shikaku[8], where the rules are simple: 1) Divide the grid into rectangles with the numbers in the cells. 2) Each rectangle is to contain only one number showing the number of cells in the rectangle. However it remains a challenging open question to design a physical ZKP for this game without revealing any information on the positions of the rectangles.

Acknowledgements. We thank the anonymous referees, whose comments have helped us to improve the presentation of the paper. This work was supported in part by JSPS KAKENHI Grant Number JP19J21153.

References

1. Bultel, X., Dreier, J., Dumas, J., Lafourcade, P.: Physical zero-knowledge proofs for Akari, Takuzu, Kakuro and KenKen. In: Demaine, E.D., Grandoni, F. (eds.) 8th International Conference on Fun with Algorithms, FUN 2016, June 8–10, 2016, La Maddalena, Italy. LIPIcs, vol. 49, pp. 8:1–8:20. Schloss Dagstuhl - Leibniz-Zentrum fuer Informatik (2016). https://doi.org/10.4230/LIPIcs.FUN.2016.8
2. Bultel, X., et al.: Physical zero-knowledge proof for Makaro. In: Izumi, T., Kuznetsov, P. (eds.) SSS 2018. LNCS, vol. 11201, pp. 111–125. Springer, Cham (2018). https://doi.org/10.1007/978-3-030-03232-6_8
3. Cramer, R., Damgård, I., Nielsen, J.B.: Multiparty computation from threshold homomorphic encryption. In: Pfitzmann, B. (ed.) EUROCRYPT 2001. LNCS, vol. 2045, pp. 280–300. Springer, Heidelberg (2001). https://doi.org/10.1007/3-540-44987-6_18
4. Damgård, I., Faust, S., Hazay, C.: Secure two-party computation with low communication. In: Cramer, R. (ed.) TCC 2012. LNCS, vol. 7194, pp. 54–74. Springer, Heidelberg (2012). https://doi.org/10.1007/978-3-642-28914-9_4
5. Dumas, J.-G., Lafourcade, P., Miyahara, D., Mizuki, T., Sasaki, T., Sone, H.: Interactive physical zero-knowledge proof for Norinori. In: Du, D.-Z., Duan, Z., Tian, C. (eds.) COCOON 2019. LNCS, vol. 11653, pp. 166–177. Springer, Cham (2019). https://doi.org/10.1007/978-3-030-26176-4_14

---

[7] http://www.nikoli.co.jp/en/puzzles/shakashaka.html.
[8] https://www.nikoli.co.jp/en/puzzles/shikaku.html.

6. Goldreich, O., Micali, S., Wigderson, A.: Proofs that yield nothing but their validity and a methodology of cryptographic protocol design. In: 27th Annual Symposium on Foundations of Computer Science (SFCS 1986), pp. 174–187, October 1986. https://doi.org/10.1109/SFCS.1986.47

7. Goldreich, O., Micali, S., Wigderson, A.: How to prove all NP statements in zero-knowledge and a methodology of cryptographic protocol design (extended abstract). In: Odlyzko, A.M. (ed.) CRYPTO 1986. LNCS, vol. 263, pp. 171–185. Springer, Heidelberg (1987). https://doi.org/10.1007/3-540-47721-7_11

8. Goldwasser, S., Micali, S., Rackoff, C.: Knowledge complexity of interactive proof-systems. In: Conference Proceedings of the Annual ACM Symposium on Theory of Computing, pp. 291–304 (1985). https://doi.org/10.1145/3335741.3335750

9. Gradwohl, R., Naor, M., Pinkas, B., Rothblum, G.N.: Cryptographic and physical zero-knowledge proof systems for solutions of sudoku puzzles. In: Crescenzi, P., Prencipe, G., Pucci, G. (eds.) FUN 2007. LNCS, vol. 4475, pp. 166–182. Springer, Heidelberg (2007). https://doi.org/10.1007/978-3-540-72914-3_16

10. Ishikawa, R., Chida, E., Mizuki, T.: Efficient card-based protocols for generating a hidden random permutation without fixed points. In: Calude, C.S., Dinneen, M.J. (eds.) UCNC 2015. LNCS, vol. 9252, pp. 215–226. Springer, Cham (2015). https://doi.org/10.1007/978-3-319-21819-9_16

11. Iwamoto, C., Haruishi, M., Ibusuki, T.: Herugolf and Makaro are NP-complete. In: Ito, H., Leonardi, S., Pagli, L., Prencipe, G. (eds.) Fun with Algorithms 2018. LIPIcs, vol. 100, pp. 24:1–24:11. Schloss Dagstuhl - Leibniz-Zentrum fuer Informatik (2018)

12. Lafourcade, P., Miyahara, D., Mizuki, T., Sasaki, T., Sone, H.: A physical ZKP for Slitherlink: how to perform physical topology-preserving computation. In: Heng, S.-H., Lopez, J. (eds.) ISPEC 2019. LNCS, vol. 11879, pp. 135–151. Springer, Cham (2019). https://doi.org/10.1007/978-3-030-34339-2_8

13. Miyahara, D., et al.: Card-based ZKP protocols for Takuzu and Juosan. In: Farach-Colton, M., Prencipe, G., Uehara, R. (eds.) Fun with Algorithms 2020. LIPIcs (2020)

14. Miyahara, D., Sasaki, T., Mizuki, T., Sone, H.: Card-based physical zero-knowledge proof for Kakuro. IEICE Trans. Fund. Electron. Commun. Comput. Sci. **E102.A**(9), 1072–1078 (2019). https://doi.org/10.1587/transfun.E102.A.1072

15. Quisquater, J.-J., et al.: How to explain zero-knowledge protocols to your children. In: Brassard, G. (ed.) CRYPTO 1989. LNCS, vol. 435, pp. 628–631. Springer, New York (1990). https://doi.org/10.1007/0-387-34805-0_60

16. Romero-Tris, C., Castellà-Roca, J., Viejo, A.: Multi-party private web search with untrusted partners. In: Rajarajan, M., Piper, F., Wang, H., Kesidis, G. (eds.) SecureComm 2011. LNICST, vol. 96, pp. 261–280. Springer, Heidelberg (2012). https://doi.org/10.1007/978-3-642-31909-9_15

17. Ruangwises, S., Itoh, T.: Physical zero-knowledge proof for Numberlink. In: Farach-Colton, M., Prencipe, G., Uehara, R. (eds.) Fun with Algorithms 2020. LIPIcs (2020)

18. Sasaki, T., Miyahara, D., Mizuki, T., Sone, H.: Efficient card-based zero-knowledge proof for Sudoku. Theor. Comput. Sci. (2020). https://doi.org/10.1016/j.tcs.2020.05.036, http://www.sciencedirect.com/science/article/pii/S0304397520303200

19. Shamir, A.: IP = PSPACE. J. ACM **39**(4), 869–877 (1992). https://doi.org/10.1145/146585.146609

Uniform Deployment of Mobile Agents in Dynamic Rings

Masahiro Shibata[1]([⊠]), Yuichi Sudo[2], Junya Nakamura[3], and Yonghwan Kim[4]

[1] Kyushu Institute of Technology, Fukuoka, Japan
shibata@cse.kyutech.ac.jp
[2] Osaka University, Osaka, Japan
y-sudou@ist.osaka-u.ac.jp
[3] Toyohashi University of Technology, Aichi, Japan
junya@imc.tut.ac.jp
[4] Nagoya Institute of Technology, Aichi, Japan
kim@nitech.ac.jp

Abstract. In this paper, we consider the uniform deployment problem of mobile agents in synchronous dynamic bidirectional rings, which requires agents to spread uniformly in the ring. So far, uniform deployment has been considered in static graphs. In this paper, we consider this problem in 1-interval connected rings, that is, one of the links may be missing at each time step. In such networks, we aim to clarify the solvability of the uniform deployment problem, focusing on global knowledge given to the agents. To the best of our knowledge, this is the first research considering uniform deployment in dynamic networks. First, we consider agents with knowledge of the number n of nodes. In this case, we show that our algorithm can solve the problem with $O(k \log n)$ memory space per agent, $O(n \log k)$ rounds, and a total number of $O(kn)$ moves, where k is the number of agents. Next, we consider agents without knowledge of n but with knowledge of k. In this case, when $k \geq 4$, we show that our algorithm can also solve the problem but requires $O(k \log n)$ memory space per agent, $O(n^2)$ rounds, and a total number of $O(n^2)$ moves. These results mean that the uniform deployment problem can be solved also in dynamic rings.

1 Introduction

1.1 Background and Related Work

A *distributed system* comprises a set of computing entities (*nodes*) connected by communication links. As a promising design paradigm of distributed systems, (mobile) agents have attracted much attention [1]. The agents can traverse the system, carrying information collected at visited nodes, and execute an action at each node using the information to achieve a task. In other words, agents can encapsulate the process code and data, which simplifies design of distributed systems [2].

© Springer Nature Switzerland AG 2020
S. Devismes and N. Mittal (Eds.): SSS 2020, LNCS 12514, pp. 248–263, 2020.
https://doi.org/10.1007/978-3-030-64348-5_20

In this paper, we consider the *uniform deployment* (or *uniform scattering*) *problem* as a fundamental problem for agents' coordination. This problem requires agents to spread uniformly in the network. Uniform deployment is useful for network management. In distributed systems, it is necessary that each node is periodically checked whether some application installed on the node works correctly or not [3]. Thus, considering agents with such services, uniformly deployed agents can visit each node at short intervals and check nodes' statuses. Uniform deployment might be useful also for load balancing. That is, when agents deploy uniformly in the network with bringing large-size database replicas, not all nodes need to store the database but each node can quickly access the database [4]. Thus, the uniform deployment problem can be seen as a particular case of the resource allocation problem (e.g., the k-server problem).

As related work, uniform deployment has been considered in rings [5,6] and grids [7,8]. All of them assumed that agents are oblivious (or memoryless) but can observe multiple nodes within its visibility range. On the other hand, Shibata et al. [9,10] considered uniform deployment in asynchronous unidirectional rings for agents that have memory but cannot observe nodes except for their currently visited nodes. They clarified the relationship between the capability of agents and the solvability of the problem [9], and they considered the relationship between the capability of agents and the memory space required per agent [10]. All of the above work on uniform deployment are considered in *static graphs*, where a network topology does not change during an execution. On the other hand, recently many problems involving agents have been studied in *dynamic graphs*, where a network topology changes during an execution. For example, the gathering problem [11], the exploration problem [12,13], the patrolling problem [14], and the dispersion problem [15] are considered in dynamic graphs.

1.2 Our Contribution

In this paper, we consider the uniform deployment problem in synchronous dynamic bidirectional rings. Similarly to [9,10], we consider agents that have memory but cannot observe nodes except for their currently visited nodes. In this paper, we consider *1-interval connected rings* [11,12,14,15], that is, one of the links may be missing at each time step. In such a network, one agent may be blocked from traversing forever during an execution of an algorithm. Hence, we say that agents solve the uniform deployment problem in a dynamic ring when all agents other than one agent spread uniformly in the ring. An example is given in Fig. 1, where n and k are the number of nodes and agents, respectively. Such a deployment can guarantee some fair resource allocation in spite of temporal link-missings. In such networks, we aim to clarify the solvability of the problem, focusing on global knowledge given to the agents. To the best of our knowledge, this is the first research considering uniform deployment in dynamic networks.

Throughout the paper, we consider agents under the following eight assumptions: (1) Agents are *anonymous*, that is, they do not have distinct IDs. (2) Agents have knowledge of k or n. (3) Agents have *chirality*, that is, they agree on the orientation of clockwise and counterclockwise direction in the ring. (4)

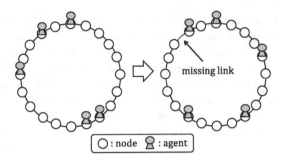

Fig. 1. An example of uniform deployment in a dynamic ring ($n = 20, k = 5$).

Agents have the *strong multiplicity detection capability*, that is, they can count the exact number of agents staying at the same node. (5) Agents cannot perform a direct message communication even when they stay at the same node. (6) Each agent initially has a *token* and can release it on the currently visited node. (7) Agents start executions of the algorithm from mutually distinct nodes. (8) Agents behave in a synchronous manner. In particular, we can simply show in Sect. 2.1 that assumptions (1), (2), (5), (6), and (7) are the weakest assumptions for solving the problem, and at this point it is possible that the other assumptions can be removed or weakened.

In Table 1, we compare our contribution with the results for agents with knowledge of k or n in [9]. We consider two problem settings. First, we consider agents with knowledge of n. In this case, we show that our algorithm can solve the problem with $O(k \log n)$ memory space per agent, $O(n \log k)$ rounds, and a total number of $O(kn)$ moves. Next, we consider agents without knowledge of n but with knowledge of k. In this case, when $k \geq 4$, we show that our algorithm can also solve the problem but requires $O(k \log n)$ memory space per agent, $O(n^2)$ rounds, and a total number of $O(n^2)$ moves. These results mean that the uniform deployment problem can be solved also in dynamic rings. In addition, it is worthwhile to mention that, while knowledge of n and knowledge of k are equivalent in static rings (that is, agents with knowledge of n can easily get to know k and vice versa [9]), this no longer holds in dynamic rings. Interestingly, in our algorithms, it requires longer time to know n using knowledge of k than to know k using knowledge of n. As a result, the algorithm with knowledge of n solves the problem faster and requires a smaller number of agent moves than the algorithm with knowledge of k. Due to page limit, we omit pseudocodes and several proofs of lemmas and theorems.

2   Preliminaries

2.1   System Model

We basically follow the model defined in [11]. A *dynamic bidirectional ring* R is defined as 2-tuple $R = (V, E)$, where $V = \{v_0, v_1, \ldots, v_{n-1}\}$ is a set of n

Table 1. Results in each model (n: #nodes, k: #agents)

	Results in [9]		Results of this paper	
	Result 1	Result 2	Result 1 (Sec. 3)	Result 2 (Sec. 4)
Static/Dynamic ring	Static	Static	Dynamic	Dynamic
Knowledge	k or n	k or n	n	k
Agent memory	$O(k \log n)$	$O(\log n)$	$O(k \log n)$	$O(k \log n)$
Time complexity	$\Theta(n)$	$O(n \log k)$	$O(n \log k)$	$O(n^2)$
Total number of agent moves	$\Theta(kn)$	$\Theta(kn)$	$\Theta(kn)$	$O(n^2)$

anonymous nodes and $E = \{e_0, e_1, \ldots, e_{n-1}\}$ ($e_i = \{v_i, v_{(i+1) \bmod n}\}$) is a set of links. For simplicity, we denote $v_{(i+j) \bmod n}$ (resp., $e_{(i+j) \bmod n}$) by v_{i+j} (resp., $e_{(i+j)}$) for any integers i and j. We define the direction from v_i to v_{i+1} (resp., v_i to v_{i-1}) as the *forward* or *clockwise* (resp., *backward* or *counterclockwise*) direction. In addition, one of the links may be missing at each time step, and which link is missing is controlled by an *adversarial scheduler*. Such a dynamic ring is known as a *1-interval connected ring*. The *distance* from node v_i to v_j is defined to be $(j - i) \bmod n$. Note that this definition of the distance is correct when there is no missing link.

Let $A = \{a_0, a_1, \ldots, a_{k-1}\}$ be a set of k ($\leq n$) agents. Agents are *anonymous*, that is, they do not have distinct IDs (assumption (1) in Sect. 1.2). Obviously, this is the weakest assumption about identification among agents. An agent is a state machine having two special states, *initial state* $s_{initial}$ and *final state* s_{final}. Once an agent changes its state to s_{final}, it never changes its state or leaves the current node thereafter. We say that an agent *terminates* when its state changes to s_{final}. We consider two problem settings: (i) agents know n, and (ii) agents know k. Note that either of knowledge n or k is necessary (assumption (2) in Sect. 1.2). This is because, it is shown in [9] that agents without knowledge of n or k cannot achieve uniform deployment even in static rings if termination detection is required (i.e., they need to change their states to s_{final}). Agents have *chirality*, that is, they agree on the orientation of clockwise and counterclockwise direction in the ring. Agents have the *strong multiplicity detection capability*, that is, they can count the exact number of agents staying at the same node. However, agents cannot perform a direct message communication even when they stay at the same node (assumption (5) in Sect. 1.2). Obviously, this is the weakest assumption about message communication. Each agent initially has a *token* and can release it on the currently visited node (assumption (6) in Sect. 1.2). After the token is released, it cannot be removed. The token on an agent can be realized by only 1-bit memory and cannot carry any additional information. Hence, even when an agent visits a node with tokens, the agent can detect only the existence of the tokens and cannot recognize the owners of the tokens[1]. Note that the assumption on tokens is necessary because, if agents are not allowed to have tokens and they move in a synchronous manner, they cannot mark nodes in any way, they cannot get any other information of other agents, and thus uniform deployment cannot be achieved. Hence, agents communicate only by the tokens and the strong multiplicity detection capability.

During execution of the algorithm, we assume that agents move instantaneously, that is, they always exist at nodes (do not exist on links). Each agent a_i executes the following three operations in an atomic action: 1) Agent a_i counts the number of agents at the same node v and the number of tokens released on v, 2) agent a_i executes some local computation, and 3) agent a_i releases its token (if it decides to do so) and leaves v (if it decides to move). When a_i tries to move to its neighboring node (e.g., from node v_j to v_{j+1}) but the corresponding link (e.g., link e_j) is missing, we say that a_i is *blocked*, and it still exists at v_j at the beginning of the next atomic action.

In an agent system, a (global) *configuration* is defined as the Cartesian product of the states of all agents, the states (the number of tokens) of all nodes, and the locations of all agents. We define C as a set of all configurations. In an initial configuration $c_0 \in C$, all agents are in the same state $s_{initial}$ and every agent has one token. Furthermore, we assume that agents are located at mutually distinct nodes in c_0 (assumption (7) in Sect. 1.2). This assumption is necessary because otherwise two agents initially located at the same node always execute the same action and they never stay at distinct nodes, which implies that the agents cannot achieve uniform deployment deterministically. The node where agent a is located in c_0 is called the *home node* of a and is denoted by $v_{HOME}(a)$.

Moreover, we define *periodic initial configurations*. First, we define the i-th forward (resp., backward) agent a' of agent a as the agent such that $i - 1$ agents exist between a and a' in a's forward (resp., backward) direction in c_0. For convenience, we define the 0-th forward agent of a as a itself. Then, in c_0, we assume that agents $a_0, a_1, \ldots, a_{k-1}$ exist in this order, that is, a_i is the i-th forward agent of a_0 in c_0. We define the *distance sequence* of agent a_i in c_0 as $D_i(c_0) = (d_0^i(c_0), \ldots, d_{k-1}^i(c_0))$, where $d_j^i(c_0)$ is the distance from the j-th forward agent of a_i to the $(j + 1)$-st forward agent of a_i in c_0. We define the distance sequence $D(c_0)$ of c_0 as the lexicographically minimum sequence among $\{D_i(c_0) \mid 0 \leq i \leq k - 1\}$. Let $shift(D, x) = (d_x, d_{x+1}, \ldots, d_{k-1}, d_0, d_1, \ldots, d_{x-1})$ for sequence $D = (d_0, d_1, \ldots, d_{k-1})$. Then, if $D(c_0) = shift(D(c_0), x)$ holds for some x $(0 < x < k)$, we say c_0 is *periodic*. Otherwise, we say c_0 is *aperiodic*.

In this paper, we consider a *synchronous execution*, that is, in each time step called *round*, all agents perform atomic actions. An *execution* from c_0 is defined as $E = c_0, c_1, \ldots$ where each c_i $(i \geq 1)$ is the configuration reached from c_{i-1} by atomic actions of all agents. An execution is infinite, or ends in a configuration where the state of every agent is s_{final}.

---

[1] In practice, obviously each node can store information more than 1-bit token, but it is sufficient to store information about tokens when considering anonymous agents.

2.2   The Uniform Deployment Problem

In [9], the uniform deployment problem in a *static* ring is defined so that all $k \, (\geq 2)$ agents spread uniformly in the ring, that is, all agents are located at distinct nodes and the *adjacent distance* between any two *adjacent agents* should be the same. Here, we say that two agents are adjacent when there exists a path (a set of consecutive links) connecting the two agents such that no other agent exists on the path, and the adjacent distance is the number of links on the path. However, if we follow this definition, agents cannot achieve uniform deployment in a dynamic ring intuitively by the following two reasons: (i) The adversarial scheduler can continue to block one agent a' at some node v_j forever during an execution by deciding that link e_j (resp., e_{j-1}) is missing when a' tries to move from v_j to v_{j+1} (resp., from v_j to v_{j-1}), and (ii) the other agents (and a') cannot detect whether a' is blocked forever and they need to determine their destinations according to the node v_j where a' is blocked, or a' can eventually leave v_j and they must not determine their destinations according to the location of v_j. Hence, the uniform deployment problem in a *dynamic* ring allows at most one agent to stay at an arbitrary node, that is, the problem requires that all agents other than one agent spread uniformly in the ring (Fig. 1). In addition, we should consider the case that n is not divisible by k. In this case, we aim to distribute agents so that the adjacent distance of any two adjacent agents should be $\lfloor n/k \rfloor$ or $\lceil n/k \rceil$. As an exception, let a' be an agent allowed to stay at an arbitrary node due to missing links. Then, when considering the configuration such that a' does not exist in the ring, the adjacent distance of one pair of adjacent agents is either $2\lfloor n/k \rfloor, 2\lfloor n/k \rfloor + 1$, or $2\lceil n/k \rceil$, and each adjacent distance of the other pairs of adjacent agents is $\lfloor n/k \rfloor$ or $\lceil n/k \rceil$. Formally, we define the uniform deployment problem in dynamic rings as follows.

Definition 1. *An algorithm solves the uniform deployment problem in a dynamic ring if any execution E satisfies the following conditions.*

- *Execution E is finite (i.e., all agents terminate in state s_{final}).*
- *When E terminates, there is an agent a' such that if we ignore a', the adjacent distance of exactly one pair of adjacent agents is either $2\lfloor n/k \rfloor, 2\lfloor n/k \rfloor + 1$, or $2\lceil n/k \rceil$, and each adjacent distance of the other pairs of adjacent agents is $\lfloor n/k \rfloor$ or $\lceil n/k \rceil$.*

In this paper, we evaluate the proposed algorithms by memory space per agent, the time complexity (the number of rounds for agents to solve the problem), and the total number of agent moves. In [9], lower bounds on the time complexity and the total number of agent moves for static rings are shown to be $\Omega(n)$ and $\Omega(kn)$, respectively. These bounds also hold in dynamic rings, because there exists an initial configuration that requires the above time and the total number of agent moves for $k - 1$ agents to spread uniformly in the ring. Thus, we have the following theorems.

Theorem 1. *When $k \leq pn$ holds for some constant $p \, (p < 1)$, a lower bound of the total number of agent moves to solve the uniform deployment problem in dynamic rings is $\Omega(kn)$.*

Theorem 2. *A lower bound of the time complexity to solve the uniform deployment problem in dynamic rings is* $\Omega(n)$.

3   Agents with Knowledge of n

In this section, for agents with knowledge of n, we propose an algorithm that solves the problem with $O(k \log n)$ memory space per agent, $O(n \log k)$ rounds, and a total number of $O(kn)$ moves. By Theorem 1, this algorithm is asymptotically optimal in terms of the total number of agent moves. For simplicity, we assume that n is divisible by k in the rest of this section since we can remove this assumption easily, but we omit the description. The basic idea for the case of static rings [9] is that each agent first travels once around the ring and then determines where it should stay at using the information obtained by the traversal. However, in dynamic rings, agents cannot travel once around the ring when one link continues to be missing. Agents treat this by additional behaviors explained in the following subsections. The proposed algorithm comprises two phases: the selection phase and the deployment phase. In the selection phase, each agent selects a *base node* as a reference node for uniform deployment. In the deployment phase, based on the base node, each agent determines and moves to its *destination node*.

3.1   Selection Phase

The aim of this phase is that each agent achieves either of the following two goals: (i) It travels once around the ring and gets the distance sequence of the initial configuration c_0, or (ii) it detects that all agents stay at the same node. We use an idea similar to [11] which considers gathering in dynamic rings, in order for agents to get information about locations of tokens and detect whether all agents stay at the same node or not. First, each agent a_i releases its token at its home node $v_{HOME}(a_i)$ and then moves forward for $3n$ rounds. During the movement, a_i measures the distance dis between every pair of adjacent token nodes, and stores dis to an array D_i for memorizing the distance sequence. After the movement, the number of nodes that a_i has visited is (a) at least n or (b) less than n due to missing links. In case (a), a_i must have completed traveling once around the ring. Thus, a_i can get the value of k and the distance sequence of c_0 (goal (i) is achieved). Then, a_i selects its base node using the obtained sequence. Let $D_i = (d_0, d_1, \ldots, d_{k-1})$ be the distance sequence observed by a_i, where d_j is the distance from the j-th token node a_i found to the $(j+1)$-st token node. We regard a_i's home node $v_{HOME}(a_i)$ as the 0-th token node. Let x_i be the minimum integer satisfying $shift(D_i, x_i) = D_{min}$, where D_{min} is the lexicographically minimum distance sequence among $\{ shift(D_i, x) \mid 0 \leq x \leq k - 1 \}$. Note that D_{min} is equal to $D(c_0)$ defined in Sect. 2.1. Then, a_i selects its base node $v_{base}(a_i)$ as the home node of its x_i-th token node. Note that the x_i-th agent has the minimum distance sequence D_{min}. If c_0 is aperiodic, all agents select the same node as a base node. If c_0 is periodic, multiple nodes are selected as base nodes. For

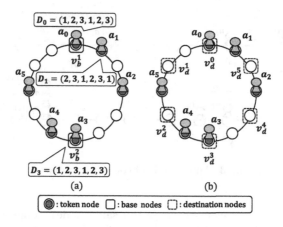

Fig. 2. Example of base nodes and destination nodes.

example, when c_0 is like Fig. 2 (a), node v_b^1 (resp., v_b^2) is selected as a base node for agents a_4, a_5, and a_0 (resp., a_1, a_2, and a_3).

In case (b) (i.e., a_i has visited less than n nodes during the $3n$ rounds), all agents must have blocked at least $2n$ times, which implies that all agents stay at the same node after the $3n$ rounds (Lemma 1), and thus goal (ii) is achieved. In this case, a_i gets the value of k by counting the number of agents at the same node using the strong multiplicity detection capability, and it selects the current node as the base node. Distance sequence D_i is not used in this case.

Concerning the selection phase, we have the following lemma.

Lemma 1. *After finishing the selection phase, each agent a_i achieves either of the following two goals: (i) It travels at least once around the ring and gets the distance sequence of the initial configuration c_0, or (ii) it detects that all agents stay at the same node.*

3.2 Deployment Phase

In this phase, based on the base nodes, each agent a_i determines its destination node and moves to and stays at the node. First, we explain how to determine the destination node of a_i. In case (a) (i.e., a_i travels at least once around the ring and has the distance sequence of c_0), a_i considers that it is the *rank*-th agent ($0 \leq rank \leq k - 1$) to its base node $v_{base}(a_i)$. Here, we say that a_i is the *rank*-th agent if $rank - 1$ tokens exist from $v_{HOME}(a_i)$ to $v_{base}(a_i)$ in c_0. Thus, *rank* is equal to x_i in Sect. 3.1. We regard the agent a_i staying at $v_{base}(a_i)$ as the 0-th agent. In Fig. 2 (a), agent a_2 (resp., a_5) is the 1-st agent since no agent exists between a_2 (resp., a_5) and $v_{base}(a_2)(= v_b^2)$ (resp., $v_{base}(a_5)(= v_b^1)$). Similarly, a_1 and a_4 are the 2-nd agents. Then, a_i determines its destination node as the node with distance $rank \times n/k$ from $v_{base}(a_i)$. In Fig. 2 (b), the destination nodes of a_0, a_1, a_2, a_3, a_4, and a_5 are $v_d^0, v_d^1, v_d^2, v_d^3, v_d^4$, and v_d^5, respectively.

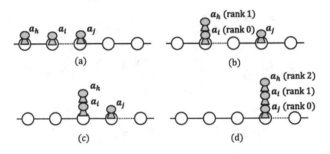

Fig. 3. An example of computing ranks.

In case (b) (i.e., all agents stay at the same node), some agent does not travel once around the ring, does not get the distance sequence of c_0, and cannot compute its rank as in case (a). To treat this, each agent computes its rank using the strong multiplicity detection capability. Concretely, since at most one agent exists at each node in c_0, we consider a synchronous execution, and so far agents have moved forward, two agents exist at the same node when (1) some agent a_i is blocked at node v_j and (2) another agent a_h visits v_j from v_{j-1}. Then, a_i and a_h can recognize which is the first agent that visited v_j using the strong multiplicity detection capability. That is, just before a_h visits v_j, a_i recognizes that no other agent exists at v_j, and it (and a_h) recognize that two agents exist at v_j just after a_h visits v_j. Thus, a_i and a_h can recognize that a_i (resp., a_h) is the 0-th (resp., 1-st) agent. Ranking when three or more agents exist at the same node is similarly computed. In Fig. 3 (a), agents a_h, a_i, and a_j try to move forward (right) but a_i is blocked. Then, the system reaches the configuration of Fig. 3 (b) and a_i (resp., a_h) recognizes that it is the 0-th (resp., 1-st) agent. From (b) to (c), agents try to move forward but a_j is blocked. From (c) to (d), a_j is still blocked, and a_h and a_i reach a_j's node. Then, a_j, a_i and a_h recognize that they are 0-th, 1-st, 2-nd agents, respectively. Using such a technique, even when all agents stay at the same node and do not have the distance sequence of c_0, each agent can compute its rank and determine its destination node as in case (a).

Next, we explain how to move to and stay at the destination node for each agent a_i. Basically, each agent continues to move forward and stays at its destination node when it reaches there. When there exists no missing link, each agent can reach its destination node even if the initial configuration is periodic. This is because the target configuration of uniform configuration is a symmetric one and agents can reach the configuration without symmetry breaking from the periodic configurations. However, if some link continues to be blocked and all agents move forward, they may continue to be blocked from traversing forever due to the same missing link and they cannot reach their destination nodes. To treat this, we introduce a technique called *splitting*. The splitting technique comprises several subphases, and each subphase comprises $12n$ rounds. In addition, the $12n$ rounds of each subphase are divided into four parts each of which comprises $3n$ rounds. For the first $3n$ rounds, each agent a_i moves forward. During the movement, if a_i reaches its destination node, it terminates an execution of the algorithm there. After the $3n$ rounds, if there exists some agent that does not reach its destination node and is still executing the algorithm, we can show by the similar discussion of agents' behaviors in Sect. 3.1 and Lemma 1 that all agents still executing the algorithm stay at the same node. Let k' be the number of agents staying at the same node. Note that the range of ranks for the k' agents are from 0 to $k' - 1$. Then, for the second $3n$ rounds, agents with rank less than $\lceil (k' - 1)/2 \rceil$ (resp., at least $\lceil (k' - 1)/2 \rceil$) try to move forward (resp., backward). By this behavior, it does not happen that all agents continue to be blocked. We call the set of agents with rank less than $\lceil (k' - 1)/2 \rceil$ (resp., at least $\lceil (k' - 1)/2 \rceil$) A_f (resp., A_b). During the movement, if a_i reaches its destination node, it terminates an execution of the algorithm there. Thereafter, for the third $3n$ rounds, each agent still executing the algorithm moves forward (regardless whether it is in A_f or A_b) and stays at its destination node when it reaches there. Then, similar to the end of the first $3n$ rounds, we can show that all agents still executing the algorithm stay at the same node. For the fourth (or last) $3n$ rounds, each agent that belonged to A_f (resp., A_b) moves forward (resp., backward) and stays at its destination node when it reaches there. By these behaviors, we can show that either of all agents in A_f or all agents in A_b can reach their destination nodes in each subphase (Lemma 2). Intuitively, this is because, if agents in some group (e.g., A_f) continue to be blocked for a long time and they have visited less than n nodes in the subphase, agents in the other group (e.g., A_b) can continue to move to the opposite direction and they can visit at least n nodes, which implies that they can reach their destination nodes within the subphase (see Fig. 4 in the proof of Lemma 2). Thus, the number of agents still executing the algorithm at least halves in each subphase. Therefore, by executing such a subphase $\lceil \log k \rceil$ times, all agents other than one agent can stay at their destination nodes.

Concerning the deployment phase, we have the following lemma.

Lemma 2. *After finishing the deployment phase, all agents other than one agent stay at their destination nodes.*

Proof. Let A_f (resp., A_b) be the set of agents that move forward (resp., backward) for the second and fourth $3n$ rounds in some subphase. Then, $-1 \leq |A_f| - |A_b| \leq 1$ holds and it is sufficient to show that either of all agents in A_f or all agents in A_b visit all n nodes of the ring in the subphase. Then, agents in A_f or agents in A_b can stay at their destination nodes and all agents other than one agent can eventually stay at their destination nodes after executing such a subphase $\lceil \log k \rceil$ times. We show the above argument by contradiction, that is, we assume that neither agents in A_f nor agents in A_b visit all n nodes in some subphase. Then, we show that, if agents in some group (e.g., A_f) continue to be blocked for a long time and they have visited less than n nodes in the subphase, agents in the other group (e.g., A_b) can continue to move to the opposite direction and they must visit at least n nodes, which is a contradiction.

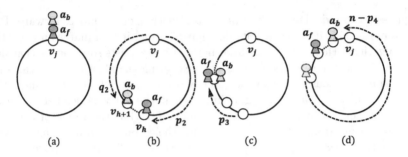

Fig. 4. An example of the splitting technique.

Now, we show the detail of the proof. Let us consider a situation such that agents completed the first $3n$ rounds in some subphase. Then, by the similar discussion in the proof of Lemma 1, all the agents still executing the algorithm stay at the same node like Fig. 4 (a). We consider the behavior of agents for the remaining second, third, and forth $3n$ rounds. By the assumption, there exist at least two agents $a_f \in A_f$ and $a_b \in A_b$ that do not reach their destination nodes during the subphase. For simplicity, in the following we consider behaviors of a_f and a_b. Let v_j be the node where agents start the second $3n$ rounds. For this $3n$ rounds, agent a_f (resp., a_b) tries to move forward (resp., backward). Since neither agents a_f nor a_b visit all n nodes by the assumption of contradiction, a_f and a_b must continue to be blocked by the same missing link before they visit n nodes. That is, when a_f is blocked at some node v_h, the link e_h is missing and a_b is blocked at node v_{h+1} (Fig. 4 (b)). Let p_2 (resp., q_2) be the number of nodes that a_f (resp., a_b) has visited from v_j by moving forward (resp., backward). Then, $0 \le p_2, q_2 < n$ and (1) $p_2 + q_2 + 1 = n$ hold. Thereafter, for the third $3n$ rounds, a_h and a_b try to move forward. Since a_f does not visit n nodes by the assumption, a_f must continue to be blocked before it visits v_j again (Fig. 4 (c)). Let p_3 be the number of nodes that a_f has visited for this $3n$ rounds. Then, $p_2 + p_3 < n$ holds and a_b has visited $p_3 - 1$ nodes for this $3n$ rounds, that is, a_b exists at the $(q_2 - (p_3 - 1))$-th backward node from v_j. Also, at the end of this $3n$ rounds, a_f and a_b stay at the same node by the similar discussion in the proof of Lemma 1. Thereafter, for the fourth $3n$ rounds, a_f (resp., a_b) tries to move forward (resp., backward). Then, a_f continues to be blocked before it visits n nodes. Let p_4 be the number of nodes that a_f visited for this $3n$ rounds. Then, (2) $p_2 + p_3 + p_4 < n$ holds. On the other hand, a_b continues to move backward and be blocked at another endpoint node of the missing link (Fig. 4 (d)). Then, a_b has visited $n - p_4$ nodes for this $3n$ rounds and the number of different nodes that a_b has visited is at least

$$q_2 - (p_3 - 1) + (n - p_4) \overset{(1)}{=} (n - p_2 - 1) - (p_3 - 1) + (n - p_4)$$
$$= 2n - (p_2 + p_3 + p_4)$$
$$\overset{(2)}{>} 2n - n = n.$$

This implies that a_b has visited at least n nodes and traveled once around the ring, which is a contradiction. Therefore, the lemma follows.   □

We have the following theorem for the proposed algorithm.

Theorem 3. *For agents with knowledge of n, the proposed algorithm solves the uniform deployment problem in dynamic rings with $O(k \log n)$ memory space per agent, $O(n \log k)$ rounds, and a total number of $O(kn)$ moves.*

4   Agents with Knowledge of k

In this section, for agents without knowledge of n but with knowledge of k, when $k \geq 4$, we propose an algorithm that solves the problem with $O(k \log n)$ memory space per agent, $O(n^2)$ rounds, and a total number of $O(n^2)$ moves. As in Sect. 3, the algorithm comprises the selection phase and the deployment phase. For simplicity, we assume that n is divisible by k in the rest of this section as in Sect. 3.

4.1   Selection Phase

Basic Idea. In this section, each agent a_i selects its base node similar to Sect. 3.1. That is, a_i first releases its token at its home node $v_{HOME}(a_i)$. Thereafter, agents try to travel once around the ring and get the distance sequence of the initial configuration c_0. However, since agents in this section have knowledge of k instead of n, each agent a_i moves until (a) it observes k tokens after leaving $v_{HOME}(a_i)$ and returns to $v_{HOME}(a_i)$ or (b) it detects that all k agents stay at the same node due to link-missing. In case (a), similar to Sect. 3.1, we can show that a_i completes traveling once around the ring. Then, a_i can get the value of n and the distance sequence of c_0, and can select a base node using the obtained sequence.

In case (b), similar to Sect. 3.1, it may be possible to select the current node as a base node. However, in this case agents do not get the value of n yet and they cannot determine their destination nodes at this stage. Hence, they need additional behaviors to get n. Concretely, when all k agents stay at the same node, we call the agent $a_{i'}$ with the maximum rank (or rank $k - 1$) *the minor agent*, and the other $k - 1$ agents *the major agents*. Then, only the minor agent $a_{i'}$ switches its direction to backward and the other agents (or the major agents) keep the forward direction. By this behavior, it does not happen that all agents continue to be blocked at the same node. In the following, we first explain the

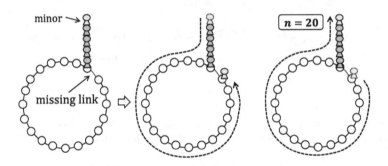

minor →

missing link ⇨

$n = 20$

Fig. 5. An execution example when a link is missing for a long time.

behavior of the minor agent $a_{i'}$ and then explain the behavior of the major agents.

Behavior of Minor Agent $a_{i'}$. We describe the case that $a_{i'}$ is not blocked when moving backward and travels once around the ring at the end of this paragraph. If $a_{i'}$ is blocked when it tries to move backward, it returns its direction to forward, and moves forward until it visits the node where the major agents exist. Then, $a_{i'}$ determines whether the major agents moved forward at least once until $a_{i'}$ comes back to them or not, by comparing the numbers of backward moves and forward moves that $a_{i'}$ made. If the major agents did not move until $a_{i'}$ comes back to them, the major agents continue to be blocked while $a_{i'}$ is moving. That is, when $a_{i'}$ is blocked when moving backward, the major agents exist at another endpoint of the missing link. Then, $a_{i'}$ can recognize that it has visited all n nodes during this traversal and get the value of n, and it can tell the major agents the value of n through the number of rounds that the major agents were blocked. An example is given in Fig. 5. Otherwise (i.e., the major agents moved at least once until $a_{i'}$ comes back to them), $a_{i'}$ executes the above behavior again. The minor agent $a_{i'}$ repeats such a behavior until (i) it continues to move to some direction, travels once around the ring, and gets the value n and the distance sequence of c_0, (ii) the major agents continue to be blocked for a long time but $a_{i'}$ can get the value of n and tell it to the major agents as mentioned above, or (iii) the major agents tell $a_{i'}$ the value of n by the method explained next.

Behavior of the Major Agents. In contrast to the minor agent $a_{i'}$, the major agents do not switch their direction and keep trying to move forward. If the major agents continue to be blocked at the same node for a long time, $a_{i'}$ gets the value of n and tells it to the major agents as mentioned above (Fig. 5). Otherwise (i.e., at every node v_j the major agents leave v_j before $a_{i'}$ comes back to them), the major agents eventually observe k tokens and get the value of n. Thereafter, the major agents try to move forward for $3n$ rounds to tell $a_{i'}$ the value of n. During the movement, if the major agents meet $a_{i'}$ at some node v_j, they tell $a_{i'}$ the value of n as follows. When major agents and $a_{i'}$ meet at round r, $a_{i'}$ stays at the current node v_j at round $r + 1$. On the other hand, the agent a_m^0 with rank

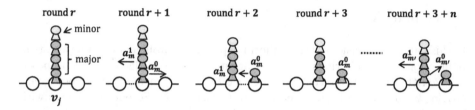

Fig. 6. An example that the major agents tell the minor agent the value of n.

0 (resp., the agent a_m^1 with rank 1) in the major agents tries to move forward (resp., backward) and the other agents in the major agents stay at v_j at round $r + 1$. By this behavior, a_m^0 or a_m^1 can leave v_j, and $a_{i'}$ can detect the change of the number of agents at v_j and the major agents know the value of n. After that, at round $r + 2$, a_m^0 and a_m^1 try to return to v_j (if a_m^0 or a_m^1 is blocked at round $r + 1$, it stays at v_j at round $r + 2$). By this behavior, at least two major agents exist at v_j at round $r + 2$ since we consider the case of $k \geq 4$, and they can tell $a_{i'}$ the value of n. That is, from rounds $r+3$ to $r+3+(n-1)$, the major agents stay at the current nodes. Then, at round $r + 3 + n$, the agent $a_{m'}^0$ with rank 0 (resp., the agent $a_{m'}^1$ with rank 1) in the major agents staying at v_j tries to move forward (resp., backward). Then, the number of agents at v_j changes at least one and $a_{i'}$ can detect that the major agents finished telling the value of n. An example is given in Fig. 6. Otherwise (i.e., the major agents do not meet $a_{i'}$ while trying to move forward for $3n$ rounds), we can show that $a_{i'}$ has already traveled once around the ring, and got the value of n and the distance sequence of c_0 (Lemma 3). Thereafter, the major agents (and $a_{i'}$) enter the deployment phase.

Concerning the selection phase, we have the following lemma.

Lemma 3. *After finishing the selection phase, each agent knows the value of n. In addition, when the minor agent $a_{i'}$ gets the value of n without being told the value by the major agents, each agent knows the distance sequence of the initial configuration c_0.*

4.2 Deployment Phase

By Lemma 3, after executing the selection phase, each agent a_i knows the value of n (and the distance sequence of c_0 when the minor agent $a_{i'}$ gets the value of n without being told the value by the major agents). Hence, a_i can determine its destination node and move to and stay at the node by the exact same way in Sect. 3.2. Thus, we have the following theorem for the proposed algorithm.

Theorem 4. *For agents with knowledge of k, when $k \geq 4$, the proposed algorithm solves the uniform deployment problem in dynamic rings with $O(k \log n)$ memory space per agent, $O(n^2)$ rounds, and a total number of $O(n^2)$ moves.*

5   Conclusion

In this paper, we considered the uniform deployment problem of mobile agents in bidirectional dynamic rings. First, for agents with knowledge of n, we proposed an algorithm that solves the problem with $O(k \log n)$ memory space per agent, $O(n \log k)$ rounds, and a total number of $O(kn)$ moves. Second, for agents with knowledge of k, when $k \geq 4$, we proposed an algorithm that solves the problem with $O(k \log n)$ memory space per agent, $O(n^2)$ rounds, and a total number of $O(n^2)$ moves. These results mean that the uniform deployment problem can be solved also in dynamic rings, and unlike [9], agents with knowledge of k cannot achieve uniform deployment by the same way as that for agents with knowledge of n.

As future works, first we will analyze the lower bound on the time complexity and consider whether or not proposed algorithms this time are time optimal. Second, we will consider whether or not some assumptions (e.g., chirality, the strong multiplicity detection capability, etc.) can be removed or weakened. Finally, we will consider agents without knowledge of n or k. In this case, it is shown in [9] that, (1) agents cannot solve the problem even in static rings when termination detection is required, and (2) if termination detection is not required and agents have communication capability, the problem is solvable. Hence, using the model like (2), we will consider whether the problem is solvable or not also in dynamic rings.

Acknowledgements. This work was partially supported by JSPS KAKENHI Grant Number 18K18029, 18K18031, and 20H04140; the Hibi Science Foundation; and Foundation of Public Interest of Tatematsu.

References

1. Gray, R.S., Kotz, D., Cybenko, G., Rus, D.: D'Agents: applications and performance of a mobile-agent system. Softw. Pract. Exp. **32**(6), 543–573 (2002)
2. Lange, D.B., Oshima, M.: Seven good reasons for mobile agents. CACM **42**(3), 88–89 (1999)
3. Kranakis, E., Krizanc, D.: An algorithmic theory of mobile agents. In: Montanari, U., Sannella, D., Bruni, R. (eds.) TGC 2006. LNCS, vol. 4661, pp. 86–97. Springer, Heidelberg (2007). https://doi.org/10.1007/978-3-540-75336-0_6
4. Cao, J., Sun, Y., Wang, X., Das, S.K.: Scalable load balancing on distributed web servers using mobile agents. JPDC **63**(10), 996–1005 (2003)
5. Flocchini, P., Prencipe, G., Santoro, N.: Self-deployment of mobile sensors on a ring. Theoret. Comput. Sci. **402**(1), 67–80 (2008)
6. Yotam, E., Alfred, B.M.: Uniform multi-agent deployment on a ring. Theoret. Comput. Sci. **412**(8), 783–795 (2011)
7. Barriere, L., Flocchini, P., Mesa-Barrameda, E., Santoro, N.: Uniform scattering of autonomous mobile robots in a grid. Int. J. Found. Comput. Sci. **22**(03), 679–697 (2011)
8. Poudel, P., Sharma, G.: Time-optimal uniform scattering in a grid. In: ICDCN, pp. 228–237 (2019)

9. Shibata, M., Mega, T., Ooshita, F., Kakugawa, H., Masuzawa, T.: Uniform deployment of mobile agents in asynchronous rings. JPDC **119**, 92–106 (2018)
10. Shibata, M., Kakugawa, H., Masuzawa, T.: Space-efficient uniform deployment of mobile agents in asynchronous unidirectional rings. Theoret. Comput. Sci. **809**, 357–371 (2020)
11. Di Luna, G.A., Flocchini, P., Pagli, L., Prencipe, G., Santoro, N., Viglietta, G.: Gathering in dynamic rings. Theoret. Comput. Sci. **811**, 79–98 (2018)
12. Di Luna, G., Dobrev, S., Flocchini, P., Santoro, N.: Distributed exploration of dynamic rings. Distrib. Comput. **33**(1), 41–67 (2018). https://doi.org/10.1007/s00446-018-0339-1
13. Gotoh, T., Sudo, Y., Ooshita, F., Kakugawa, H., Masuzawa, T.: Group exploration of dynamic tori. In: ICDCS, pp. 775–785 (2018)
14. Das, S., Di Luna, G.A., Gasieniec, L.A.: Patrolling on dynamic ring networks. In: Catania, B., Královič, R., Nawrocki, J., Pighizzini, G. (eds.) SOFSEM 2019. LNCS, vol. 11376, pp. 150–163. Springer, Cham (2019). https://doi.org/10.1007/978-3-030-10801-4_13
15. Agarwalla, A., Augustine, J., Moses Jr., W.K., Madhav, S.K., Sridhar, A.K.: Deterministic dispersion of mobile robots in dynamic rings. In: ICDCN, pp. 1–4 (2018)

Partial Gathering of Mobile Robots from Multiplicity-Allowed Configurations in Rings

Masahiro Shibata[1(✉)] and Sébastien Tixeuil[2]

[1] Kyushu Institute of Technology, Fukuoka, Japan
shibata@cse.kyutech.ac.jp
[2] Sorbonne University, CNRS, LIP6, Paris, France
Sebastien.Tixeuil@lip6.fr

Abstract. In this paper, we consider the problem of partial gathering for mobile robots in ring topologies. The partial gathering problem is a generalization of the (well-investigated) total gathering problem, which requires that all k robots distributed in the network terminate at a non-predetermined single node. The partial gathering problem requires, for a given positive integer $g (< k)$, that each robot moves to a node and terminates, so that at least g robots, or no robot, exists at any node. We consider the solvability of the problem for anonymous and oblivious mobile robots from initial configurations where several robots may share the same node. If an algorithm can solve the problem from any such initial configuration, the algorithm is called self-stabilizing. In addition, since the requirement for the partial gathering problem is weaker than that for the total gathering problem, we aim to clarify whether partial gathering is *(i)* solvable in weaker models than those for total gathering, and *(ii)* solvable in a strictly smaller total number of moves than that for total gathering.

First, we show that no probabilistically self-stabilizing partial gathering algorithm exists in the asynchronous (ASYNC) model, or if only local-strong and global-weak multiplicity detection is available. Surprisingly, these impossibility results are the same as the case of total gathering. Next, in the semi-synchronous (SSYNC) and global-strong multiplicity detection model (*i.e.*, the weakest system assumption that may allow a solution), we consider deterministic and probabilistic algorithms. In the deterministic case, we first show that unsolvable initial configurations exist, and then propose an algorithm to solve the problem using $O(gn)$ moves from any solvable initial configuration, where n is the number of nodes. In the probabilistic case, we propose an algorithm to solve the problem using $O(gn)$ moves in expectation from any initial configuration (*i.e.,* the algorithm is probabilistically self-stabilizing). Note that $g < k$ holds, and the partial (resp., total) gathering problem requires $\Omega(gn)$ (resp., $\Omega(kn)$) moves. Thus, our results show that the proposed algorithms achieve partial gathering in a strictly smaller total number of moves than that for total gathering.

Keywords: Mobile robot · Partial gathering · Self-stabilizing

© Springer Nature Switzerland AG 2020
S. Devismes and N. Mittal (Eds.): SSS 2020, LNCS 12514, pp. 264–279, 2020.
https://doi.org/10.1007/978-3-030-64348-5_21

1   Introduction

Background. Studies for mobile robot networks have emerged recently in the field of Distributed Computing. Their goal is to achieve some tasks by a team of mobile robots with weak capabilities. Most studies assume that robots are identical (they execute the same algorithm and cannot be distinguished by their appearance) and oblivious (they cannot remember their past actions). In addition, it is assumed that robots cannot communicate with other robots explicitly. Instead, the communication is done implicitly by having each robot observe the positions of others.

Since Suzuki and Yamashita presented a pioneering work [19], many problems were studied in various settings [7]. In this paper, we focus on unoriented anonymous ring-shaped networks, since such topologies proved the hardest to cope with, particularly with respect to symmetries [2,13,14]. The main concern of previous works in ring networks is to characterize the minimum assumptions allowing deterministic algorithms to solve a particular problem. Many algorithms for fundamental problems such as the total gathering problem, which requires all k robots to meet at a non-predetermined single node, have been proposed using various settings [2].

Most previous works considering robots in a discrete model (*i.e.*, a graph) assume that initial robot positions are unique, that is, in the initial configuration, no two robots share the same location [2]. A notable exception is due to Ooshita et al. [14], where all initial configurations are possible (including those with multiplicity points), yet their probabilistic algorithm yields a solution to the total gathering problem. Such an algorithm is called *probabilistically self-stabilizing* since it solves the problem from any initial configuration in a probabilistic way.

Our Contribution. In this paper, we explore the possibility to obtain self-stabilizing mobile robot algorithms (*i.e.,* algorithms that allow any initial configuration, including those with multiplicity points) for the *g-partial gathering* *problem.* Considering a set of k robots, the g-partial gathering requires, for a given positive integer $g\,(< k)$, that each robot moves to a node and terminates, so that every node is either empty or hosting at least g robots when all robots terminate, (*e.g.*, Fig. 1).

Let us observe that the requirement for the g-partial gathering problem is weaker than that for the (well-investigated) total gathering problem, as a solution to the total gathering problem is also a solution to the g-partial gathering problem (obvious, the converse is not true). Thus, we aim to clarify whether g-partial gathering is *(i)* solvable in weaker models (*i.e.*, assuming fewer hypotheses) than those for total gathering, and *(ii)* solvable in a strictly smaller total number of moves than that for total gathering.

We summarize our results in Table 1. We consider various hypotheses such as synchrony assumptions (the network could be asynchronous or semi-synchronous) and the ability for the robots to detect multiplicity (that is, detecting several robots occupy the same node). For multiplicity, we consider the global (a robot can sense multiplicity in the entire network) and the local (a robot can sense multiplicity only at its host node) variants, as well as the strong (the exact

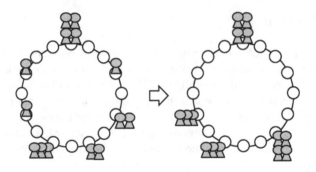

Fig. 1. An example of a partial gathering solution ($g = 3$)

number of robots on a node can be sensed) and the weak (only the "multiple robots" information is sensed) variants.

On the negative side, we show that no probabilistic self-stabilizing g-partial gathering algorithm exists in the asynchronous (ASYNC) model, or if only local-strong and global-weak multiplicity detection is available. Surprisingly, these impossibility results are the same as the case of the total gathering problem [14] despite the studied problem (g-partial gathering) being "simpler" than total gathering. Next, in the semi-synchronous (SSYNC) model, and assuming global-strong multiplicity detection model (*i.e.*, the weakest system assumption that may allow a solution), we consider deterministic and probabilistic algorithms. In the deterministic case, we first show that unsolvable initial configurations remain, and then propose an *almost-stabilizing algorithm*, that is, an algorithm to solve the problem from *any* solvable initial configuration. This algorithm requires $O(n)$ rounds, and a total number of $O(gn)$ moves, where n is the number of nodes. So, there exist initial configurations with multiplicity points that allow deterministic g-partial gathering solvability. By contrast, deterministic results for total gathering only consider initial distinct configurations (*i.e.*, initial configurations where at most one robot is located on a given node). In the probabilistic case, we propose an algorithm to solve the problem in $O(n)$ rounds in expectation, and a total number of $O(gn)$ moves in expectation, from *any* initial configuration. Hence, our probabilistic algorithm is probabilistically self-stabilizing. Note that, since $g < k$ holds and the g-partial (resp., the total) gathering problem requires $\Omega(gn)$ (resp., $\Omega(kn)$) moves [15], our algorithms solve the g-partial gathering problem with an asymptotically optimal number of moves, and this number is strictly smaller than that for total gathering.

Related Works. The majority of works for mobile robots that have been published since the pioneering work of Suzuki and Yamashita [19] considers a two-dimensional Euclidean space (*a.k.a.* the continuous model). The original paper [19] demonstrates that when robots are not fully synchronous (that is, in the SSYNC or the ASYNC model), the deterministic gathering of two robots is impossible. This impossibility result was generalized to an even number of robots initially located evenly at two positions by Courtieu et al. [5]

Table 1. Results for the g-partial gathering problem (n: #nodes)

	Model 1	Model 2	Model 3	Model 4
Deterministic/Probabilistic	Probabilistic	Probabilistic	Deterministic	Probabilistic
Synchronicity	ASYNC	SSYNC	SSYNC	SSYNC
Multiplicity	Global-strong	Local-strong and global-weak	Global-strong	Global-strong
Solvable self-stabilizingly	No	No	No	Yes
Solvable almost-stabilizingly	?	?	Yes	Yes
Rounds	-	-	$O(n)$	$O(n)$
Total number of moves	-	-	$O(gn)$	$O(gn)$

(those configurations are known as *bivalent* configurations). By contrast, the case of an odd number of robots permits to withstand any initial position in a deterministic manner, as demonstrated by Dieudonné et al. [6], assuming global-strong multiplicity detection. Alternatively, probabilistic approaches permit gathering without assumptions on the number of robots or their initial configuration [3,8]. Clement et al. [3] (resp., Izumi et al. [8]) show the trade-off between the time complexity and the availability of global-weak multiplicity detection or global-strong multiplicity detection (resp., local-weak multiplicity detection or local-strong multiplicity detection). Finally, weakening synchrony assumptions down to the fully synchronous model (FSYNC) permits to solve gathering [4], even from arbitrary initial configurations [1].

Recently, a number of works consider the discrete model (*a.k.a.* graph model), where robots may only be located at nodes of the graph, and move from one position to the next if there is an edge in the graph linking the two corresponding nodes. For ring networks, Klasing et al. [12] propose deterministic total gathering algorithms in the ASYNC model with global-weak multiplicity detection. They also show that there exist some initial configurations where no deterministic algorithm can achieve total gathering. Izumi et al. [9] provide a deterministic total gathering algorithm with local-weak multiplicity detection. Their algorithms assume that initial configurations are non-symmetric or non-periodic, and that the number of robots is less than half of the number of nodes. For odd number of robots or odd number of nodes in the same model, Kamei et al. [10,11] propose total gathering algorithms that also work when started from symmetric configurations. Note that all of the above works assume that locations of robots in the initial configuration are distinct, and thus they are not self-stabilizing.

Going probabilistic proved useful: the work of Ooshita et al. [14] demonstrated that probabilistic algorithm can ensure gathering (in the probabilistic sense) starting from any initial configuration (including those with multiplicity points).

To the best of our knowledge, the g-partial gathering problem was considered only for the so called *agent model* [15–18], where mobile entities have persistent

memory but cannot observe others' positions unless they are located on the exact
same node. Also, they assume that no two mobile entities may share the same
position in the initial configuration. So, this paper is the first to investigate the
g-partial gathering problem in the classical look-compute-move model on graphs.

2 Model

System Models. We use almost the same model as that in Ooshita et al.'s
paper [14]. The system comprises n nodes and k mobile robots. The nodes
$v_0, v_1, \ldots, v_{n-1}$ construct a directed ring in this order. For simplicity, we consider
mathematical operations to indices of nodes as operations modulo n. We consider two types of rings: oriented and unoriented rings. Neither nodes nor links
have any labels, and consequently robots cannot distinguish nodes and links.
Robots occupy some nodes of the ring.

Robots considered here have the following characteristics and assumptions.
Robots are *identical*, that is, robots execute the same algorithm and cannot be
distinguished by their appearance. Robots are *oblivious*, that is, robots have no
persistent memory and cannot remember the history of their execution. Robots
cannot communicate with other robots directly, however they can observe the
positions of other robots. This means that robots can communicate implicitly by
their positions. We assume each robot has some multiplicity detection capability.
We consider two types of multiplicity detection: *global-strong multiplicity detection*, and *local-strong and global-weak multiplicity detection*. When each robot
has global-strong multiplicity detection capability, robots can detect the exact
number of robots at each node. When each robot has local-strong and global-
weak multiplicity detection capability, robots can detect the exact number of
robots only at its current node and detect whether the number of robots at
every other node is zero, one, or more than one.

Each robot executes the algorithm by repeating cycles. At the beginning of
each cycle, the robot observes the environment and the positions of other robots
(look phase). According to the observation, the robot computes whether it moves
to its adjacent node or stays idle (compute phase). If the robot decides to move,
it moves to the node by the end of the cycle (move phase). Such an execution
model is called the *Look-Compute-Move model*. To analyze the asynchronous
behavior of robots, we introduce the notion of a *scheduler* that decides when
each robot executes phases. When the scheduler makes robot r execute some
phase, we say the scheduler *activates* r. We consider two types of synchronicity:
the SSYNC (semi-synchronous) and the ASYNC (asynchronous) model. In the
SSYNC model, a set of robots is selected by the scheduler, and the selected
robots execute cycles synchronously. In the *ASYNC model*, cycles of robots are
executed asynchronously. Note that in the ASYNC model each robot can move
based on the outdated view that the robot observed before. On the other hand,
in the SSYNC model robots can move based on the latest view. For both models,
the scheduler is *fair*, that is, each robot is activated infinitely often. When we
analyze the worst-case performance of algorithms, we consider the scheduler as

an adversary. That is, we assume that the scheduler knows all information, such as positions and decisions of all robots, and activates robots to degrade the performance of algorithms as much as possible.

A *configuration* of the system is defined as the number of robots at each node. If some robots occupy a node, the node is called a *robot node*. If exactly one robot occupies a node, the node is called a *solo node*. If m robots $(m \geq 2)$ occupy a node, the node is called a *m-robot node* or a *tower node*. If no robot occupies a node, the node is called a *free node*.

When a robot observes the environment, it gets a *view* of the system. Consider a configuration such that nodes $v_{i_0}, v_{i_1}, \ldots, v_{i_{w-1}} (i_0 < i_1 < \cdots < i_{w-1})$ are robot nodes and each robot node v_{i_x} is occupied by m_{i_x} robots. A robot obtains two views: a forward view and a backward view. When robots have global-strong multiplicity detection capability, let M_y be the number of robots at its y-th robot node in the forward direction (*i.e.*, $M_y = m_{i_{(x+y)}}$) and D_y be the distance from its y-th robot node to $(y+1)$-th robot node (*i.e.*, $D_y = (i_{(x+y+1)} - i_{(x+y)})$). Then, the forward view and the backward view of a robot at node v_{i_x} are defined as $V_f = (M_0, D_0, M_1, D_1, \ldots, M_{w-1}, D_{w-1})$ and $V_b = (M_0, D_{w-1}, M_{w-1}, \ldots, M_1, D_0)$, respectively. When robots have local-strong and global-weak multiplicity detection capability, views are defined similarly except that $M_y(0 < y < w)$ is one or two: $M_y = 1$ implies $m_{i_{(x+y)}} = 1$ and $M_y = 2$ implies $m_{i_{(x+y)}} > 1$ (note that $M_0 = m_{i_x}$). Figure 2 shows an example of configurations. When robots have global-strong multiplicity detection capability, robot r at v_0 obtains two views $(4, 3, 4, 2, 3, 1, 1, 2)$ and $(4, 2, 1, 1, 3, 2, 4, 3)$. When robots have local-strong and global-weak multiplicity detection capability, r obtains two views $(4, 3, 2, 2, 2, 1, 1, 2)$ and $(4, 2, 1, 1, 2, 2, 2, 3)$. When we assume unoriented rings (or robots have no chirality) and some robot has the same forward and backward view, we say that the ring is *symmetric*. In such cases, we assume the scheduler decides which direction each robot moves to.

We evaluate the algorithm by the (expected) number of asynchronous rounds and (expected) total number of moves. A round is defined as the shortest fragment of an execution where each robot executes at least one complete cycle. The total number of moves is the sum of moves each agent makes.

Problem to be Solved. We define the g-gathering problem as follows: at least g robots remains at nodes where at least one robot is present, and remain there thereafter, as in Fig. 1.

Definition 1. *An algorithm A solves the g-partial gathering problem if and only if the system reaches a configuration where each node is a robot node with at least g robots or a free node.*

For the g-partial gathering problem, a lower bound on the total number of moves for the agent model is shown by Shibata et al. [15]. The result also holds for the robot model we consider in this paper, and thus we have the following theorem.

Theorem 1. [15] *The lower bound of the total number of agents moves to solve the g-partial ring networks is $\Omega(gn)$ if $g \geq 2$.*

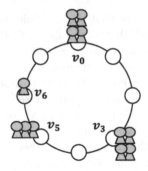

Fig. 2. An example of configurations

3   Impossibility Results

In this section, we show the impossibility of probabilistically self-stabilizing g-Partial gathering algorithms in weaker models. First, we show the impossibility with a weaker synchronicity model, *i.e.*, the ASYNC model. Recall that cycles of robots are executed asynchronously in the ASYNC model. We use the technique by Ooshita et al. [14], which shows that the total gathering problem is unsolvable in the ASYNC model (even probabilistically). Intuitively, Ooshita et al. [14] shows that, from the configuration that every node is occupied by exactly one robot (so, $k = n$), with some probability, there exists the following asynchronous execution: When the scheduler activates some robot r at node v_x and r decides to move forward to v_{x+1}, the scheduler activates robot r' at v_{x+1}. If r' decides to move backward to v_x, robots r and r' move in the ring. Then, views of all robots do not change. If r' decides to move forward to v_{x+2}, the scheduler activates robot r'' at v_{x+2}. The scheduler repeats such activations until all robots decide to move forward. After the movement, views of all robots do not change. The case that robot r at v_x decides to move backward to v_{x-1} is treated similarly. As a result, views of all robots do not change during the execution and robots cannot solve the problem. The above idea can be directly applied to the g-partial gathering problem and we obtain the following theorem.

Theorem 2. *No probabilistically self-stabilizing g-partial gathering algorithm with global-strong multiplicity detection exists in the ASYNC model even if rings are oriented.*

Next, we consider a model with weaker multiplicity detection, *i.e.*, local-strong and global-weak multiplicity detection. Similarly, we use the technique by Ooshita et al. [14]. Intuitively, from a configuration that each node is occupied by three robots, the proof of Ooshita et al. [14] demonstrates that, with some probability, there exists a following execution: When robot r at v_x is activated and it moves to v_{x+1} or v_{x-1}, the scheduler continues to activate only r until it returns to v_x. After that, the scheduler continues to activate another robot r' until it moves in the ring and returns to its initial node. The scheduler repeats such activations until all robots are activated, move in the ring, and return to their initial nodes. During such an execution, the numbers of robots at each node are between 2 and 4. Hence, the above technique can be also applied to the g-partial gathering problem for the case of $g \geq 5$. We thus have the following theorem.

Theorem 3. *For the case of $g \geq 5$, no probabilistically self-stabilizing g-gathering algorithm with local-strong and global-weak multiplicity detection exists in the SSYNC model even if rings are oriented.*

By the above theorems, in the following sections we consider the semi-synchronous and global-strong multiplicity detection model.

4   Almost-Stabilizing Deterministic g-Partial Gathering

In this section, we consider a almost-stabilizing deterministic g-partial gathering algorithm. First, we show that unsolvable initial configurations exist, and then propose a deterministic algorithm to solve the problem from any solvable initial configuration.

4.1   Unsolvable Initial Configurations

For explanation, we define the periodicity of the initial configuration. For a forward view $V = (M_0, D_0, M_1, D_1, \ldots, M_{w-1}, D_{w-1})$, let $shift\ (V, t) = (M_{w-t}, D_{w-t}, M_{w-t+1}, D_{w-t+1}, \ldots, M_{w-1}, D_{w-1}, M_0, D_0, \ldots, M_{w-t-1}, D_{w-t-1})$. Then, when $shift(V, t) = V$ holds for some t $(0 < t < w - 1)$, we say that the configuration is *periodic*. Otherwise, we say that the configuration is *aperiodic*. Moreover, for the minimum (positive) integer t satisfying $shift(V, t) = V$, we define the *number of robots in one period* (*nOnePeri*) as $\sum_{i=0}^{t-1} M_i$. Note that $nOnePeri = k$ holds in aperiodic configurations. Then, we have the following theorem

Theorem 4. *If the initial configuration is periodic and $nOnePeri < g$ holds, the g-partial gathering problem is not deterministically solvable.*

Proof. We assume that robots move in a synchronous manner. Let V_i be a view of robot r_i in the initial configuration, and let t be the minimum integer satisfying $shift(V_i, t) = V_i$. Then $\sum_{i=0}^{t-1} M_i = nOnePeri$ holds. In this case, robots with either of views $V_i, shift(V_i, t), \ldots, shift(V_i, pt - 1)$ $(p = k/nOnePeri)$ have the same view and behave in the exactly same way in a synchronous execution. Thus, they cannot break the periodicity of the initial configuration and any pair of them cannot gather at the same node. Then, in the final configuration, there exist at least $p = k/nOnePeri$ robot nodes and the number of robots at each of p robot nodes is at most $k/p = nOnePeri < g$. This means that the g-partial gathering problem is not solvable by a deterministic algorithm. □

4.2   Proposed Algorithm

In this section, we propose the g-partial gathering algorithm in $O(n)$ rounds and the total number of $O(gn)$ moves from any solvable initial configuration. The ring is unoriented, that is, robots have no common chirality. The basic idea is to partition robots into groups each containing at least g robots. Robots realize this using views. Robots determine their behaviors following the three cases: (i) Case 1 is when no tower node exists, (ii) Case 2 is when at least one tower node and at least one solo node exist, and (iii) Case 3 is when no solo node exists. Recall that a tower node represents a node where at least two robots exist. At first, we describe the abstract behavior of robots following the above three cases. Note that, since we allow initial configurations with multiplicity points, robots may start the algorithm from any of the above cases.

In Case 1, several robots move in the ring to construct a configuration such that at least one tower node exists (Case 2). To do this, we define a *min robot* r_{min} as the robot with the lexicographically minimum view among all forward and backward views robots observe. We also define a *min node* v_{min} as the node where r_{min} exists. Note that in periodic configurations several min nodes (or min robots) exist. Then, robots closer to v_{min} move to v_{min} to make a tower node. Concretely, for each robot r_i, if its nearest backward or forward robot node is v_{min}, r_i moves to v_{min}. If both of r_i's backward and forward robot nodes are min nodes, it moves to its backward min node. If neither of r_i's backward nor forward robot node is v_{min}, it remains at the current node. By this behavior, the system eventually reaches the configuration where at least one tower node exists. Note that, when the ring is periodic, several min nodes exist. In this case, since each moving robot gets closer to its nearest min node, at least one min node v_{min} keeps its view the minimum and some robot eventually reaches v_{min}.

In Case 2, that is, when at least one tower node and at least one solo node exist, robots at solo nodes move in the ring to construct a configuration such that no solo node exists (Case 3). To do this, let $V_{max} = (M_0^{max}, D_0^{max}, \ldots, M_{w-1}^{max}, M_{w-1}^{max})$ be the lexicographically maximum view among all forward and backward views robots observe, and r_{max} be a robot whose view is V_{max}. Then, in Case 2 and 3, robots determine the direction from M_0^{max}-robot node (i.e., the node r_{max} is staying) to M_1^{max}-robot node as the global forward direction of the ring. In Fig. 2, robots regard the direction from v_0 to v_3 as the forward direction. If the ring is symmetric, each robot r_i except for r_{max} regards the direction from its nearest node r_{max} exists to r_i as its local forward direction.

Then, each robot r_i computes its ordinal number from the nearest backward tower node v_t. Concretely, when $\ell - 1$ solo nodes exist from v_t and r_i's node in the forward direction, r_i recognizes that it stays at the ℓ-th robot node from v_t. Let m_t be the number of robots at v_t. Then, r_i decides its behavior depending on values of ℓ and m_t. Concretely, if $(\min\{m_t, g\} + \ell \mod g = 1)$ holds, it remains at the current node. Otherwise, it moves backward. Intuitively, this behavior partitions robots into groups each containing at least g robots, except for a fraction. For example, in Fig. 3 (a) $(g = 4)$, v_t^1 is a $3 (< g)$-tower node. Then, since r_2 stays at the 2-nd robot node and $((3+2) \mod g = 1)$ holds, it remains

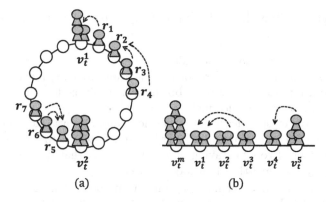

Fig. 3. (a): An example of robots' behaviors in Case 2 ($g = 4$). (b): An example of robots' behaviors in Case 3 ($g = 5$).

at the current node. On the other hand, r_1 (resp., r_3 and r_4) moves backward to v_t^1 (resp., the node where r_2 stays). Meanwhile, v_t^2 is a 4 ($\geq g$)-robot node. Then, since r_5 stays at the 1-st robot node from v_t^2 and $((4 + 1) \mod g = 1)$ holds, it remains at the current node. Oh the other hand, r_6 and r_7 move backward to the node where r_5 stays. By this behavior, the system eventually reaches the configuration such that no tower node exists.

In Case 3, that is, when no solo node exists, let V_g^{more} (resp., V_g^{less}) be the set of nodes each containing at least (resp., less than) g robots. Then, robots staying at nodes in V_g^{less} move in the ring to achieve g-partial gathering. To do this, we assume that v_t^m is a robot node in V_g^{more} and there exist tower nodes $v_t^1, v_t^2, \ldots, v_t^\ell$ each containing less than g robots in v_t^m's forward direction in this order. Then, robots at v_t^1 remain at v_t^1 and robots at $v_t^2, v_t^3, \ldots v_t^{\ell'}$, such that the total number of robots at $v_t^1, v_t^2, \ldots, v_t^{\ell'-1}$ (resp., $v_t^1, v_t^2, \ldots, v_t^{\ell'}$) is less than (resp., at least) g, move backward to v_t^1. At the same time, robots at $v_t^{\ell'+1}$ remain at $v_t^{\ell'+1}$ and robots at the next several consecutive robot nodes move backward to $v_t^{\ell'+1}$, so that at least g robots meet at $v_t^{\ell'+1}$. For example, in Fig. 3(b) ($g = 5$), since the total number of robots at v_t^1 and v_t^2 (resp., v_t^1, v_t^2, and v_t^3) is $4 < g$ (resp., $6 > g$), robots at v_t^2 and v_t^3 move to v_t^1. Similarly, since the number of robots at v_t^4 (resp., v_t^4 and v_t^5) is less than (resp., at least) g, robots at v_t^5 move backward to v_t^4. Robots execute such a behavior until all robots reach their destination nodes. Note that, if robots at the same tower node move at different timings, the system reaches a configuration in Case 2. However, even in this case the destination node of all the robots that have stayed at the same node does not change. Thus, by continuing such behaviors, robots are partitioned into groups each containing at least g robots and robots eventually achieve g-partial gathering.

In the following, we describe the detailed behavior of robots. The goal of g-partial gathering is to form a configuration such that at least g robots exist at each node where robots exist. We denote by C_g a set of such configurations.

Then, the behavior of each robot r_i in configuration C ($C \neq C_g$) is described as follows:

- **Case 1**: There exists no tower node.
  - If its nearest backward robot node is a min node, r_i moves there.
  - Else if its nearest forward robot node is a min node, r_i moves there.
  - Otherwise, r_i does not move.
- **Case 2**: There exist at least one tower node and at least one solo node. Let v_t (resp., $v_{t'}$) be r_i's nearest backward (resp., forward) tower node. Let m_t be the number of robots staying at v_t. We assume that if r_i stays at a solo node, it stays at the ℓ-th robot node from v_t.
  - If r_i stays at a tower node, r_i remains at the current node.
  - If $\ell = 1$ holds (i.e., r_i's nearest backward robot node is v_t),
    - If ($m_t < g$) or (the nearest forward robot node is $v_{t'}$) or (the ring is symmetric and there exists exactly one solo node between r_i and $v_{t'}$), r_i moves backward to v_t.
    - Otherwise, r_i does not move.
  - If ($\ell \neq 1$) holds,
    - If $(\min\{m_t, g\} + \ell) \mod g = 1$, r_i remains at the current node.
    - If $(\min\{m_t, g\} + \ell) \mod g = 2$, r_i moves backward.
    - Otherwise, if the neighboring backward node is a free node, r_i moves backward.
  - * If r_i decides to move backward, it moves only when the global forward direction does not change after the movement.
- **Case 3**: There exists no solo node. We assume that if there exists at least one tower node in V_g^{more} and r_i stays at a tower node $v_t^\ell \in V_g^{less}$, there exist $\ell - 1$ tower nodes $v_t^1, v_t^2, \ldots v_t^{\ell-1}$ in V_g^{less} in this order between the nearest backward tower node $v_t^m \in V_g^{more}$ and v_t^ℓ. Let $v_{t'}^m$ be r_i's nearest forward tower node in V_g^{more}. Let m_j ($1 \leq j \leq \ell$) be the number of robots staying at v_t^j, and let $f(1) = m_1$ and $f(j) = f(j-1) + m_j$ if $f(j-1) + m_j < g$ holds and $f(j) = 0$ otherwise.
  - If $V_g^{more} = \emptyset$ holds,
    - If its nearest backward robot node is v_{min}, r_i moves to v_{min}.
    - Otherwise, r_i does not move.
  - If r_i stays at a node in V_g^{more}, r_i remains at the current node.
  - If $\ell = 1$ holds, (i.e., r_i's nearest backward robot node is v_t^m),
    - If (the nearest forward robot node is a $v_{t'}^m$) or (the ring is symmetric and there exists exactly one tower node in V_g^{less} between r_i and $v_{t'}^m$), r_i moves backward to v_t^m.
    - Otherwise, r_i does not move.
  - If $\ell \neq 1$ holds,
    - If $f(\ell - 1) = 0$ holds, r_i remains at the current node.
    - If $f(\ell - 2) = 0$ holds, or r_i stays at the 2-nd robot node from v_t^m (i.e., $\ell = 2$), r_i moves backward.
    - Otherwise, if the neighboring backward node is a free node, r_i moves backward.

* If r_i decides to move backward, it moves only when the global forward direction does not change after the movement.

Note that, for the case of $\ell = 1$ in Case 2, if r_i stays at a solo node and both of its nearest backward and forward robot nodes are tower nodes v_t and $v_{t'}$, or if the ring is symmetric and exactly one solo node exists between r_i and $v_{t'}$, regardless of m_t, r_i moves backward to v_t to construct a configuration in Case 3. In addition, for the case of $V_g^{more} = \emptyset$ in Case 3, if no tower node in V_g^{more} exists, only robots nearest to v_{min} move to the node.

We have the following lemmas for each case.

Lemma 1. *From any configuration in Case 1, the system reaches a configuration in Case 2 in $O(n)$ rounds and the total number of $O(n)$ moves.*

Proof. Let R_{next} be the set of robots each of whose nearest forward or backward robot node is a min node. Then, since each robot in R_{next} moves to its nearest min node, at least one min robot keeps its view the lexicographically minimum. Hence, after at most n rounds, at least one robot in R_{next} reaches a min node and system reaches a configuration in Case 2. Since each link is passed by at most once during execution in Case 1, the lemma holds.   □

Lemma 2. *From any configuration in Case 2 or 3, the system reaches a configuration in C_g in $O(n)$ rounds and the total number of $O(gn)$ moves.*

By Lemma 1 and 2, we obtain the following theorem.

Theorem 5. *The proposed deterministic algorithm solves the g-partial gathering problem in $O(n)$ rounds and the total number of $O(gn)$ moves from any solvable configuration.*

Proof. We assume that the ring is not symmetric, and we can similarly show the proof for the case that the ring is symmetric. At first, we show that from any configuration in Case 2 the system reaches a configuration in Case 3 or C_g in $O(n)$ rounds and the total number of $O(gn)$ moves. We assume that v_t is a m_t-robot node and there exist p solo nodes $v_s^1, v_s^2, \ldots v_s^p$ in this order between v_t and the nearest forward tower node v_t'. If $p = 1$ holds, there exists only one robot r_i between v_t and v_t'. In this case, r_i moves backward to v_t, which clearly requires $O(n)$ rounds and the total number of $O(n)$ moves. When $(p > 1)$ and $(m_t < g)$, the robots at $v_s^1, v_s^2, \ldots v_s^{g-m_t}$ move backward to v_t. At the same time, robots at $v_s^{g-m_t+1}, v_s^{2g-m_t+1}, \ldots$ stay at the current nodes and the other robots moves backward to the nearest node among $v_s^{jg-m_t+1}$ $(j \geq 1)$. Robots execute such a behavior until there exists no solo robot node. Also, for the case of $m_t \geq g$, robots behave similarly to the above behavior. Thus, after at most n rounds, the system eventually reaches a configuration in Case 3 or C_g. Since each links is passed by $O(g)$ times, the total number of moves during execution of this case is $O(gn)$.

Next, we show that from any configuration in Case 3 the system eventually reaches a configuration in C_g in $O(n)$ rounds and the total number of $O(gn)$ moves. If there exists no robot node in V_g^{more}, the robots nearest to v_{min} move backward to v_{min}. Hence, the system eventually reaches a configuration such that there exists at least one robot node in V_g^{more}. We assume that v_t^m is a robot node in V_g^{more} and there exist p tower nodes $v_t^1, v_t^2, \ldots, v_t^p$ each containing less than g robots in this order between v_t^m and the nearest forward tower node $v_{t'}^m \in V_g^{more}$. Let m_j $(1 \leq j \leq p)$ be the number of robots at v_t^j. Then, from the algorithm robots staying at nodes $v_t^{g_1} (= v_t^1), v_t^{g_2}, v_t^{g_3}, \ldots$ $(g_1 < g_2 < g_3 < \cdots)$ such that for each $j > 0$ $(\sum_{i=g_j}^{g_{j+1}-2} m_i < g) \wedge (\sum_{i=g_j}^{g_{j+1}-1} m_i \geq g)$ hold, remain at the current nodes and the other robots move backward to the nearest node among $v_t^{g_j}$ (≥ 1). After finishing the movement, the system reaches a configuration in C_g. Note that, if robots at the same tower node move at different timings, the system may reach a configuration in Case 2. However, even in this case the destination node of all the robots that have stayed at the same node does not change. Thus, after at most n rounds, the system eventually reaches a configuration in C_g. Since each link is passed by $O(g)$ times, the total of agent moves during execution of this case is $O(gn)$. Therefore, lemma holds.   □

5   Probabilistically Self-stabilizing g-Partial Gathering

In this section, we propose a probabilistic algorithm to solve the problem in $O(n)$ rounds and the total number of $O(gn)$ moves in expectation starting from any initial configuration. Hence, the proposed algorithm is probabilistically self-stabilizing. The idea of the algorithm is as follows. From Sect. 4.2, the proposed deterministic algorithm achieves g-partial gathering from any solvable configuration. Hence, in this section we use randomization if the current configuration is not solvable deterministically, and in the other cases robots behave in the exact same way as in the previous section. Thus, from a deterministically unsolvable configuration, if robots can reach a deterministically solvable configuration in $O(n)$ rounds and the total number of $O(gn)$ moves in expectation, they can solve the problem in $O(n)$ rounds and the total number of $O(gn)$ moves in expectation from any initial configuration.

The behavior of each robot r_i in configuration C $(C \neq C_g)$ is described as follows:

- **Case 0**: C is deterministically unsolvable. Let R_{max} be the set of robots having V_{max}.
    - If $r_i \in R_{max}$, r_i moves backward with probability $1/|R_{max}|$.
    - Otherwise, r_i does not move.
- **Case 0'**: C is deterministically solvable (*i.e.*, C is either Case 1, 2, or 3 in Sect. 4.2). Robot r_i behaves in the exact same way as in Sect. 4.2.

We have the following theorem for the proposed algorithm.

Theorem 6. *The proposed probabilistic algorithm solves the g-partial gathering problem in $O(n)$ rounds and the total number of $O(gn)$ moves in expectation, starting from any initial configuration.*

Proof. If the current configuration is deterministically solvable, from the algorithm in Sect. 4.2 and Theorem 5, robots can solve the problem in $O(n)$ rounds and the total number of $O(gn)$ moves. Hence, in the following we consider deterministically unsolvable configurations. In this case, each robot in R_{max} moves backward with probability $1/|R_{max}|$. Then, unless any of the robots in R_{max} does not move or all of them move, the system reaches a deterministically solvable configuration. The probability is $1 - (1 - \frac{1}{|R_{max}|})^{|R_{max}|} - (\frac{1}{|R_{max}|})^{|R_{max}|} \geq 1 - (\frac{1}{|R_{max}|})^{|R_{max}|} \geq 3/4$. The first deformation comes from Bernoulli's inequality and the second deformation comes from the fact that $|R_{max}| \geq 2$. Hence, after constant rounds robots can reaches a deterministically solvable configuration. Thus, the theorem holds. □

6   Conclusion

In this paper, we considered g-partial gathering algorithms for mobile robots starting from multiplicity-allowed configurations in rings. First, we proved that no probabilistic (or deterministic) self-stabilizing g-partial gathering algorithm can exist in the ASYNC model, or if only local-strong and global-weak multiplicity detection is available. Second, in the SSYNC and global-strong multiplicity detection model, we considered deterministic and probabilistic algorithms. In the deterministic case, we first showed that unsolvable initial configurations exist, and then proposed an algorithm to solve the problem in $O(n)$ rounds and the total number of $O(gn)$ moves starting from any solvable initial configuration (*i.e.*, the algorithm is almost-stabilizing). In the probabilistic case, we proposed an algorithm to solve the problem in $O(n)$ rounds and the total number of $O(gn)$ moves in expectation from *any* initial configuration (*i.e.*, the algorithm is probabilistically self-stabilizing).

Future researches are as follows. First, we plan to extend the impossibility in the local-strong and global-weak multiplicity detection model for the case of $2 \leq g \leq 4$. We conjecture that global-strong multiplicity detection is also necessary to solve the problem in the total number of $O(gn)$ moves. Next, we plan to consider existence of almost-stabilizing algorithms in weaker models (*e.g..*, ASYNC model, local-strong and global-weak multiplicity detection model, etc.). Finally, we plan to characterize the problems that are actually solvable in a self-stabilizing setting using weaker models.

Acknowledgement. This work was partially supported by JSPS KAKENHI Grant Number 18K18031. This work was partially funded by the ANR project SAPPORO, ref. 2019-CE25-0005-1.

References

1. Balabonski, T., Delga, A., Rieg, L., Tixeuil, S., Urbain, X.: Synchronous gathering without multiplicity detection: a certified algorithm. Theory Comput. Syst. **63**(2), 200–218 (2019)
2. Cicerone, S., Di Stefano, G., Navarra, A.: Asynchronous robots on graphs: gathering. In: Distributed Computing by Mobile Entities, Current Research in Moving and Computing, vol. 11340, pp. 184–217. Springer (2019). https://doi.org/10.1007/978-3-030-11072-7_8
3. Clement, J., Défago, X., Potop-Butucaru, M.G., Izumi, T., Messika, S.: The cost of probabilistic agreement in oblivious robot networks. Inf. Process. Lett. **110**(11), 431–438 (2010)
4. Cohen, R., Peleg, D.: Convergence properties of the gravitational algorithm in asynchronous robot systems. SIAM J. Comput. **34**(6), 1516–1528 (2005)
5. Courtieu, P., Rieg, L., Tixeuil, S., Urbain, X.: Impossibility of gathering, a certification. Inf. Process. Lett. **115**(3), 447–452 (2015)
6. Dieudonné, Y., Petit, F.: Self-stabilizing gathering with strong multiplicity detection. Theoret. Comput. Sci. **428**, 47–57 (2012)
7. Flocchini, P., Prencipe, G., Santoro, N. (eds.): Distributed Computing by Mobile Entities, Current Research in Moving and Computing. Lecture Notes in Computer Science, vol. 11340. Springer (2019). https://doi.org/10.1007/978-3-030-11072-7_1
8. Izumi, T., Izumi, T., Kamei, S., Ooshita, F.: Feasibility of polynomial-time randomized gathering for oblivious mobile robots. IEEE Trans. Parallel Distrib. Syst. **24**(4), 716–723 (2013)
9. Izumi, T., Izumi, T., Kamei, S., Ooshita, F.: Time-optimal gathering algorithm of mobile robots with local weak multiplicity detection in rings. IEICE Trans. Fundam. Electr. Commun. Comput. Sci. **96**(6), 1072–1080 (2013)
10. Kamei, S., Lamani, A., Ooshita, F., Tixeuil, S.: Asynchronous mobile robot gathering from symmetric configurations without global multiplicity detection. In: Kosowski, A., Yamashita, M. (eds.) SIROCCO 2011. LNCS, vol. 6796, pp. 150–161. Springer, Heidelberg (2011). https://doi.org/10.1007/978-3-642-22212-2_14
11. Kamei, S., Lamani, A., Ooshita, F., Tixeuil, S.: Gathering an even number of robots in an odd ring without global multiplicity detection. In: Rovan, B., Sassone, V., Widmayer, P. (eds.) MFCS 2012. LNCS, vol. 7464, pp. 542–553. Springer, Heidelberg (2012). https://doi.org/10.1007/978-3-642-32589-2_48
12. Klasing, R., Kosowski, A., Navarra, A.: Taking advantage of symmetries: gathering of many asynchronous oblivious robots on a ring. Theoret. Comput. Sci. **411**(34–36), 3235–3246 (2010)
13. Kranakis, E., Krizanc, D., Markou, E.: The mobile agent rendezvous problem in the ring. Synth. Lect. Distrib. Comput. Theory **1**(1), 1–122 (2010)
14. Ooshita, F., Tixeuil, S.: On the self-stabilization of mobile oblivious robots in uniform rings. Theoret. Comput. Sci. **568**, 84–96 (2015)
15. Shibata, M., Kawai, S., Ooshita, F., Kakugawa, H., Masuzawa, T.: Partial gathering of mobile agents in asynchronous unidirectional rings. Theoret. Comput. Sci. **617**, 1–11 (2016)
16. Shibata, M., Kawata, N., Sudo, Y., Ooshita, F., Kakugawa, H., Masuzawa, T.: Move-optimal partial gathering of mobile agents without identifiers or global knowledge in asynchronous unidirectional rings. Theoret. Comput. Sci. **822**, 92–109 (2020)

17. Shibata, M., Nakamura, D., Ooshita, F., Kakugawa, H., Masuzawa, T.: Partial gathering of mobile agents in arbitrary networks. IEICE Trans. Inf. Syst. **102**(3), 444–453 (2019)
18. Shibata, M., Ooshita, F., Kakugawa, H., Masuzawa, T.: Move-optimal partial gathering of mobile agents in asynchronous trees. Theoret. Comput. Sci. **705**, 9–30 (2018)
19. Suzuki, I., Yamashita, M.: Distributed anonymous mobile robots: formation of geometric patterns. SIAM J. Comput. **28**(4), 1347–1363 (1999)

Efficient Dispersion of Mobile Agents without Global Knowledge

Takahiro Shintaku[1], Yuichi Sudo[1(✉)], Hirotsugu Kakugawa[2],
and Toshimitsu Masuzawa[1]

[1] Osaka University, Suita, Japan
{t-shintaku,y-sudou,masuzawa}@ist.osaka-u.ac.jp
[2] Ryukoku University, Kyoto, Japan
kakugawa@rins.ryukoku.ac.jp

Abstract. We consider the dispersion problem for mobile agents. Initially, k agents are located at arbitrary nodes in an undirected graph. Agents can migrate from node to node via an edge in the graph synchronously. Our goal is to let the k agents be located at different k nodes while minimizing the number of steps before dispersion is completed and the working memory space used by the agents. Kshemkalyani, Molla, and Sharma [ALGOSENSORS, 2019] present a fast and space-efficient dispersion algorithm with the assumption that each agent has global knowledge such as the number of edges and the maximum degree of a graph. In this paper, we present a dispersion algorithm that does not require such global knowledge but keeps the asymptotically same running time and slightly smaller memory space.

Keywords: Dispersion · Mobile agents · Autonomous robots

1 Introduction

We consider the dispersion problem of mobile robots, which we call mobile agents (or just *agents*) in this paper. At the beginning of an execution, k agents are arbitrarily placed in an undirected graph at the beginning of an execution. The goal of this problem is to let al.l agents be located at different nodes. This problem was originally formulated by Augustine and Moses Jr. [1] in 2018. The most interesting point of this problem is the uniqueness of the computation model. Unlike many problems of mobile agents on graphs, we cannot access the identifiers of the nodes and cannot use a local memory at each node, usually called whiteboard. In this setting, an agent cannot get/store any information from/on a node when it visits the node. Instead, k agents have unique identifiers and can communicate with each other when they visit the same node in a graph. The agents must solve a common task in a coordinated manner via direct communication with each other.

This work was supported by JSPS KAKENHI Grant Numbers 19H04085, 19K11826, and 20H04140 and JST SICORP Grant Number JPMJSC1606.

S. Devismes and N. Mittal (Eds.): SSS 2020, LNCS 12514, pp. 280–294, 2020.
https://doi.org/10.1007/978-3-030-64348-5_22

Table 1. Dispersion algorithm for arbitrary undirected graphs ($m' = \min(m, k\delta_{\max}/2,$ $\binom{k}{2})$). The information about Sync./Async. and/or knowledge is shown in the setting that we do not require termination.

	Memory space	Running time	Sync./Async	Knowledge
[1]	$O(k\log(\delta_{\max} + k))$	$O(m')$	Asynchronous	
[2]	$O(\log n)$ bits	$O(m'\ell)$ steps	Asynchronous	
[4]	$O(\ell\log(\delta_{\max} + k))$ bits	$O(m')$ steps	Asynchronous	
[4]	$O(d\log\delta_{\max})$ bits	$O(\delta_{\max}^d)$ steps	Asynchronous	
[4]	$O(\log(\delta_{\max} + k))$ bits	$O(m'\ell)$ steps	Asynchronous	
[5]	$O(\log n)$ bits	$O(m'\log\ell)$ steps	Synchronous	$m, k, \delta_{\max}$
This work	$O(\log(\delta_{\max} + k))$ bits	$O(m'\log\ell)$ steps	Synchronous	

Several algorithms were presented for the dispersion problem of mobile agents in the literature. Let $m' = \min(m, k\delta_{\max}/2, \binom{k}{2})$, where m and $\delta_{\max}$ are the number of edges and the maximum degree of a graph, respectively. The value m' is a upper bound on the number of edges between two nodes each with at least one robot. Thus m' appears in the time complexities of many algorithms. Augustine and Moses Jr. [1] gave an algorithm that achieves dispersion within $O(m')$ steps. In this algorithm, each agent uses $O(k\log(\delta_{\max} + k))$ bits of memory space. In the arXiv version [2], they also gave a dispersion algorithm with much smaller space at the cost of increased execution time, *i.e.*, $O(\log n)$ bits per agent and $O(m'\ell)$ steps, where n is the number of nodes in a graph and ℓ is the number of nodes at which at least one agent is located at the beginning of an execution. Kshemkalyani and Ali [4] presents three dispersion algorithms. Compared to the algorithm of [1], the first one requires the same execution time and slightly smaller memory space, *i.e.*, $O(\ell\log(\delta_{\max} + k))$ bits. The second one uses $O(d\log\delta_{\max})$ bits per agent and solves the dispersion problem within $O(\delta_{\max}^d)$ steps, where d is the diameter of a graph. The third one achieves dispersion with much smaller space: $O(\log(\delta_{\max} + k))$ bits per agent. However, it requires $O(m'\ell)$ time steps before dispersion is achieved. The current state of art algorithm was given by Kshemkalyani, Molla, and Sharma [5]. This algorithm is both time and space efficient. The running time is $O(m'\log\ell)$ steps and the memory space used by each agent is $O(\log n)$ bits.

The above algorithms are summarized in Table 1. We say that the agents are *synchronous* if they have a common global clock, *i.e.*, they can move simultaneously at each time step. If the agents may move at different pace, we say that they are *asynchronous*. Note that the information of Table 1 are shown for the case that we do not require the agents to terminate. If termination is required, all the above algorithms except for the fourth algorithm require the synchronousness of the agents and some global knowledge to decide whether a sufficient number of steps has passed or not. For example, the agents can terminate in $O(m')$ steps in an execution of the first algorithm only if the agents are synchronous and they know an asymptotically tight upper bound on m'. If we do not require termination, all of the first five algorithms correctly work in the asynchronous setting and without any global knowledge as shown in Table 1.

The sixth algorithm [5] essentially requires the synchronous setting and global knowledge, *i.e.*, m, k, and δ_{max}, even if termination is not required.[1] To the best of our understanding, the assumption for global knowledge can be reasonably relaxed: the algorithm correctly works if the agents know common upper bounds M, K, and Δ on m, k, and δ_{max}, respectively. However, time and space complexities may increase depending on how large those upper bounds are. Indeed, given those upper bounds, the algorithm achieves dispersion within $O(M' \log \ell)$ steps and uses $O(\log M')$ bits of memory space, where $M' = \min(M, K\Delta, K^2)$.

1.1   Our Contribution

The main contribution of this paper is removing the requirement of global knowledge of the algorithm given by [5]. Specifically, we give a dispersion algorithm whose running time is $O(m' \log \ell)$ steps and uses $O(\log(\delta_{max} + k))$ bits of the memory space of each agent. The proposed algorithm does not require any global knowledge such as m, k, and δ_{max}. In addition, the space complexity is slightly smaller than the algorithm given by [5], while the running times of both algorithms are asymptotically the same.

As with the aforementioned algorithms, the proposed algorithm works on an arbitrary simple, connected, and undirected graph. This algorithm solves the dispersion problem regardless of the initial locations of the agent in a graph. We require that the agents are synchronous, as in the algorithm given by [5].

1.2   Other Related Work

The dispersion problem has been studied not only for arbitrary undirected graphs, but for graphs of restricted topology. Augustine and Moses Jr. [1] addressed this problem for paths, rings, and trees. Kshemkalyani, Molla, and Sharma [5] studied the dispersion problem also in grid networks. Very recently, the same authors introduced *the global communication model* [6]. Unlike the above setting, all agents can always communicate with each other regardless of their current locations. They studied the dispersion problem for arbitrary graphs and trees under this communication model.

The exploration problem of mobile agents is closely related to the dispersion problem. This problem requires that each node (or each edge) of a graph is visited at least once by an agent. If the unique node-identifiers are available, a single agent can easily visit all nodes within $2m$ steps in a simple depth first search traversal. Panaite and Pelc [7] gave a faster algorithm, whose cover time is $m + 3n$ steps. Their algorithm uses $O(m \log n)$ bits in the agent-memory, while it does not use whiteboards, *i.e.*, local memories of the nodes. Sudo, Baba,

---

[1] This algorithm also uses the knowledge of n in [5], but it can keep the same time and space complexities without knowing n.

Nakamura, Ooshita, Kakugawa, and Masuzawa [9] gave another implementation of this algorithm: they removed the assumption of the unique identifiers and reduced the space complexity on the agent-memory from $O(m \log n)$ bits to $O(n)$ bits by using $O(n)$ bits in each whiteboard. The algorithm given by Priezzhev, Dhar, Dhar, and Krishnamurthy [8], which is well known as the *rotor-router*, also solves the exploration problem efficiently. The agent uses $O(\log \delta_v)$ bits in the whiteboard of each node $v \in V$ and the agent itself is oblivious, *i.e.*, it does not use its memory space at all. The rotor-router algorithm is self-stabilizing, *i.e.*, it guarantees that starting from any (possibly corrupted) configuration, the agent visits all nodes within $O(mD)$ steps [10].

1.3   Organization

In Sect. 2, we define the model of computation and the problem specification. In Sect. 3, we briefly explain the existing techniques used for the dispersion problem in the literature. This section may help the readers to clarify what difficulties we addresses and the novelty of the techniques that we introduce to design the proposed protocol in this paper. In Sect. 4, we present the proposed protocol. In Sect. 5, we conclude this paper with short discussion for an open problem.

2   Preliminaries

Let $G = (V, E)$ be any simple, undirected, and connected graph. Define $n = |V|$ and $m = |E|$. We define the degree of a node v as $\delta_v = |\{u \in V \mid (u, v) \in E\}|$. Define $\delta_{\max} = \max_{v \in V} \delta_v$, *i.e.*, $\delta_{\max}$ is the maximum degree of G. The nodes are anonymous, *i.e.*, they do not have unique identifiers. However, the edges incident to a node v are locally labeled at v so that a robot located at v can distinguish those edges. Specifically, those edges have distinct labels $0, 1, \ldots, \delta_v - 1$ at node v. We call these local labels *port numbers*. We denote the port number assigned at v for edge $\{v, u\}$ by $p_v(u)$. Each edge $\{v, u\}$ has two endpoints, thus has labels $p_u(v)$ and $p_v(u)$. Note that these two labels are not necessarily the same, *i.e.*, $p_u(v) \neq p_v(u)$ may hold. We say that an agent moves *via port* p from a node v when the agent moves from v to the node u such that $p_v(u) = p$.

We consider that k agents exist in graph G, where $k \leq n$. The set of all agents is denoted by R. Each agent is always located at some node in G, *i.e.*, the move of an agent is *atomic* and an agent is never located at an edge at any time step (or just *step*). The agents have unique identifiers, *i.e.*, each agent a has a positive identifier $a.\mathrm{ID}$ such that $a.\mathrm{ID} \neq b.\mathrm{ID}$ for any $b \in R \setminus \{a\}$. The agents know a common upper bound $id_{\max} = O(poly(k))$ such that $id_{\max} \geq \max_{a \in R} a.\mathrm{ID}$, thus the agents can store the identifier of any agent on $O(\log k)$ space.[2] Each agent has

---

[2] Strictly speaking, this means that the agent knows a common integer $id_{\max} = O(k^c)$, where c is some constant. We do not assume that the agents know the degree c. Thus, the knowledge of $id_{\max}$ does not necessarily give the agents an upper bound on k. Therefore, this setting does not contradict the main claim of this paper: the proposed algorithm does not require any global knowledge.

a read-only variable $a.\text{p\_in} \in \{-1, 0, 1, \ldots, \delta_v - 1\}$. At time step 0, $a.\text{p\_in} = -1$ holds. For any $t \geq 1$, if a moves from u to v at step $t - 1$, $a.\text{p\_in}$ is set to $p_v(u)$ at the beginning of step t. If a does not move at step $t - 1$, $a.\text{p\_in}$ is set to -1.

The agents are synchronous. All k agents are given a common algorithm $\mathcal{A}$. Let $R(v, t) \subseteq R$ be the set of the agents located at a node v. We define $\ell = |\{v \in V \mid R(v, 0) \geq 1\}|$, i.e., ℓ is the number of nodes with at least one agent in step 0. At each step $t \geq 0$, the agents in $R(v, t)$ first communicate with each other and agrees on how each agent $a \in R(v, t)$ updates the variables in its memory space in step t, including a variable $a.\text{p\_out} \in \{-1, 0, 1, \ldots, \delta_v - 1\}$, according to algorithm $\mathcal{A}$. The agents next update the variables according to the agreement. Finally, each agent $a \in R(v, t)$ with $a.\text{p\_out} \neq -1$ moves via port $a.\text{p\_out}$. If $a.\text{p\_out} = -1$, agent a does not move and stays in v in step t.

A node does not have any local memory accessible by the agents. Thus, the agents have to coordinate only by communicating with each other. No agents are given a priori any global knowledge such as m, $\delta_{\max}$, and k.

The values of all variables in agent a constitute the state of a. Let $\mathcal{M}_{\mathcal{A}}$ be the set of all possible agent-states for algorithm $\mathcal{A}$. ($\mathcal{M}$ may be an infinite set.) Algorithm must specify the initial state s_{init}, which is common to all agents in R. A global state of the network or a *configuration* is defined as a function $C : R \to (\mathcal{M}, V)$ that specifies the state and the location of each agent $a \in R$. In this paper, we consider only deterministic algorithms. Thus, if the network is in a configuration C at a step, a configuration C' in the next step is uniquely determined. We denote this configuration C' by $next_{\mathcal{A}}(C)$. The execution $\Xi_{\mathcal{A}}(C_0)$ of $\mathcal{A}$ starting from a configuration C_0 is defined as an infinite sequence $C_0, C_1, \ldots$ of configurations such that $C_{t+1} = next_{\mathcal{A}}(C_t)$ for all $t = 0, 1, \ldots$.

Definition 1 (Dispersion Problem). A configuration C of an algorithm $\mathcal{A}$ is called *legitimate* if (i) all agents in R are located in different nodes in C, and (ii) no agent changes its location in execution $\Xi_{\mathcal{A}}(C)$. We say that $\mathcal{A}$ solves the dispersion problem if execution $\Xi_{\mathcal{A}}(C_0)$ reaches a legitimate configuration for any configuration C_0 where all agents are in state s_{init}.

We evaluate the *running time* of algorithm $\mathcal{A}$ as the maximum number of steps until $\Xi_{\mathcal{A}}(C_0)$ reaches a legitimate configuration, where the maximum is taken over all configurations C_0 in which all agents are in state s_{init}.

We define $m' = \min(m, k\delta_{\max}/2, \binom{k}{2})$, which is frequently used in this paper. We simply write $R(v)$ for $R(v, t)$ when time step t is clear from the context.

3   Existing Techniques

3.1   Simple DFS

If we assume that all k agents are initially located at the same node $v_{\text{st}} \in V$, the dispersion problem can be solved by a simple depth-first search (DFS). The pseudocode of the simple DFS is shown in Algorithm 1. Each agent $a \in R$ maintains a variable $\text{mode} \in \{settled, unsettled\}$. We say that agent a is *settled*

Algorithm 1: Simple DFS

```
   /* the actions of all agents in R(v) for any v ∈ V                    */
1  a.p_out ← −1 for all a ∈ R(v)                          // Initialize p_out.
2  if R(v) ≥ 2 then
3  |   if there is no settled agent in R(v) then
4  |   |   The agent with the smallest identifier in R(v) becomes settled.
5  |   |   a_v.last ← a_v.p_in + 1 mod δ_v
6  |   |   a.p_out ← a_v.last for all a ∈ R(v) \ {a_v}
7  |   else
8  |   |   Let p be the common a.p_in for a ∈ R(v) \ {a_v}
9  |   |   if a.last ≠ p then
10 |   |   |   a.p_out ← p for all a ∈ R(v) \ {a_v}                    // Backtrack
11 |   |   else
12 |   |   |   a_v.last ← a_v.last + 1 mod δ_v
13 |   |   |   a.p_out ← a_v.last for all a ∈ R(v) \ {a_v}

14 Every agent a ∈ R(v) such that a.p_out ≠ −1 moves via a.p_out
```

---

if $a.\mathtt{mode} = settled$, and *unsettled* otherwise. All agents are initially unsettled. Once an agent becomes settled at a node, it never becomes unsettled nor moves to another node. We say that a node v is *settled* if an settled agent is located at v; otherwise, v is unsettled. We denote by a_v the settled agent located at v when v is settled. (No two agents become settled at the same node.) In this model, no node provides a local memory accessible by agents. However, when a node v is settled, unsettled agents can use the memory of a_v like the local memory of v since those agents can communicate with a_v at v.

Unsettled agents always move together in a DFS fashion. Each time they find an unsettled node v, the agent with the smallest identifier becomes settled at v (Line 4). An settled agent a_v maintains a variable $a_v.\mathtt{last} \in \{0, 1, \ldots, \delta_v - 1\}$ to remember which port was used for the last time by unsettled agents to leave v. At step 0, all agents are unsettled and located at node v_{st}. First, after one agent becomes settled at v_{st}, the other $k - 1$ agents move via port 0. Thereafter, the unsettled agents basically migrates between nodes by the following simple rule: each time they move from u to v, it moves via port $p_v(u) + 1 \mod \delta_v$ (Lines 5, 6, 12, and 13). Only exception is the case that node v has already been settled when they move from u to v and $p_v(u) \neq a_v.\mathtt{last}$. At this time, the unsettled agents immediately backtracks from v to u and this backtracking does not update $a_v.\mathtt{last}$ (Line 10).

If $k > n$, this well-known DFS traversal guarantees that the unsettled agents visit all nodes within $4m$ steps, during which the agents move through each edge at most four times. Since we assume $k \leq n$, the unsettled agents move through at most $m' = \min(m, k\delta_{\max}/2, \binom{k}{2})$ different edges. Thus, all agents become settled[5] within $4m' = O(m')$ steps, at which point the dispersion is achieved.[3] The space complexity is $O(\log \delta_{\max})$ bits per agent because each agent maintains only one non-constant variable $a.\texttt{last}$ in its working memory.

3.2   Parallel DFS

Kshemkalyani and Ali [4] generalized the above simple DFS to handle the case that the agents may be initially located at multiple nodes, using $O(\ell \log(\delta + k))$ bits per agent. (Remember that ℓ is the number of nodes at which one or more agents are located in step 0.) That is, ℓ groups of unsettled agents perform DFS in parallel. Specifically, in step 0, the agents located at each node v compute $\max\{a.\texttt{ID} \mid a \in A(v, 0)\}$ and store it in a variable $\texttt{group}$. Thereafter, the agents use the values of $\texttt{group}$ as their *group identifier*. Settled agents can distinguish each group by group identifiers, thus they can maintain ℓ slots of individual memory space such that unsettled agents of each group can dominantly access one slot of the space. Since simple DFS requires $O(\log(\delta_{\max}))$ bits, This implementation of parallel DFS requires $O(\ell \log \delta_{\max} + \ell \log k)) = O(\ell \log(\delta_{\max} + k))$ bits. The running time is still $O(m')$ steps.

3.3   Zombie Algorithm

We can solve the dispersion with memory space of $O(\log(\delta_{\max} + k))$ bits per agent at the cost of increasing the running time to $O(m'k)$, regardless the initial locations of k agents. In step 0, all agents compute its group identifier in the same way as in Sect. 3.2. However, due to the memory constraints, each settled agent cannot maintain one slot of memory space for each of ℓ groups. Instead, it provides only one group with memory space of $O(\log \delta_{\max})$ bits. Each settled agent memorizes the largest group identifier it observes: each time unsettled agent visits a node v, $a_v.\texttt{group}$ is updated to $\max\{a.\texttt{group} \mid a \in R(v)\}$. Thus, at least one group of agents can perform its DFS by using the memory space of settled agents exclusively. If an unsettle agent a visits a node v such that $a.\texttt{group} < a_v.\texttt{group}$, a becomes a *zombie*, which always chase the agent whose identifier is equal to $a_v.\texttt{group}$, which we call the *leader* of the group. Specifically, a zombie z chases a leader by moving via port $a_u.\texttt{last}$ each time z visits any node u. When z catch up with the leader, z always follow the leader thereafter until z becomes settled, which occurs when z reaches an unsettled node v and z has the smallest identifier among $A(v)$. Every agent moves at most $4m'$ times until it observes larger group identifier than any identifier it has observed so far. Thus, the running time of this algorithm is $O(m'\ell)$ steps.

We call this algorithm the *zombie algorithm*. The zombie algorithm is almost the same as the $O(\log n)$ space algorithm of [2] and the tree-switching algorithm of [4], while the zombie algorithm is faster only by a constant factor than those algorithms.

---

[3] Strictly speaking, according to Algorithm 1, the last one agent never becomes settled even if it visits an unsettled node. However, this does not matter because thereafter the last agent never moves nor change its state.

3.4   Dispersion with $O(m' \log \ell)$ Steps and $O(\log n)$ Bits per Agent

Kshemkalyani, Molla, and Sharma [5] gave the current state of art algorithm that achieves dispersion on arbitrary graphs regardless of the initial locations of the k agents. Their algorithm requires that each agent knows *a priori* upper bounds M, K, and Δ on m, k, and $\delta_{\max}$, respectively. Then, the running time is $O(\min(M, K\Delta, K^2) \log \ell)$ steps and the required memory space per agent is $O(\log M)$ bits. In particular, if those upper bounds are asymptotically tight, i.e., $N = O(n)$, $M = O(m)$, $K = O(k)$, and $\Delta = O(\delta_{\max})$, those complexities are $O(m' \log \ell)$ steps and $O(\log n)$ bits, respectively. In this section, we briefly explain the key idea of their algorithm to clarify how this algorithm requires global knowledge M, K, and Δ. The implementation in the following explanation is slightly different from the original implementation in [5], but the difference is not essential.

Let T be a sufficiently large $O(\min(M, K\Delta, K^2))$ value. Each unsettled agent maintains a timer variable and counts how many steps have passed since an execution began modulo T. Agents switches from stage 1 to stage 2 and from stage 2 to stage 1 in every T steps. At the switch from stage 1 to stage 2 and from stage 2 to stage 1, all settled agents reset their group identifiers to -1, which is smaller than any group identifier of unsettled agents.

In stage 1, all unsettled agents at each node v first compare their identifiers and adopts the largest one as their group identifier. Then, unsettled agents perform DFS in parallel like the zombie algorithm. The difference arises when an unsettled agent finds a settled agent with a larger group identifier. Then, instead of becoming a zombie, it stops until the end of stage 1. Moreover, unlike the zombie algorithm, an unsettled agent located at a node v becomes settled even when $|R(v)| = 1$.

At the beginning of stage 2, one or more unsettled agents may be located at settled nodes. Let G' be the subgraph induced by all settled nodes in the beginning of stage 2 and $G'(v)$ the component that includes a node v in G'. The goal of stage 2 is to collect all agents in $G'(v)$ and locate them to the same node in $G'(v)$ for each $v \in V$. Specifically, each unsettled agents located at any node u tries to perform DFS in $G'(u)$ twice. Since each settled agent b has only $O(\log M)$ bits space, the memory space of b can be used only by unsettled agents in the group with the largest identifier that b has observed in stage 2. An settled agents with smaller group identifier stops when it observes a larger group identifier. Then, all unsettled agents in the group with the largest group identifier in $G'(u)$ can perform DFS and visit all nodes in $G(u')$ twice within T steps. During the period, they pick up all stopped and unsettled agents in the component and goes back to the node that they are located at in the beginning of stage 2.

Hence, each iteration of stages 1 and 2 decreases the number of nodes with at least one unsettled agent at least by half. This yields that all agents become settled and the dispersion is achieved within $O(T \log \ell) = O(\min(M, K\Delta, K^2) \log \ell)$ steps.

The knowledge of the global knowledge M, K, and Δ is inherent to the key idea of this algorithm. Without those upper bounds, the agents may switch between two stages before all agents complete a stage so that the correctness is no longer guaranteed or the agents may stay in one stage too long so that the running time is much larger than, for example, $\Omega(mn)$. To the best of our knowledge, no simple modification removes the requirement of the knowledge keeping the same running time asymptotically.

4   Proposed Algorithm

In this section, we give an algorithm $\mathcal{A}_{svl}$ that solves a dispersion within $O(m' \log \ell)$ steps and uses $O(\log(k + \delta_{\max}))$ bits of memory space per agent. Algorithm $\mathcal{A}_{svl}$ requires no global knowledge.

4.1   Overview

Algorithm $\mathcal{A}_{svl}$ is based on the zombie algorithm explained in Sect. 3.3 but it has more sophisticated mechanism to achieve dispersion within $O(m' \log \ell)$ steps. An agent a maintains its *mode* on a variable $a.\mathtt{mode} \in \{L, Z, S\}$. We say that an agent a is a leader (resp. a zombie, a settled agent) when $a.\mathtt{mode} = L$ (resp. $a.\mathtt{mode} = Z$, $a.\mathtt{mode} = S$). A leader may become a zombie, and a zombie eventually becomes a settled agent. However, a zombie never becomes a leader again, and a settled agent never changes its mode. An agent a also maintains its *level* on a variable $a.\mathtt{lv}$, while a stores the identifier of some leader on a variable $a.\mathtt{leader}$. As long as an agent l is a leader, $l.\mathtt{ID} = l.\mathtt{leader}$ holds. These two variables, $\mathtt{lv}$ and $\mathtt{leader}$, determine the strength of agent a. We say that an agent a is stronger than an agent b if $a.\mathtt{lv} > b.\mathtt{lv}$ or $a.\mathtt{lv} = b.\mathtt{lv} \wedge a.\mathtt{leader} > b.\mathtt{leader}$ holds.

All agents are leaders at the beginning of an execution of $\mathcal{A}_{svl}$. Each time two or more leaders visit the same node, the strongest leader *kills* all other leaders, and the killed leaders become zombies. Zombies always follow the strongest leader that it has ever observed. A leader always tries to perform a DFS. Each time a leader visits an unsettled node v, the leader picks an arbitrary one zombie $z \in R(v)$ and makes z settled. If there is no zombie at v, the leader suspends its DFS until zombies visits v. As in Sect. 3, we denote the settled agent located at a node v by a_v if it exists. We say that a settled agent s is a *minion* of a leader l when $s.\mathtt{lv} = l.\mathtt{lv}$ and $s.\mathtt{leader} = l.\mathtt{ID}$ hold. As in the zombie algorithm, a leader l performs a DFS by using the memory space of its minions, in particular, by using the values of their $\mathtt{lasts}$. When a leader l visits a settled node v such that a_v is weaker than l, a_v becomes a minion of l by executing $(a_v.\mathtt{lv}, a_v.\mathtt{leader}) \leftarrow (l.\mathtt{lv}, l.\mathtt{ID})$. Conversely, the leader l becomes a zombie if a_v is stronger than l. At this time, the stronger leader may not be located at node v. Then, this new zombie begins to chase the stronger leader by moving via $a_v.\mathtt{last}$ repeatedly.

Table 2. Variables of $\mathcal{A}_{svl}$

Variables	Description	Init. value
$a.\texttt{mode} \in \{L, Z, S\}$:	the mode of an agent a	L
$a.\texttt{slot} \in \{0, 1, 2, 3\}$:	the current timeslot	0
$a.\texttt{lv} \in \mathbb{N}$:	the level of an agent a	0
$a.\texttt{leader} \in \mathbb{N}$:	the identifier of the strongest leader that a observed	$a.\texttt{ID}$
$a.\texttt{last} \in \mathbb{N}$:	the pointer to the strongest leader that a observed	0
$a.\texttt{port} \in \mathbb{N} \cup \{-1\}$:	the last non-negative value of $\texttt{p\_in}$	-1

Each leader maintains its level like the famous minimum spanning tree algorithm given by Gallager, Humblet, and Spira [3]. Initially, the level of every agent is zero. A leader increases its level by one each time it kills another leader with the same level or meets a zombie with the same level. We have the following lemma.

Lemma 1. *No agent reaches a level larger than* $\log_2 \ell + 1$.

Proof. Let ℓ_2 be the nodes at which two or more leaders are located in the initial configuration, *i.e.*, $\ell_2 = |\{v \in V \mid |R(v, 0)| \geq 2\}|$. In step 0, (i) ℓ_2 leaders changes their levels from 0 to 1, (ii) the levels of $\ell - \ell_2$ leaders remain 0, and (iii) other $k - \ell$ leaders become zombies with level 0. Thus, at most $\ell_2 + (\ell - \ell_2) = \ell$ leaders reaches level 1. Since a zombie never increases its level until it becomes a settled agent, for any $i \geq 1$, at most $\ell/2^{i-1}$ leaders reaches level i. Thus, no agent reaches a level larger than $\log_2 \ell + 1$. $\qquad\square$

By definition of minions, a leader l loses all its minions when l increases its level by one. As we will see in Sect. 4.2, when a leader visits a node v, the leader considers that it has already visited a node v in its DFS if and only if a_v is a minion at that time. This means that a leader l restarts a new DFS each time it increases its level. In every DFS, a leader l moves at most $4m'$ times. However, this does not necessarily mean that l completes its DFS or increases its level within $4m'$ steps because l suspends its DFS while no other agent is located at the same node. Thus, we require a mechanism to bound the running time by $O(m' \log \ell)$ steps. As we will see in Sects. 4.2 and 4.3, we achieve this by differentiating the moving speed of agents according to various conditions.

4.2 Detailed Description

The list of variables and the pseudocode of $\mathcal{A}_{svl}$ are given in Table 2 and Algorithm 2, respectively.

First, we introduce some terminologies and notations. Let t be any time step. We denote the set of leaders, the set of zombies, and the set of settled agents at step t by $R_L(t)$, $R_Z(t)$, and $R_S(t)$, respectively. In addition, we define $R_L(v, t) = R(v, t) \cap R_L(t)$, $R_Z(v, t) = R(v, t) \cap R_Z(t)$, and $R_S(v, t) = R(v, t) \cap R_S(t)$ for any

---

Algorithm 2: $\mathcal{A}_{svl}$

/* the actions of all agents in $R(v)$ for any $v \in V$ */
1 Let $a_{\max}$ be the unique leader in $R_{\max}(v)$ if it exists. Otherwise, let $a_{\max} = a_v$.
2 $a.\text{p\_out} \leftarrow -1$ for all $a \in R(v)$ // Initialize p_out.
3 $a.\text{mode} \leftarrow Z$ for all $a \in R_L(v) \setminus \{a_{\max}\}$ // $a_{\max}$ kills the other leaders.

4 $a_{\max}.\text{port} \leftarrow \begin{cases} a_{\max}.\text{p\_in} & a_{\max}.\text{mode} = L \wedge a_{\max}.\text{p\_in} \neq -1 \\ a_{\max}.\text{port} & \text{otherwise} \end{cases}$

5 **if** $|R(v)| \geq 2$ **then**
6   **if** $a_{\max}.\text{mode} = L$ **then**
7          $a_{\max}.\text{lv} \leftarrow \begin{cases} a_{\max}.\text{lv} + 1 & \exists z \in R_Z(v) : z.\text{lv} = a_{\max}.\text{lv} \\ a_{\max}.\text{lv} & \text{otherwise} \end{cases}$
8   **if** $R_S(v) = \emptyset$ **then**
9   Choose any zombie in $R(v) \setminus \{a_{\max}\}$ and makes it settled.
10             $a_v.\text{lv} \leftarrow a_{\max}.\text{lv}$; $a_v.\text{leader} \leftarrow a_{\max}.\text{leader}$; $a_v.\text{last} \leftarrow a_{\max}.\text{port}$
11   **else if** a_v *is not a minion of* $a_{\max}$ **then**
12             $a_v.\text{lv} \leftarrow a_{\max}.\text{lv}$; $a_v.\text{leader} \leftarrow a_{\max}.\text{leader}$; $a_v.\text{last} \leftarrow a_{\max}.\text{port}$
13   **else if** a_v *is a minion of* $a_{\max}$ **and** $a_{\max}.\text{port} \neq a_v.\text{last}$ **then**
14             $a.\text{p\_out} \leftarrow a_{\max}.\text{port}$ for all $a \in R(v) \setminus \{a_v\}$ // Backtrack
15   **else if** $a_{\max}.\text{slot} = 0$ **then**
16             $a.\text{p\_out} \leftarrow a_{\max}.\text{port} + 1 \bmod \delta_v$ for all $a \in R(v) \setminus \{a_v\}$
17             $a_v.\text{last} \leftarrow a_{\max}.\text{p\_out}$
18   **else if** $\left(\begin{array}{l} \max\limits_{z \in R_Z(v)} z.\text{lv} < a_v.\text{lv} \wedge a_v.\text{slot} \in \{2, 3\} \\ \vee \max\limits_{z \in R_Z(v)} z.\text{lv} = a_v.\text{lv} \wedge a_v.\text{slot} = 2 \end{array} \right)$ **then**
19         $a.\text{p\_out} \leftarrow a_v.\text{last}$ for all $a \in R(v) \setminus \{a_v\}$

20 $a.\text{slot} \leftarrow a.\text{slot} + 1 \bmod 4$ for all $a \in R(v)$
21 Every agent $a \in R(v)$ such that $a.\text{p\_out} \neq -1$ moves via $a.\text{p\_out}$

---

node $v \in V$. Define $R_{\max}(v, t)$ as the set of strongest agents in $R(v, t)$. We omit time step t from those notation, *e.g.*, simply write $R_L(v)$ for $R_L(v, t)$, when time step t is clear from the context. A leader l located at a node v is called an *active leader* when $|R(v)| \geq 2$. When $|R(v)| = 1$, l is called a *waiting* leader. A zombie z located at a node v is called a *strong zombie* if $z.\text{lv} = a_v.\text{lv}$. Otherwise, z is called a *weak zombie*. In the pseudocode, we use notation $a_{\max}$ for simplicity. We define $a_{\max}$ as follows[6]: If there is a leader in $R_{\max}(v)$, $a_{\max}$ denotes the (unique) leader in $R_{\max}(v)$; Otherwise, let $a_{\max} = a_v$.[4]

---

[4] This definition assumes that there is at most one leader in $R_{\max}(v)$ and there is no zombie in $R_{\max}(v)$ at any time step. The former proposition holds because for any two leaders l_1 and l_2, l_1 is stronger than l_2, or l_2 is stronger than l_1. The latter holds because (i) a leader becomes a zombie only if it observes a stronger agent, (ii) a zombie never becomes stronger, *i.e.*, never change its lv or leader, unless it becomes settled, and (iii) a zombie moves by chasing a stronger leader or moves together with a stronger leader.

As mentioned in Sect. 4.1, we differentiate the moving speed of agents. We implement the differentiation with a variable slot $\in \{0, 1, 2, 3\}$. Each agent a counts how many steps have passed since an execution of $\mathcal{A}_{svl}$ began modulo 4 and stores it on a.slot (Line 20). Thus, all agents always have the same value for this variable. This variable represents timeslots indicating which kind of agents are allowed to move at each step. For example, an active leader is allowed to move at slots 0 and 1, while a strong zombie is allowed to move only at slot 2.

The pseudocode (Algorithm 2) specifies how the agents located at a node v updates their variables including port p_out, via which they move to the next destination. A leader is immediately killed, *i.e.*, becomes a zombie, if it meets a stronger leader or a stronger settled agent (Line 3). A surviving leader increases its level by one each time it meets a (weaker) leader or zombie with the same level (Line 7). A leader l performs a DFS using a variable last in its minions. It basically moves in timeslot 0, while backtracking occurs in timeslot 1. Since the information of incoming port in l.p_in gets lost in timeslots 2 and 3, the leader l remembers p_in $= p_v(u)$ in a variable l.port each time l moves from u to v (Line 4). Specifically, a leader l performs a DFS as follows until it is killed by a stronger leader or a stronger settled agent:

- When l moves from u to v such that a_v is not a minion of l, the leader l considers that it visits v for the first time in the current DFS. Then, l changes a_v to a minion of l (Line 12) and waits for the next timeslot 0.
- When l moves from u to v such that v is an unsettled node, l considers that it visits v for the first time in the current DFS. If there is at least one zombie at v, l changes arbitrary one zombie to a settled agent and make it a minion of l (Lines 9 and 10). Otherwise, it suspends the current DFS until a zombie or a (weaker) leader a visits v (the condition in Line 5 implements this suspension). Then, l does the same thing for a, *i.e.*, make a settled and a minion of l. Thereafter, l waits for timeslot 0.
- When l moves from u to v such that a_v is a minion and a_v.last $= l$.port, the leader l considers that it has just backtracked from u to v. Then, it just waits for the next time slot 0.
- In timeslot 0, l leaves the current node v via port l.port $+ 1 = p_v(u) + 1$ after storing this port on a_v.last (Lines 16 and 17).
- When l moves from u to v such that a_v is a minion and a_v.last $\neq l$.port, the leader l considers that it has already visited v before in the current DFS. Thus, it immediately backtracks from v to u without updating a_v.last (Line 15). Since a leader makes a non-backtracking move only in timeslot 0, this backtracking occurs only in timeslot 1.

In the first and second case, the leader l simultaneously substitute l.port for a_v.last to avoid triggering backtracking mistakenly. Thus, each leader can perform a DFS in the same way as a simple DFS explained in Sect. 3.1 until it is killed and becomes a zombi.e.

A zombie always tries to follow a leader. If a zombie z is located at a node v without a leader, it moves via port a_v.last in timeslot 2 if z is a strong zombie or some strong strong zombie is located at the same node v (Line 19). Otherwise,

z moves via port a_v.last in timeslots 2 and 3 (Line 19). If z is located at a node with a leader, it just moves with the leader (Lines 14 and 16) or becomes settled (Line 9).

4.3   Correctness and Complexities

In this section, we prove that an execution $\Xi_{\mathcal{A}_{svl}}(C)$ of $\mathcal{A}_{svl}$ starting from any configuration C where all agents are in the initial state achieves dispersion within $O(m' \log \ell)$ steps, and each agent uses $O(\log(k + \delta_{\max}))$ bits of memory space.

First, we gave an upper bound on the space complexity.

Lemma 2. *Each agent uses at most $O(\log(k + \delta_{\max}))$ bits of its memory space in an execution of $\mathcal{A}_{svl}$.*

Proof. Each agent a maintains six non-constant variables: a.lv, a.leader, a.last, a.port, a.p_in, and a.p_out. The first variable a.lv uses only $O(\log \log \ell)$ bits by Lemma 1. Each of the other five variables just stores the identifier of some agent or the port numbers of some node. Thus, they uses only $O(\log(k + \delta_{\max}))$ bits. □

In an execution of $\mathcal{A}_{svl}$, agents located at a node v leaves v only when there are two or more agents at v. Therefore, we have the following lemma.

Lemma 3. *In an execution of $\mathcal{A}_{svl}$, once all agents are located at different nodes, no agent leaves the current location thereafter.*

Thus, it suffices to show that all agents will be located at different nodes within $O(m' \log \ell)$ steps in an execution of $\mathcal{A}_{svl}$. To prove this, we introduce some terminologies and notations. Let t be any time step and let a be any agent in $R(v,t)$ for any v. Then, define the *virtual level* of agent a at step t as $VL(a,t) = \max_{b \in R(v)} b$.lv. Define $\mathcal{L}_{\min}(t)$ be the minimum virtual level of all active leaders and all zombies in R at step t. For simplicity, we define $\mathcal{L}_{\min}(t) = \infty$ if there is no active leader and no zombie in R at step t. Again, we simply write $VL(a)$ and $\mathcal{L}_{\min}$ for $VL(a,t)$ and $\mathcal{L}_{\min}(t)$, respectively, when time step t is clear from the context. Note that for any agent a, the virtual level and the level of a differ if and only if a is a weak zombi.e.

By definition of $\mathcal{A}_{svl}$, we have the following lemma.

Lemma 4. *In an execution of $\mathcal{A}_{svl}$, for any agent a, the virtual level of a never decreases.*

Lemma 5. *In an execution of $\mathcal{A}_{svl}$, from any time step, each active leader l increases its level at least by one or becomes a waiting leader, a zombie, or a settled agent within $O(m')$ steps.*

Proof. The active leader l performs a DFS correctly unless it meets a leader/zombie with the same level or finds a stronger agent than l. Thus, it becomes a waiting leader if such event does not occur for sufficiently large $O(m')$ steps, Otherwise, it increases its level by one or becomes a zombie or a settled agent within $O(m')$ steps. □

Lemma 6. *In an execution of $\mathcal{A}_{svl}$, from any time step, each weak zombie z increases its virtual level or catches up a leader or a strong zombie within $O(k)$ steps.*

Proof. Suppose that now z is located at a node v and has a virtual level i. By definition of $\mathcal{A}_{svl}$, we must have a path $w = v_1, v_2, \ldots, v_s$ such that $s \leq k$, $v = v_1$, $p_{v_i}(v_{i+1}) = a_{v_i}.\texttt{last}$ for $i = 1, 2, \ldots, s-1$, and a leader with at least level i is located at v_s. Let j be the smallest integer such that a leader, a strong zombie, or an agent with virtual level $i' \geq i+1$ is located at v_j. Then, sub-path w' of w is defined as $w' = v_1, v_2, \ldots, v_j$. This sub-path w' changes as time passes. Zombie z moves forward in w' two times in every four steps because it moves in timeslots 2 and 3. However, a leader moves only once or moves and backtracks in every four steps, while a strong zombie moves only once in every four steps. Therefore, by Lemma 4, the length of the w' decreases at least by one in every four steps, from which the lemma follows. $\qquad\square$

Lemma 7. *In an execution of $\Xi = \Xi_{\mathcal{A}_{svl}}(C_0) = C_0, C_1, \ldots$, from any time step such that $\mathcal{L}_{\min} < \infty$, $\mathcal{L}_{\min}$ increases at least by one in $O(m')$ steps.*

Proof. Let C_t be any configuration where $\mathcal{L}_{\min} = i$ holds. It suffices to show that the suffix Ξ' of Ξ after C_t, i.e., $\Xi' = C_t, C_{t+1}, \ldots$ reaches a configuration where $\mathcal{L}_{\min} \geq i+1$ holds within $O(m')$ steps. Let R_γ be the set of all weak zombies with virtual level i that are not located at a node with a leader or a strong zombi.e. No agent in $R \setminus R_\gamma$ becomes an agent in R_γ in Ξ' because all leaders and all strong zombies must have levels no less than i, and thus they become weak zombies only after their virtual levels reach $i+1$. Therefore, by Lemma 6, $R_\gamma = \emptyset$ holds within $O(k')$ steps in Ξ', and $R_\gamma = \emptyset$ always hold thereafter. If there is no agent in R_γ, no waiting leader with at most level i goes back to an active leader without increasing its level. Thus, by Lemma 5, within $O(m')$ steps, Ξ reaches a configuration $C_{t'}$ from which there is always no active leader with level i.

Let Ξ'' be the suffix $C_{t'}, C_{t'+1}, \ldots$ of Ξ'. By definition, in Ξ'', there is no active leader with level at most i. In addition, the virtual level of a zombie is i only if it is located at a node with a strong zombie with level i. Thus, it suffices to show that every strong zombie z with level i increases its virtual level at least by one within $O(m')$ steps. Suppose that z is located at a node v. By definition of $\mathcal{A}_{svl}$, we must have a path $w = v_1, v_2, \ldots, v_s$ such that $s \leq k$, $v = v_1$, $p_{v_i}(v_{i+1}) = a_{v_i}.\texttt{last}$ for $i = 1, 2, \ldots, s-1$, and a leader l is located at v_s. In Ξ'', there is no active leader with level i. Thus, if l is active, its level is at least $i+1$. Even if l is waiting, its level is at least i, and the virtual level of a_{v_s} immediately becomes at least $i+1$ after an active leader or a zombie visits v_s. Therefore, if none of the agents $a_{v_1}, a_{v_2}, \ldots, a_{v_{s-1}}$ changes the value of its $\texttt{last}$ during the next $4s \leq 4k$ steps, the virtual level of z reaches at least $i+1$. Thus, suppose that some a_{v_j} ($1 \leq j \leq s-1$) changes the value of its $\texttt{last}$ during the $4s$ steps. This yields that an active leader must visit v_j during the period. Moreover, in Ξ'', every active leader has level at least $i+1$. Thus, the virtual level of agent a_{v_j} must reach at least $i + 1$ at that time. Therefore, in any case, zombie z increases its virtual level at least by one within $O(k)$ steps in Ξ''. $\square$

Theorem 1. *Algorithm $\mathcal{A}_{svl}$ solves a dispersion problem within $O(m' \log \ell)$ steps. It uses $O(\log(k + \delta_{\max}))$ bits of memory space per agent.*

Proof. By Lemmas 1 and 7, $\mathcal{L}_{\min} = \infty$ holds within $O(m' \log \ell)$ steps. Since $\mathcal{L}_{\min} = \infty$ yields that no active leader and no zombie exists, all agents are located at different nodes at that time. Therefore, the theorem immediately follows from Lemmas 2 and 3. $\square$

5   Conclusion

In this paper, we presented a both time and space efficient algorithm for the dispersion problem. This algorithm does not require any global knowledge. However, we require that all agents compute and move synchronously. The proposed algorithm inherently requires the synchronous assumption: active leaders, strong zombies, and weak zombies move in different speeds.

References

1. Augustine, J., Moses Jr., W.K.: Dispersion of mobile robots. In: Proceedings of the 19th International Conference on Distributed Computing and Networking, January 2018
2. Augustine, J., Moses Jr., W.K.: Dispersion of mobile robots: A study of memory-time trade-offs. arXiv preprint arXiv:1707.05629 (2018)
3. Gallager, R.G., Humblet, P.A., Spira, P.M.: A distributed algorithm for minimum-weight spanning trees. ACM Trans. Program. Lang. Syst. (TOPLAS) **5**(1), 66–77 (1983)
4. Kshemkalyani, A.D., Ali, F.: Efficient dispersion of mobile robots on graphs. In: Proceedings of the 20th International Conference on Distributed Computing and Networking, pp. 218–227 (2019)
5. Kshemkalyani, A.D., Molla, A.R., Sharma, G.: Fast dispersion of mobile robots on arbitrary graphs. In: Dressler, F., Scheideler, C. (eds.) ALGOSENSORS 2019. LNCS, vol. 11931, pp. 23–40. Springer, Cham (2019). https://doi.org/10.1007/978-3-030-34405-4_2
6. Kshemkalyani, A.D., Molla, A.R., Sharma, G.: Dispersion of mobile robots in the global communication model. In: Proceedings of the 21st International Conference on Distributed Computing and Networking, pp. 1–10 (2020)
7. Panaite, P., Pelc, A.: Exploring unknown undirected graphs. J. Algorithms **33**(2), 281–295 (1999)
8. Priezzhev, V.B., Dhar, D., Dhar, A., Krishnamurthy, S.: Eulerian walkers as a model of self-organized criticality. Phys. Rev. Lett. **77**(25), 5079 (1996)
9. Sudo, Y., Baba, D., Nakamura, J., Ooshita, F., Kakugawa, H., Masuzawa, T.: A single agent exploration in unknown undirected graphs with whiteboards. IEICE Trans. Fundam. Electron. Commun. Comput. Sci. **98**(10), 2117–2128 (2015)
10. Yanovski, V., Wagner, I.A., Bruckstein, A.M.: A distributed ant algorithm for efficiently patrolling a network. Algorithmica **37**(3), 165–186 (2003)

Brief Announcement: TRIX: Low-Skew Pulse Propagation for Fault-Tolerant Hardware

Christoph Lenzen and Ben Wiederhake[✉]

Saarbrücken Graduate School of Computer Science,
Max Planck Institute for Informatics, Saarland Informatics Campus,
Saarbrücken, Germany
{clenzen,bwiederh}@mpi-inf.mpg.de

Abstract. We present a simple grid structure to use in a fault-tolerant clock propagation method and study it by means of simulation experiments. A key question is how well neighboring grid nodes are synchronized, even without faults. Our statistical approach provides substantial evidence that this system performs surprisingly well. In a grid of height H, the standard deviation of the delay seems to be $O(H^{1/4})$ ($\approx$2.7 link delay uncertainties for $H = 2000$) and the standard deviation of the skew to be $o(\log \log H)$ ($\approx$0.77 link delay uncertainties for $H = 2000$).

1 Introduction

Traditionally, clocking of synchronous systems is performed by clock trees or other structures that cannot sustain faulty components [12]. This imposes limits on scalability on the physical size of clock domains. To the best of our knowledge, work on fault-tolerant clocking schemes started in earnest in the last decade, with an upsurge of interest in single event upsets of the clocking subsystem [1,2,8,10]. Larger systems and smaller components require going beyond these techniques.

There is a significant body of work on fault-tolerant synchronization from the area of distributed systems considering Byzantine faults [9,11]. A line of works culminating in [6] additionally consider *self-stabilization*, the ability of a system to recover from an unbounded number of transient faults. These highly desirable properties come at a high price, usually in the form of high connectivity [4].

A suitable relaxation of requirements is proposed in [3], requiring that Byzantine faults are distributed across the system not in a worst-case fashion, but more "spread out". Distributing a clock signal through a grid-like network called HEX is proposed, which tolerates one out of each node's four in-neighbors being faulty. Unfortunately, HEX has poor synchronization performance: a crashed node causes a "detour" resulting in a clock skew between neighbors of at least one maximum node-to-node communication delay d. This is much larger than the *uncertainty* u in the node-to-node delay, which is engineered to be small ($u \ll d$).

© Springer Nature Switzerland AG 2020
S. Devismes and N. Mittal (Eds.): SSS 2020, LNCS 12514, pp. 295–300, 2020.
https://doi.org/10.1007/978-3-030-64348-5_23

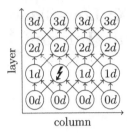

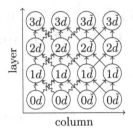

Fig. 1. Zoom-in on a part of a larger TRIX grid with one crashing node. Observe that the fault causes no significant additional skew.

Fig. 2. Worst-case assignment of wire delays causing large skew for TRIX. Squiggly lines indicate slow wires, straight lines indicate fast wires. $\check{d} := d - u$.

We propose a novel clock distribution topology that overcomes the above shortcoming of HEX, in particular the high skew between neighboring nodes. Similar to HEX, the clock signal is propagated through layers, but for each node, all of its three in-neighbors are on the preceding layer. If at most one in-neighbor is faulty, each node still has two correct in-neighbors on the preceding layer, as demonstrated in Fig. 1. Hence, we can now focus on fault-free executions, because single isolated faults only introduce an additional uncertainty of at most $u \ll d$. Predictions in the fault-free model are therefore still meaningful for systems with rare and non-malicious faults.

The TRIX topology is acyclic, which conveniently means that self-stabilization is trivial to achieve, as any incorrect state is "flushed out" from the system.

Despite its apparent attractiveness and even greater simplicity, we note that this choice of topology should not be obvious. The fact that nodes do not check in with their neighbors on the same layer implies that the worst-case clock skew between neighbors grows as uH, where H is the number of layers and (for the sake of simplicity) we assume that the skew on the first layer (which can be seen as the "clock input") is 0, see Fig. 2. However, reaching the skew of d between neighbors on the same layer, which is necessary to give purpose to any link between them, takes many layers, at least $d/u \gg 1$ many. This is in contrast to HEX, where the worst-case skew is bounded, but more easily attained.

While the worst-case behavior is easy to understand, it originates from a very unlikely configuration, where one side of the grid is entirely slow and the other is fast, see Fig. 2. In contrast, correlated but gradual changes will also result in spreading out clock skews. Any change that affects an entire region in the same way will not affect local timing differences at all. This motivates to study the extreme case of independent noise on each link in the TRIX grid. Moreover, we assume "perfect" input, i.e., each node on the initial layer signals a clock pulse at time 0, and that the grid is infinitely wide. We argue that this simplistic abstraction captures the essence of (independent) noise on the channels.

We provide evidence that TRIX behaves better than conventional concentration bounds might suggest. The full version [7] argues in-depth that these results are not just artifacts of the simulation, the model, or due to various biases.

We point out the open problem of analyzing the stochastic process we use as an abstraction for TRIX. Understanding of the underlying cause would allow making qualitative and quantitative predictions beyond the considered setting.

2 Model

The network topology is a grid of height H and width W. To simplify, we choose $W = \infty$, because we aim to focus on the behavior in large systems. We refer to the grid nodes by integer coordinates (x, y), where $x \in \mathbb{Z}$ and $y \in \mathbb{N}_0$. Layer $0 \le \ell \le H$ consists of the nodes (x, ℓ), $x \in \mathbb{Z}$.

Nodes in layer 0 represent the clock source. Note that for the purposes of this paper, we assume that the problem of fault-tolerant clock signal *generation* has already been sufficiently addressed (e.g. using [5]), but the signal still needs to be *distributed*. All other nodes (x, ℓ) for $\ell > 0$ are TRIX nodes. Each TRIX node propagates the clock signal to the three nodes "above" it, i.e., the vertices $(x+c, y+1)$, $c \in \{-1, 0, +1\}$. Each of the wire delays is modeled as i.i.d. random variables $w_c^{x,y}$ (or w_c for short) that are fair coin flips, i.e., attain the values 0 or 1 with probability $1/2$ each. This reflects that any absolute delay does not matter, as the number of wires is the same for any path from layer 0 (the clock generation layer) to layer $\ell > 0$; also, this normalizes the uncertainty from u to 1.

Let $d(x, y)$ be the time at which node (x, y) fires. Clock generation provides us with $d(x, 0) = 0$. Each TRIX node fires when receiving the second signal from its predecessors: Define $t_c := d(x - c, y) + w_c^{x-c,y}$ as the time at which node $(x, y + 1)$ receives each clock pulse. Then node $(x, y + 1)$ fires a clock pulse at the median time $t := \text{median}\{t_{-1}, t_0, t_{+1}\}$.

We concentrate on two important metrics to analyze this system: absolute delay and relative skew. Our main interests are the random variables $d(H) := d(0, H)$, i.e. the total delay at the top, and $s(H) := d(1, H) - d(0, H)$, i.e. the relative skew between neighboring nodes.

3 Delay is Tightly Concentrated

We examine $d(2000)$, the delay at layer 2000. The estimated probability mass function of $d(2000)$ looks like a binomial distribution. The empiric standard deviation is only 2.741, i.e. less than three delay uncertainties. The full version [7] explains the statistic methods in detail, contains more figures, and proves all following lemmas. The peak of the probability mass function falls in the middle of the support $[0, H]$:

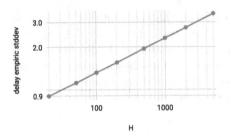

Fig. 3. Log-log plot of the empiric standard deviation of $d(H)$.

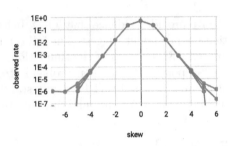

Fig. 4. Estimated probability mass function for $s(2000)$ with logarithmic Y-axis. Error bounds are only visible at fringes.

Lemma 1. $\mathbb{E}[d(H)] = H/2$.

The behavior at $H = 2000$ is similar for other heights and changes slowly with increasing H.

The empiric standard deviation for various values of H can be seen in the data plotted in Fig. 3 as a log-log plot. This suggests a polynomial relationship between standard deviation σ and grid height H. The slope of the line is close to $1/4$, which suggests $\sigma \sim H^\beta$ with $\beta \approx 1/4$. This is a quadratic improvement over standard concentration bounds, which would predict $\beta \approx 1/2$.

4   Skew is Tightly Concentrated

We examine $s(2000)$, the skew at layer 2000 between neighboring nodes. As expected, we see a high concentration around 0 in Fig. 4, with roughly half of the probability mass at 0.

Observe that the skew does not follow a normal distribution at all: The probability mass seems to drop off exponentially like $e^{-\lambda|x|}$ for $\lambda \approx 2.9$ (where x is the skew), and not quadratic-exponentially like $e^{-x^2/(2\sigma^2)}$, as it would happen in the normal distribution. The probability mass for 0 is a notable exception, not matching this behavior.

We observe that the skew seems to be symmetric with mean 0.

Corollary 1. $s(H)$ is symmetric with $\mathbb{E}[s(H)] = 0$.

Furthermore, the worst-case skew on layer H is indeed H, c.f. Fig. 2.

Lemma 2. There is an assignment for all c_w such that $s(H) = H$.

We conjecture that the probability mass of high-skew assignments is very low.

Again, the behavior at $H = 2000$ is similar for other heights and changes extremely slowly with increasing H. Figure 5 shows that the skew remains small even for large values of H. Note that the X-axis is doubly logarithmic. This suggests that the standard-deviation of $s(H)$ grows strongly sub-logarithmically, possibly even converges to a finite value. In fact, the plot suggests that $s(H) \in O(\log \log H)$.

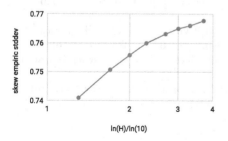

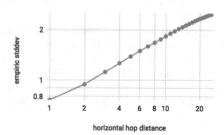

Fig. 5. Empiric standard deviation of $s(H)$ as a function of H, as a loglog-lin plot.

Fig. 6. Empiric standard deviation of $d(\delta, 500) - d(0, 500)$ as a function of horizontal distance δ in a log-log plot.

Note that if we pretended that adjacent nodes exhibit independent delays, the skew would have the same concentration as the delay. In contrast, we see that adjacent nodes are tightly synchronized; this is ideal for clock propagation.

So far, we have limited our attention to the skew between neighboring nodes. In the other extreme end, at horizontal distances $\delta \geq 2H$, node delays are independent, as they do not share any wires on any path to any clock generator. In Fig. 6, we see that the skew grows steadily with increasing $\delta \ll 2H$. The plot suggests that the standard deviation increases roughly proportional to δ^{γ} for $\gamma \approx 1/3$. This is noticeably less steep than the naive guess $\gamma \approx 1/2$ for small δ. It is not surprising that the slope falls off towards larger values, as it must become constant for $\delta \geq 2H = 1000$.

References

1. Chipana, R., Kastensmidt, F.L.: SET susceptibility analysis of clock tree and clock mesh topologies. In: ISVLSI, pp. 559–564 (2014)
2. Chipana, R., Kastensmidt, F.L., Tonfat, J., Reis, R., Guthaus, M.: SET susceptibility analysis in buffered tree clock distribution networks. In: RADECS, pp. 256–261 (2011)
3. Dolev, D., Függer, M., Lenzen, C., Perner, M., Schmid, U.: HEX: scaling honeycombs is easier than scaling clock trees. J. Comput. Syst. Sci. **82**(5), 929–956 (2016)
4. Dolev, D., Halpern, J.Y., Strong, H.R.: On the possibility and impossibility of achieving clock synchronization. J. Comput. Syst. Sci. **32**(2), 230–250 (1986)
5. Dolev, D., Függer, M., Lenzen, C., Schmid, U.: Fault-tolerant algorithms for tick-generation in asynchronous logic. J. ACM **61**(5), 30:1–30:74 (2014)
6. Lenzen, C., Rybicki, J.: Self-stabilising byzantine clock synchronisation is almost as easy as consensus. J. ACM **66**(5), 32:1–32:56 (2019)

7. Lenzen, C., Wiederhake, B.: TRIX: Low-skew pulse propagation for fault-tolerant hardware (2020). https://arxiv.org/abs/2010.01415
8. Malherbe, V., Gasiot, G., Clerc, S., Abouzeid, F., Autran, J.L., Roche, P.: Investigating the single-event-transient sensitivity of 65 nm clock trees with heavy ion irradiation and Monte-Carlo simulation. In: IRPS, pp. SE-3-1–SE-3-5 (2016)
9. Srikanth, T.K., Toueg, S.: Optimal clock synchronization. J. ACM **34**(3), 626–645 (1987)
10. Wang, H.B., et al.: Single-event transient sensitivity evaluation of clock networks at 28-nm CMOS technology. IEEE Trans. Nucl. Sci. **63**(1), 385–391 (2016)
11. Welch, J.L., Lynch, N.A.: A new fault-tolerant algorithm for clock synchronization. Inf. Comput. **77**(1), 1–36 (1988)
12. Xanthopoulos, T. (ed.): Clocking in Modern VLSI Systems. ICIR. Springer, Boston (2009). https://doi.org/10.1007/978-1-4419-0261-0

Time-Optimal Self-stabilizing Leader Election on Rings in Population Protocols

Daisuke Yokota[✉], Yuichi Sudo, and Toshimitsu Masuzawa

Osaka University, Suita, Japan
{d-yokota,y-sudou,masuzawa}@ist.osaka-u.ac.jp

Abstract. We propose a self-stabilizing leader election protocol on directed rings in the model of population protocols. Given an upper bound N on the population size n, the proposed protocol elects a unique leader within $O(nN)$ expected steps starting from any configuration and uses $O(N)$ states. This convergence time is optimal if a given upper bound N is asymptotically tight, *i.e.*, $N = O(n)$.

1 Introduction

We consider the *population protocol* (PP) model [2] in this paper. A network called *population* consists of a large number of finite-state automata, called *agents*. Agents make *interactions* (*i.e.*, pairwise communication) with each other to update their states. The interactions are opportunistic, *i.e.*, they are unpredictable for the agents. A population is modeled by a graph $G = (V, E)$, where V represents the set of agents, and E indicates which pair of agents can interact. Each pair of agents $(u, v) \in E$ has interactions infinitely often, while each pair of agents $(u', v') \notin E$ never has an interaction. At each time step, one pair of agents chosen uniformly at random from all pairs in E has an interaction. This assumption enables us to evaluate time complexities of population protocols. Almost all studies in the population protocol model make this assumption when they evaluate time complexities of population protocols. In the field of population protocols, many efforts have been devoted to devising protocols for a complete graph, *i.e.*, a population where every pair of agents interacts infinitely often. In addition, several studies [1,2,4,7,8,10,14,15,17,19,20] have investigated populations forming graphs other than complete graphs.

Self-stabilization [11] is a fault-tolerant property whereby, even when any number and kinds of faults occur, the network can autonomously recover from the faults. Formally, self-stabilization is defined as follows: (i) starting from an arbitrary configuration, a network eventually reaches a *safe configuration* (*convergence*), and (ii) once a network reaches a safe configuration, it maintains its specification forever (*closure*). Self-stabilization is of great importance in the PP model because self-stabilization tolerates any finite number of transient faults,

This work was supported by JAPS KAKENHI Grant No. 19H04085 and 20H04140.

© Springer Nature Switzerland AG 2020
S. Devismes and N. Mittal (Eds.): SSS 2020, LNCS 12514, pp. 301–316, 2020.
https://doi.org/10.1007/978-3-030-64348-5_24

and this is a necessary property in a network consisting of a large number of inexpensive and unreliable nodes.

Consequently, many studies have been devoted to self-stabilizing population protocols [1,4,6–8,12,13,15–21]. For example, Angluin et al. [1] proposed self-stabilizing protocols for a variety of problems, i.e., leader election in rings, token circulation in rings with a pre-selected leader, 2-hop coloring in degree-bounded graphs, consistent global orientation in undirected rings, and spanning tree construction in regular graphs. Sudo et al. [17,20] gave a self-stabilizing 2-hop coloring protocol that uses a much smaller memory space of agents. Sudo et al. [22] investigates the possibility of self-stabilizing protocols for leader election, ranking, degree recognition, and neighbor recognition on arbitrary graphs.

Many of the above studies on self-stabilizing population protocols have focused on self-stabilizing leader election (SS-LE) because leader election is one of the most fundamental problems in the PP model. Several important protocols [1–3] require a pre-selected unique leader. In particular, Angluin et al. [3] show that all semi-linear predicates can be solved very quickly if we have a unique leader. The goal of the leader election problem is electing exactly one agent as a leader in the population. Unfortunately, SS-LE is impossible to solve without an additional assumption even if we focus only on complete graphs [1,6,22]. The studies to overcome this impossibility in the literature are roughly classified into four categories. (Some of the studies belong to multiple categories.)

The first category [5,6,22] assumes that every agent knows the exact number of agents. With this assumption, Cai et al. [6] gave an SS-LE protocol for complete graphs, Burman et al. [5] gave faster protocols in the same setting, and Sudo et al. [22] gave an SS-LE protocol for arbitrary graphs.

The second category [4,7,12] employs *oracles*, a kind of failure detectors. Fischer and Jiang [12] introduced an oracle Ω? that eventually tells all agents whether or not at least one leader exists. They proposed two SS-LE protocols using Ω?, one for complete graphs and the other for rings. Canepa et al. [7] proposed two SS-LE protocols that use Ω?, i.e., a deterministic protocol for trees and a randomized protocol for arbitrary graphs. Beauquier et al. [4] presented a deterministic SS-LE protocol for arbitrary graphs that uses two copies of Ω?.

The third category [13,15–21] slightly relaxes the requirement of the original self-stabilization and gave *loosely-stabilizing* leader election protocols. Specifically, the studies of this category allow a population to deviate from the specification of the problem (i.e., a unique leader) after the population satisfies the specification for an extremely long time. This concept was introduced by [18]. The protocols given by [13,16,18,21] work for complete graphs and those given by [15,17,19,20] work for arbitrary graphs. Recently, Sudo et al. [16] gave a time-optimal loosely-stabilizing leader election protocol for complete graphs: given a parameter $\tau \geq 1$, an execution of their protocol reaches a configuration with a unique leader within $O(\tau n \log n)$ expected steps starting from any configuration, and thereafter, it keeps the unique leader for $\Omega(n^\tau)$ expected steps, where n is the number of agents in the population.

The forth category [1,8,9,12] restricts the topology of a graph to avoid the impossibility of SS-LE. A class $\mathcal{G}$ of graphs is called *simple* if there does not exist a graph in $\mathcal{G}$ which contains two disjoint subgraphs that are also in $\mathcal{G}$. Angluin *et al.* [1] proves that there exists no SS-LE protocol that works for all the graphs in any non-simple class. Thus, if we focus on a simple class of graphs, there may exist an SS-LE protocol for all graphs in the class. As a typical example, the class of *rings* is simple. Angluin *et al.* [1] gave an SS-LE protocol that works for all rings whose sizes are not multiples of a given integer k (in particular, rings of odd size). They posed a question whether SS-LE is solvable or not for general rings (*i.e.*, rings of any size) without any oracle or knowledge such as the exact number of agents in the population, while Fischer and Jiang [12] solves SS-LE for general rings using oracle $\Omega?$. This question had been open for a decade until Chen and Chen [8] recently gave an SS-LE protocol for general rings. These three protocols given by [1,8,12] use only a constant number of states per agent. The expected convergence times (*i.e.*, the expected numbers of steps required to elect a unique leader starting from any configuration) of the protocols proposed by [1,12] are $\Theta(n^3)$, while the protocol given by [8] requires an exponentially long convergence time. The oracle $\Omega?$ only guarantees that it *eventually* reports to each agent whether there exists a leader in the population. Here, the convergence time of the protocol of [12] is bounded by $\Theta(n^3)$ assuming that the oracle immediately reports the absence of the leader to each agent. All of the three protocols assume that the rings are oriented or *directed*. However, this assumption is not essential because Angluin *et al.* [1] also presented a self-stabilizing ring orientation protocol, which gave a common sense of direction to all agents in the ring. In this paper, we also consider directed rings. Very recently, Chen and Chen [9] generalized their work on the rings for regular graphs.

Table 1. Self-stabilizing leader election on rings

	Assumption	Convergence Time	#states
[1]	n is not multiple of a given k	$\Theta(n^3)$	$O(1)$
[12]	oracle $\Omega?$	$\Theta(n^3)$	$O(1)$
[8]	none	exponential	$O(1)$
This work	$n \leq N$ for a given N	$O(nN)$	$O(N)$

Our Contribution. This paper belongs to the fourth category. We propose an SS-LE protocol P_{RL} for directed rings. Specifically, given an upper bound N on the population size n, P_{RL} elects a unique leader in $O(nN)$ expected steps for all directed rings (whose size is at most N). One can easily prove that no protocol can solve SS-LE in $o(n^2)$ expected steps. Thus, P_{RL} is time-optimal if a given upper bound N is asymptotically tight, *i.e.*, $N = O(n)$. The results are summarized in Table 1.

The main contribution of this paper is a novel mechanism that largely improves the number of steps required to decrease the number of leaders to one when there are multiple leaders in the population. The mechanism requires only $O(n^2)$ expected steps, while the existing three SS-LE protocols for rings [1,8,12] requires $\Omega(n^3)$ expected steps to decrease the number of leaders to one. Our mechanism requires only $O(1)$ states, which is the same as the existing three protocols [1,8,12]. (Protocol P_{RL} requires $O(N)$ states only to detect the absence of a leader.) Thus, if we assume an oracle that reports to each leader an absence of a leader within $O(n^2)$ expected steps, we immediately obtain an SS-LE protocol with $O(n^2)$ expected convergence time and constant space per agent by using the proposed mechanism to remove leaders. We leave open an interesting question whether or not this oracle can be implemented with $o(N)$ states.

2   Preliminaries

In this section, we describe the formal definitions of our computation model.

A *population* is a simple and weakly connected digraph $G(V, E)$, where V ($|V| \geq 2$) is a set of *agents* and $E \subseteq V \times V$ is a set of arcs. Each arc represents a possible *interaction* (or communication between two agents): If $(u, v) \in E$, agents u and v can interact with each other, where u serves as an *initiator* and v serves as a *responder*. If $(u, v) \notin E$, agents u and v never have an interaction. In this paper, we consider only a population represented by a *directed ring*, *i.e.*, we assume that $V = \{u_0, u_1, \ldots, u_{n-1}\}$ and $E = \{(u_i, u_{i+1 \bmod n}) \mid i = 0, 1, \ldots, n - 1\}$. Here, we use the indices of the agents only for simplicity of description. The agents are *anonymous*, *i.e.*, they do not have unique identifiers. We call $u_{i-1 \bmod n}$ and $u_{i+1 \bmod n}$ the left neighbor and the right neighbor of u_i, respectively. We omit "modulo by n" (*i.e.*, mod n) in the index of agents when no confusion occurs.

A *protocol* $P(Q, Y, T, \pi_{out})$ consists of a finite set Q of states, a finite set Y of output symbols, transition function $T : Q \times Q \to Q \times Q$, and an output function $\pi_{out} : Q \to Y$. When an interaction between two agents occurs, T determines the next states of the two agents based on their current states. The *output of an agent* is determined by π_{out}: the output of agent v with state $q \in Q$ is $\pi_{out}(q)$. We assume that all agents have a common knowledge N on n such that $n \leq N = O(poly(n))$. Thus, the parameters Q, Y, T, and π_{out} can depend on the knowledge N. However, for simplicity, we do not explicitly write protocol P as parameterized with N, *e.g.*, $P_N = (Q_N, Y_N, T_N, \pi_{out,N})$.

A *configuration* is a mapping $C : V \to Q$ that specifies the states of all the agents. We denote the set of all configurations of protocol P by $\mathcal{C}_{all}(P)$. We say that configuration C changes to C' by an interaction $e = (u_i, u_{i+1})$, denoted by $C \xrightarrow{e} C'$ if we have $(C'(u_i), C'(u_{i+1})) = T(C(u_i), C(u_{i+1}))$ and $C'(v) = C(v)$ for all $v \in V \setminus \{u_i, u_{i+1}\}$. We simply write $C \to C'$ if there exits $e \in E$ such that $C \xrightarrow{e} C'$. We say that a configuration C' is reachable from C if there exists a sequence of configurations $C_0, C_1, \ldots, C_k$ such that $C = C_0$, $C' = C_k$, and

$C_i \rightarrow C_{i+1}$ for all $i = 0, 1, \ldots, k - 1$. We also say that a set $\mathcal{C}$ of configurations is *closed* if no configuration outside $\mathcal{C}$ is reachable from a configuration in $\mathcal{C}$.

A scheduler determines which interaction occurs at each time step (or just *step*). In this paper, we consider a *uniformly random scheduler* $\Gamma = \Gamma_0, \Gamma_1, \ldots$: each $\Gamma_t \in E$ is a random variable such that $\Pr(\Gamma_t = (u_i, u_{i+1})) = 1/n$ for any $t \geq 0$ and $i = 0, 1, \ldots, n - 1$. Each Γ_t represents the interaction that occurs at step t. Given an initial configuration C_0, the *execution* of protocol P under Γ is defined as $\Xi_P(C_0, \Gamma) = C_0, C_1, \ldots$ such that $C_t \xrightarrow{\Gamma_t} C_{t+1}$ for all $t \geq 0$. We denote $\Xi_P(C_0, \Gamma)$ simply by $\Xi_P(C_0)$ when no confusion occurs.

We address the self-stabilizing leader election problem in this paper. For simplicity, we give the definition of a self-stabilizing leader election protocol instead of giving the definitions of self-stabilization and the leader election problem separately.

Definition 1 (Self-stabilizing Leader Election). For any protocol P, we say that a configuration C of P is *safe* if (i) exactly one agent outputs L (leader) and all other agents output F (follower) in C, and (ii) at every configuration reachable from C, all agents keep the same outputs as those in C. A protocol P is a *self-stabilizing leader election (SS-LE) protocol* if $\Xi_P(C_0, \Gamma)$ reaches a safe configuration with probability 1.

We evaluate a SS-LE protocol P with two metrics: *the expected convergence time* and *the number of states*. For a configuration $C \in \mathcal{C}_{\text{all}}(P)$, let $t_{P,C}$ be the expected number of steps until $\Xi_P(C_0, \Gamma)$ reaches a safe configuration. The expected convergence time of P is defined as $\max_{C \in \mathcal{C}_{\text{all}}(P)} t_{P,C}$. The number of states of $P = (Q, Y, T, O)$ is simply $|Q|$.

3   Self-stabilizing Leader Election Protocol

In this section, we propose a SS-LE protocol P_{RL} that works in any directed ring consisting of N agents or less. The expected convergence time is $O(nN)$, and the number of states is $O(N)$. Thus, P_{RL} is time-optimal if a given upper bound N of n is asymptotically tight, *i.e.*, $N = O(n)$.

The pseudocode of P_{RL} is given in Algorithm 1, which describes how two agents l and r update their states, *i.e.*, their variables, when they have an interaction. Here, l and r represents the initiator and the responder in the interaction, respectively. That is, l is the left neighbor of r, and r is the right neighbor of l. We denote the value of the variable var at agent $v \in V$ by $v.\text{var}$. Similarly, we denote the variable var in state $q \in Q$ by $q.\text{var}$. In this algorithm, each agent $v \in V$ maintains an *output* variable $v.\text{leader} \in \{0, 1\}$, according to which it determines its output. Agent v outputs L when $v.\text{leader} = 1$ and outputs F when $v.\text{leader} = 0$. We say that v is a *leader* if $v.\text{leader} = 1$; otherwise v is called a *follower*. For each $u_i \in V$, we define the distance to *the nearest left leader* and the distance to *the nearest right leader* of u_i as $d_L(i) = \min\{j \geq 0 \mid u_{i-j}.\text{leader} = 1\}$ and $d_R(i) = \min\{j \geq 0 \mid u_{i+j}.\text{leader} = 1\}$, respectively. When there is no leader in the ring, we define $d_L(i) = d_R(i) = \infty$.

Algorithm P_{RL} consists of two parts: the leader creation part (Lines 1–5) and the leader elimination part (Lines 6–19). Since P_{RL} is a self-stabilizing protocol, it has to handle any initial configuration, where there may be no leader or multiple leaders. The leader creation part creates a new leader when there is no leader, while the leader elimination part decreases the number of leaders to one when there are two or more leaders.

3.1   Leader Elimination

We are inspired by the algorithm of [12] to design the leader elimination part (Lines 6–19). Roughly speaking, the strategy of [12] can be described as follows.

- Each agent may have a *bullet* and/or a *shield*.
- A leader always fires a bullet: each time a leader u_{i+1} having no bullet interacts with an agent u_i, the leader u_{i+1} makes a bullet.
- Bullets move from left to right in the ring: each time u_i having a bullet interacts with u_{i+1}, the bullet moves from u_i to u_{i+1}.
- Conversely, shields move from right to left: each time u_{i+1} having a shield interacts with u_i, the shield moves from u_{i+1} to u_i.
- Each time two agents both with bullets (resp. shields) have an interaction, the left bullet (resp. the right shield) disappears.
- When a bullet and a shield pass each other, *i.e.*, u_i with a bullet and u_{i+1} with a shield have an interaction, the bullet disappears.
- When a bullet moves to a leader without a shield, the leader is *killed* (*i.e.*, becomes a follower).

The algorithm of [12] assumes an oracle, called an *eventual leader detector* Ω?, which detects and tells each agent whether a leader exists or not, when there is continuously a leader or there is continuously no leader. A follower becomes a leader when it is reported by Ω? that there is no leader in the population. At this time, the new leader simultaneously generates both a shield and a bullet. One can easily observe that by the above strategy together with oracle Ω?, the population eventually reaches a configuration after which there is always one fixed leader. However, the algorithm of [12] requires $\Omega(n^3)$ steps to elect one leader in the worst case even if oracle Ω? can immediately report to each agent whether there is a leader in the population.

We drastically modify the above strategy of [12] for the leader elimination part of P_{RL} to decrease the number of leaders to one within $O(n^2)$ steps. First, a shield never moves in our algorithm. Only leaders have shields. A leader sometimes generates a shield and sometimes breaks a shield. Second, a leader does not always fire a bullet. Instead, a leader fires a new bullet only after it detects that the last bullet it fired reaches a (possibly different) leader. Third, we have two kinds of bullets: *live bullets* and *dummy bullets*. A live bullet kills a leader without a shield. However, a dummy bullet does not have capability to kill a leader. When a leader decides to fire a new bullet, the bullet becomes live or dummy with the probability 1/2 each. When a leader fires a live bullet, it simultaneously generates a shield (if it does not have a shield). When a leader fires a

Algorithm 1. P_{RL}

Interaction between initiator l and responder r:

1: $l.\text{distL} \leftarrow \begin{cases} 0 & l.\text{leader} = 1 \\ l.\text{distL} & \text{otherwise} \end{cases}$

2: $r.\text{distL} \leftarrow \begin{cases} 0 & r.\text{leader} = 1 \\ \min(l.\text{distL} + 1, N) & r.\text{leader} = r.\text{bullet} = 0 \\ r.\text{distL} & \text{otherwise} \end{cases}$

3: **if** $r.\text{distL} = N$ **then**
4:     $r.\text{leader} \leftarrow 1$; $r.\text{bullet} \leftarrow 2$; $r.\text{shield} \leftarrow 1$; $r.\text{signal} \leftarrow 0$; $r.\text{distL} \leftarrow 0$;
5: **end if**

6: **if** $l.\text{leader} = l.\text{signal} = 1$ **then**
7:     $l.\text{bullet} \leftarrow 2$; $l.\text{shield} \leftarrow 1$; $l.\text{signal} \leftarrow 0$
8: **end if**
9: **if** $r.\text{leader} = r.\text{signal} = 1$ **then**
10:     $r.\text{bullet} \leftarrow 1$; $r.\text{shield} \leftarrow 0$; $r.\text{signal} \leftarrow 0$
11: **end if**
12: **if** $l.\text{bullet} > 0 \wedge r.\text{leader} = 1$ **then**
13:     $r.\text{leader} \leftarrow \begin{cases} 0 & l.\text{bullet} = 2 \wedge r.\text{shield} = 0 \\ 1 & \text{otherwise} \end{cases}$
14:     $l.\text{bullet} \leftarrow 0$;
15: **else if** $l.\text{bullet} > 0 \wedge r.\text{leader} = 0$ **then**
16:     $r.\text{bullet} \leftarrow \begin{cases} l.\text{bullet} & r.\text{bullet} = 0 \\ r.\text{bullet} & r.\text{bullet} > 0 \end{cases}$
17:     $l.\text{bullet} \leftarrow 0$; $r.\text{signal} \leftarrow 0$;
18: **end if**
19: $l.\text{signal} \leftarrow \max(l.\text{signal}, r.\text{signal}, r.\text{leader})$

dummy bullet, it breaks the shield if it has a shield. Thus, roughly speaking, each leader is *shielded* (*i.e.,* has a shield) with probability $1/2$ at each step. Therefore, when a live bullet reaches a leader, the leader is killed with probability $1/2$. This strategy is well designed: not all leaders kill each other simultaneously because a leader must be shielded if it fired a live bullet in the last shot. As a result, the number of leaders eventually decreases to one.

In what follows, we explain how we implement this strategy. Each agent v maintains variables $v.\text{bullet} \in \{0, 1, 2\}$, $v.\text{shield} \in \{0, 1\}$, and $v.\text{signal} \in \{0, 1\}$. As their names imply, $v.\text{bullet} = 0$ (resp. $v.\text{bullet} = 1$, $v.\text{bullet} = 2$) indicates that v is now conveying no bullet (resp. a dummy bullet, a live bullet), while $v.\text{shield} = 1$ indicates that v is shielded. Unlike the protocol of [12], we ignore the value of $v.\text{shield}$ for any follower v. A variable signal is used by a leader to detect that the last bullet it fired already disappeared. Specifically, $v.\text{signal} = 1$ indicates that v is propagating a *bullet-absence signal*. A leader always generates a bullet-absence signal in its left neighbor when it interacts with its left neighbor (Line 19). This signal propagates from right to left (Line 19), while a bullet moves from left to right (Lines 16–17). A bullet disables a bullet-absence signal regardless of whether it is live or dummy, *i.e.*, u_{i+1}.signal is reset to 0 when two agents u_i and u_{i+1} such that u_i.bullet > 0 and u_{i+1}.signal $= 1$ have an interaction (Lines 16 and 17). Thus, a bullet-absence signal propagates to a leader only after the last bullet fired by the leader disappears.

When a leader u_i receives a bullet-absence signal from its right neighbor u_{i+1}, u_i waits for its next interaction to extract randomness from the uniformly random scheduler. At the next interaction, by the definition of the uniformly random scheduler, u_i meets its right neighbor u_{i+1} with probability $1/2$ and its left neighbor u_{i-1} with probability $1/2$. In the former case, u_i fires a live bullet and becomes shielded (Lines 6–8). In the latter case, u_i fires a dummy bullet and becomes unshielded (Lines 9–11). In both cases, the received signal is deleted (Lines 7 and 10). The fired bullet moves from left to right each time the agent with the bullet, say u_i, interacts with its right neighbor u_{i+1} (Lines 16 and 17). However, the bullet disappears without moving to u_{i+1} if u_{i+1} already has another bullet at this time. Suppose that the bullet now reaches a leader. If the bullet is live and the leader is not shielded at that time, the leader is killed by the bullet (Line 13). The bullet disappears at this time regardless of whether the bullet is alive and/or the leader is shielded (Line 14).

3.2   Leader Creation

The leader creation part is simple (Lines 1–5). Each agent u_i estimates $d_L(i)$ and stores the estimated value on variable u_i.distL $\in \{0, 1, \ldots, N\}$. Specifically, at each interaction (u_i, u_{i+1}), agents u_i and u_{i+1} update their distL as follows (Lines 1 and 2): (i) u_i (resp. u_{i+1}) resets its distL to zero if u_i (resp. u_{i+1}) is a leader, and (ii) if u_{i+1} is not a leader and does not have a bullet, $\min(l.\text{distL} + 1, N)$ is substituted for u_{i+1}.distL. Thus, if there is no leader in the population, some agent v eventually increases v.distL to N, and at that time, the agent decides that there is no leader. Then, this agent becomes a leader, executing v.leader $\leftarrow 1$ and v.distL $\leftarrow 0$ (Line 4). At the same time, v fires a live bullet, generates a shield, and disables a bullet-absence signal (Line 4). This live bullet prevents the new leader from being killed for a while: the leader becomes unshielded only after it receives a bullet-absence signal, and the live bullet prevents the leader from receiving a bullet-absence signal before another shielded leader appears.

As mentioned above, at interaction (u_i, u_{i+1}), the distance propagation does not occur if u_{i+1} is a leader. This exception helps us to simplify the analysis of the convergence time, *i.e.*, we can easily get an upper bound on the expected number of steps before each bullet disappears. Note that there are two cases that a bullet disappears: (i) when it reaches a leader, and (ii) when it reaches another bullet. The first case includes an interaction (u_i, u_{i+1}) where u_i.distL $\geq N-1$ holds and u_{i+1} becomes a leader by Line 4. Formally, at an interaction (u_i, u_{i+1}) such that u_i.bullet ≥ 1, we say that a bullet located at u_i disappears if u_{i+1}.leader $= 1$, u_{i+1}.bullet ≥ 1, or u_i.distL $\geq N - 1$. We have the following lemma thanks to the above exception.

Lemma 1. *Every bullet disappears before it moves (from left to right) N times.*

We should note that the leader creation part may create a leader even when there is one or more leaders, thus this part may prevent the leader elimination part from decreasing the number of leaders to one. Fortunately, as we will see in Sect. 4, within $O(nN)$ steps in expectation, the population reaches a configuration after which no new leader is created.

4 Correctness and Time Complexity

In this section, we prove that P_{RL} is a SS-LE protocol on directed rings of any size $n(\leq N)$ and that the expected convergence time of P_{RL} is $O(nN)$. In Sect. 4.1, we define a set $\mathcal{S}_{RL}$ of configurations and prove that every configuration in $\mathcal{S}_{RL}$ is safe. In Sect. 4.2, we prove that the population starting from any configuration reaches a configuration in $\mathcal{S}_{RL}$ within $O(nN)$ steps in expectation.

Due to lack of space, we omit proofs for Lemmas 3, 5, and 10, while we give only a proof sketch for Lemma 12. See a preprint paper [23] for the complete proofs of these four lemmas.

4.1 Safe Configurations

In this paper, we use several functions whose return values depend on a configuration, such as $d_L(i)$ and $d_R(i)$. When a configuration should be specified, we explicitly write a configuration as the first argument of those functions. For example, we write $d_L(C, i)$ and $d_R(C, i)$ to denote $d_L(i)$ and $d_R(i)$ in a configuration C, respectively.

In protocol P_{RL}, leaders kill each other by firing live bullets to decrease the number of leaders to one. However, it is undesirable that all leaders are killed and the number of leaders becomes zero. Therefore, a live bullet should not kill a leader if it is the last leader (*i.e.*, the unique leader) in the population. We say that a live bullet located at agent u_i is *peaceful* when the following predicate holds:

$$Peaceful(i) \stackrel{\text{def}}{\equiv} \left(\begin{array}{l} u_{i-d_L(i)}.\texttt{shield} = 1 \\ \wedge \, \forall j = 0, 1, \ldots, d_L(i) : u_{i-j}.\texttt{signal} = 0 \end{array} \right).$$

A peaceful bullet never kills the last leader in the population because its nearest left leader is shielded. A peaceful bullet never becomes non-peaceful; because letting u_i be the agent at which the bullet is located, the agents $u_{i-d_L(i)}, u_{i-d_L(i)+1}, \ldots, u_i$ will never have a bullet-absence signal thus $u_{i-d_L(i)}$ never becomes unshielded before the bullet disappears. At the beginning of an execution, there may be one or more non-peaceful live bullets. However, every newly-fired live bullet is peaceful because a leader becomes shielded and disables the bullet-absence signal when it fires a live bullet. Thus, once the population reaches a configuration where every live bullet is peaceful and there is one or more leaders, the number of leaders never becomes zero. Formally, we define the set of such configurations as follows:

$$\mathcal{C}_{\text{PB}} = \left\{ C \in \mathcal{C}_{\text{all}}(P_{RL}) \;\middle|\; \begin{array}{l} \exists u_i \in V : C(u_i).\texttt{leader} = 1 \\ \wedge \, \forall u_j \in V : C(u_j).\texttt{bullet} = 2 \Rightarrow \textit{Peaceful}(C, j) \end{array} \right\}.$$

The following lemma holds from the above discussion.

Lemma 2. $\mathcal{C}_{\text{PB}}$ *is closed.*

Thus, once the population reaches a configuration in $\mathcal{C}_{\text{PB}}$, there is always one or more leaders.

In protocol P_{RL}, a new leader is created when $\texttt{distL}$ of some agent reaches N. We require this mechanism to create a new leader when there is no leader. However, it is undesirable that a new leader is created when there is already one or more leaders. We say that an agent u_i is *secure* when the following predicate holds:

$$Secure(i) \stackrel{\text{def}}{\equiv} \begin{cases} u_i.\texttt{distL} = 0 & u_i.\texttt{leader} = 1 \\ u_i.\texttt{distL} \leq N - d_R(i) & \text{otherwise} \end{cases}.$$

One may think that no leader is created once the population reaches a configuration in $\mathcal{C}_{\text{PB}}$ such that all agents are secure. Unfortunately, this does not hold. For example, consider the case $n = N = 100$ and a configuration $C \in \mathcal{C}_{\text{PB}}$ where

- only two agents u_0 and u_{50} are leaders,
- $u_0.\texttt{distL} = u_{50}.\texttt{distL} = 0$,
- $u_i.\texttt{distL} = 100 - d_R(i)$ for all $i = 1, 2, \ldots, 49, 51, 52, \ldots, 100$,
- u_{49} carries a live (and peaceful) bullet in C, *i.e.*, $u_{49}.\texttt{bullet} = 2$, and
- u_{50} is not shielded, *i.e.*, $u_{50}.\texttt{shield} = 0$.

Note that the above condition does not contradict the assumption $C \in \mathcal{C}_{\text{PB}}$. In this configuration, all agents are secure. However, starting from this configuration, the population may create a new leader even when another leader exists. In configuration C, $u_{49}.\texttt{distL} = 99$. If u_{49} and u_{50} have two interactions in a row, then u_{50} becomes a follower in the first interaction, and $u_{49}.\texttt{distL} + 1 = 100$ is substituted for $u_{50}.\texttt{distL}$ and u_{50} becomes a leader again in the second interaction (even though u_0 is a leader during this period).

We introduce the definition of *modest bullets* to clarify the condition by which a new leader is no longer created. A live bullet located at u_i is said to be *modest* when the following predicate holds:

$$Modest(i) \stackrel{\text{def}}{\equiv} Peaceful(i) \wedge \forall j = 0, 1, \ldots, d_L(i) : u_{i-j}.\texttt{distL} \leq d_L(i - j).$$

As we will see soon, a new leader is no longer created in an execution starting from a configuration in $\mathcal{C}_{\text{PB}}$ where all agents are secure and all live bullets are modest. Note that in the above example, a live bullet located at u_{49} in C is not modest. We define a set $\mathcal{C}_{\text{NI}}$ of configurations as follows:

$$\mathcal{C}_{\text{NI}} = \left\{ C \in \mathcal{C}_{\text{PB}} \;\middle|\; \begin{array}{l} \forall u_i \in V : Secure(C, i) \\ \wedge \, (C(u_i).\texttt{bullet} = 2 \Rightarrow Modest(C, i)) \end{array} \right\}.$$

Lemma 3. *A modest bullet never becomes non-modest.*

Lemma 4. *A newly-fired live bullet is modest.*

Proof. Assume that a leader u_i fires a live bullet b at interaction (u_i, u_{i+1}) in $C \to C'$. Bullet b immediately disappears by Lines 14 and 16 if u_{i+1} is a leader or has a bullet in C. Otherwise, b moves to u_{i+1}. Then, $u_i.\texttt{shield} = 1$, $u_i.\texttt{distL} = 0$, $u_{i+1}.\texttt{distL} = 1$, and $u_i.\texttt{signal} = u_{i+1}.\texttt{signal} = 0$ must hold in C', which yields that b is modest in C'. □

Lemma 5. *Let C be any configuration where all live bullets are modest and C' any configuration such that $C \to C'$. Then, a secure agent u_i becomes insecure in $C \to C'$ only if u_i interacts with u_{i-1} in $C \to C'$ and u_{i-1} is insecure in C.*

Lemma 6. $\mathcal{C}_{\mathrm{NI}}$ *is closed.*

Proof. Immediately follows from Lemmas 2, 3, 4, and 5. □

Lemma 7. *No new leader is created in any execution starting from any configuration in $\mathcal{C}_{\mathrm{NI}}$.*

Proof. Since $\mathcal{C}_{\mathrm{NI}}$ is closed by Lemma 6, every configuration that appears in an execution starting from a configuration in $\mathcal{C}_{\mathrm{NI}}$ is also in $\mathcal{C}_{\mathrm{NI}}$. All agents are secure in a configuration in $\mathcal{C}_{\mathrm{NI}}$. Thus, $u_i.\texttt{distL} \leq N - 1$ holds for all $u_i \in V$ and $u_i.\texttt{distL} = N - 1$ holds only if u_{i+1} is a leader. Therefore, in a configuration in $\mathcal{C}_{\mathrm{NI}}$, no agent increases its $\texttt{distL}$ to N, thus no new leader is created. □

Finally, we define $\mathcal{S}_{\mathrm{RL}}$ as the set of all configurations included in $\mathcal{C}_{\mathrm{NI}}$ where there is exactly one leader.

Lemma 8. $\mathcal{S}_{\mathrm{RL}}$ *is closed and includes only safe configurations.*

Proof. Let C be any configuration in $\mathcal{S}_{\mathrm{RL}}$ and C' any configuration such that $C \to C'$. Since $\mathcal{C}_{\mathrm{NI}}$ is closed by Lemma 6 and exactly one agent is a leader in C, it suffices to show that no one changes its output (*i.e.*, the value of variable $\texttt{leader}$) in $C \to C'$. Since $C \in \mathcal{C}_{\mathrm{PB}}$, the unique leader in C is never killed in $C \to C'$. By Lemma 7, no other agent becomes a leader in $C \to C'$. Thus, no agent changes its output in $C \to C'$. □

4.2   Convergence

In this subsection, we prove that an execution of P_{RL} starting from any configuration in $\mathcal{C}_{\mathrm{all}}(P_{RL})$ reaches a configuration in $\mathcal{S}_{\mathrm{RL}}$ within $O(nN)$ steps in expectation. Formally, for any $C \in \mathcal{C}_{\mathrm{all}}(P_{RL})$ and $\mathcal{S} \subseteq \mathcal{C}_{\mathrm{all}}(P_{RL})$, we define $ECT(C, \mathcal{S})$ as the expected number of steps that execution $\Xi_{P_{RL}}(C, \Gamma)$ requires to reach a configuration in $\mathcal{S}$. The goal of this subsection is to prove $\max_{C \in \mathcal{C}_{\mathrm{all}}(P_{RL})} ECT(C, \mathcal{S}_{\mathrm{RL}}) = O(nN)$. We give this upper bound by showing $\max_{C \in \mathcal{C}_{\mathrm{all}}(P_{RL})} ECT(C, \mathcal{C}_{\mathrm{NI}}) = O(nN)$ and $\max_{C \in \mathcal{C}_{\mathrm{all}}(\mathcal{C}_{\mathrm{NI}})} ECT(C, \mathcal{S}_{\mathrm{RL}}) = O(n^2)$ in Lemmas 11 and 13, respectively.

In this subsection, we denote interaction (u_i, u_{i+1}) by e_i. In addition, for any two sequences of interactions $s = e_{k_0}, e_{k_1}, \ldots, e_{k_h}$ and $s' = e_{k'_0}, e_{k'_1}, \ldots, e_{k'_j}$, we define $s \cdot s' = e_{k_0}, e_{k_1}, \ldots, e_{k_h}, e_{k'_0}, e_{k'_1}, \ldots, e_{k'_j}$. That is, we use "$\cdot$" for the concatenation operator. For any sequence s of interactions and integer $i \geq 1$, we define s^i by induction: $s^1 = s$ and $s^i = s \cdot s^{i-1}$. For any $i, j \in \{0, 1, \ldots, n-1\}$, we define $seq_R(i, j) = e_i, e_{i+1}, \ldots, e_j$ and $seq_L(i, j) = e_i, e_{i-1}, \ldots, e_j$.

Definition 2. Let $\gamma = e_{k_1}, e_{k_2}, \ldots, e_{k_h}$ be a sequence of interactions. We say that γ occurs within l steps when $e_{k_1}, e_{k_2}, \ldots, e_{k_h}$ occurs in this order (not necessarily in a row) within l steps. Formally, the event "γ occurs within l steps from a time step t" is defined as the following event: $\Gamma_{t_i} = e_{k_i}$ holds for all $i = 1, 2, \ldots, h$ for some sequence of integers $t \leq t_1 < t_2 < \cdots < t_h \leq t + l - 1$. We say that from step t, γ completes at step $t + l$ if γ occurs within l steps but does not occur within $l - 1$ steps. When t is clear from the context, we write "γ occurs within l steps" and "γ completes at step l", for simplicity.

Lemma 9. *From any time step, a sequence* $\gamma = e_{k_1}, e_{k_2}, \ldots, e_{k_h}$ *with length h occurs within nh steps in expectation.*

Proof. For any interaction e_i, at each step, e_i occurs with probability $1/n$. Thus, e_i occurs within n steps in expectation. Therefore, γ occurs within nh steps in expectation. □

Lemma 10. *Let C be a configuration where no leader exists. In execution $\Xi = \Xi_{P_{RL}}(C, \Gamma)$, a leader is created within $O(nN)$ steps in expectation.*

Lemma 11. $\max_{C \in \mathcal{C}_{\mathrm{all}}(P_{RL})} ECT(C, \mathcal{C}_{\mathrm{NI}}) = O(nN)$.

Proof. Let C_0 be any configuration in $\mathcal{C}_{\mathrm{all}}(P_{RL})$ and consider $\Xi = \Xi_{P_{RL}}(C_0, \Gamma)$. All bullets that exist in C_0 disappear before $\gamma = (seq_R(0, n-1))^{\lceil N/n \rceil + 1}$ completes by Lemma 1, while γ occurs within $O(nN)$ steps in expectation by Lemma 9. Thus, by Lemmas 3, 4 and 10, within $O(nN)$ steps in expectation, Ξ reaches a configuration C' where all live bullets are modest, there is at least one leader, and every leader u_i satisfies $u_i.\mathtt{distL} = 0$.

Let Ξ' be the suffix of Ξ after Ξ reaches C'. In the rest of this proof, we show that Ξ' reaches a configuration in $\mathcal{C}_{\mathrm{NI}}$ within $O(n^2)$ steps. Since $\mathcal{C}_{\mathrm{PB}}$ is closed (Lemma 2) and $C' \in \mathcal{C}_{\mathrm{PB}}$, there is always at least one leader in Ξ'. Thus, by Lemmas 3 and 4, it suffices to show that Ξ' reaches a configuration where all agents are secure within $O(n^2)$ steps in expectation.

Here, we have the following two claims.

Claim 1. *In Ξ', once an agent u_i becomes a leader, u_i is always secure thereafter (even after it becomes a follower).*

Proof. Suppose that u_i is a leader in some point of Ξ'. At this time, $u_i.\mathtt{distL} = 0$. As long as u_i is a leader, u_i is secure. Agent u_i becomes a follower only when a live bullet reaches u_i. In Ξ', all live bullets are modest. This yields that, when u_i becomes a follower, all agents $u_i, u_{i-1}, \ldots, u_{i-d_L(i)}$ are secure. Thus, letting

$u_j = u_{i-d_L(i)}$, agent u_i is secure as long as u_j is secure. Similarly, u_j is secure as long as u_j is a leader. Even if u_j becomes a follower, there is a leader u_k such that u_j is secure as long as u_k is a leader, and so on. Therefore, u_i never becomes insecure. □

Claim 2. *In Ξ', an insecure agent u_i becomes secure if it interacts with u_{i-1} when u_{i-1} is secure.*

Proof. Let $C \to C'$ be any transition (u_{i-1}, u_i) that appears in Ξ such that u_{i-1} is secure in C. If u_{i-1} is a leader in C (thus in C'), $C'(u_i).\texttt{distL} \le 1 \le N - d_R(C', i)$. Otherwise, $d_R(C', i) = d_R(C', i-1) + 1$ must hold. By Lemma 5, u_{i-1} is still secure in C', hence $C'(u_i).\texttt{distL} \le C'(u_{i-1}).\texttt{distL} + 1 \le N - d_R(C', i-1) + 1 = N - d_R(C', i)$. Thus, u_i is secure in C' in both cases. □

Let u_i be a leader in C'. By Lemma 5 and Claims 1 and 2, all agents are secure when $seq_R(i, i-2)$ completes. This requires $O(n^2)$ expected steps by Lemma 9. □

As mentioned above, we give only a proof sketch for the following lemma. The sketch gives the reader the intuition why this lemma holds, but it lacks a strict analysis on probability. See a preprint [23] for the complete proof.

Lemma 12. *Let C_0 be a configuration in $\mathcal{C}_{\text{NI}}$ where there are at least two leaders. Let u_i, u_j, and u_k be the leaders in C_0 such that $i = j - d_L(C_0, j)$ and $j = k - d_L(C_0, k)$, i.e., u_i is the nearest left leader of u_j and u_j is the nearest left leader of u_k in C_0. ($u_i = u_k$ may hold.) Let $d_1 = d_L(C_0, j)$ and $d_2 = d_L(C_0, k)$. Then, in an execution $\Xi = \Xi_{PRL}(C_0, \Gamma) = C_0, C_1, \dots,$ the event that one of the three leaders becomes a follower occurs within $O(n(d_1 + d_2))$ steps in expectation.*

Proof Sketch. We show that u_j is killed within $O(n(d_1 + d_2))$ steps as long as both u_i and u_j remain leaders. Leader u_j fires a bullet at least once in Ξ' before or when $\gamma = seq_R(j, k-1) \cdot seq_L(k-1, j) \cdot e_j$ completes. Thereafter, at any step, u_j is unshielded with probability $1/2$ as long as u_k remains a leader. The event u_i fires a bullet at least once and the bullet reaches u_j before or when $\gamma' = seq_R(i, j-1) \cdot seq_L(j-1, i) \cdot seq_R(i, j-1)$ completes. Thus, each time $\gamma \cdot \gamma'$ completes, the event that a live bullet reaches u_j at the time u_j is shielded occurs with probability at least $(1/2) \cdot (1/2) = 1/4$. By Lemma 9, u_j is killed in $O(n(d_1 + d_2))$ expected steps as long as both u_i and u_j remain leaders. □

Lemma 13. $\max_{C \in \mathcal{C}_{\text{all}}(\mathcal{C}_{\text{NI}})} ECT(C, \mathcal{S}_{\text{RL}}) = O(n^2)$.

Proof. Let C_0 be any configuration in $\mathcal{C}_{\text{NI}}$ and let $\Xi = \Xi_{PRL}(C_0, \Gamma)$. By Lemmas 6 and 7, the number of leaders is monotonically non-increasing and never becomes zero in Ξ. For any real number x, define $\mathcal{L}_x$ as the set of configurations where the number of leaders is at most x. First, we prove the following claim.

Claim 3. *Let $\alpha = 12/11$. For any sufficiently large integer $k = O(1)$, if $C_0 \in (\mathcal{L}_{\alpha^{k+1}} \setminus \mathcal{L}_{\alpha^k}) \cap \mathcal{C}_{\text{NI}}$, execution Ξ reaches a configuration in $\mathcal{L}_{\alpha^k} \cap \mathcal{C}_{\text{NI}}$ within $O(n^2/\alpha^k)$ steps in expectation.*

Proof. Let $l_0, l_1, \ldots, l_{s-1} = u_{\pi_0}, u_{\pi_1}, \ldots, u_{\pi_{s-1}}$ be the leaders in C, where $\pi_0 < \pi_1 < \cdots < \pi_{s-1}$. We say that l_j and $l_{j+1 \bmod s}$ are neighboring leaders for each $j = 0, 1, \ldots, s-1$. Since there are s leaders in C_0, there are at least $3s/4$ leaders $l_i = u_{\pi_i}$ such that $d_L(\pi_{i+1 \bmod s}) \leq 4n/s$. Thus, there are at least $n/2$ leaders $l_j = u_{\pi_j}$ such that $d_L(\pi_{j+1 \bmod s}) \leq 4n/s$ and $d_L(\pi_{j+2} \bmod s) \leq 4n/s$ in C_0. Let S_L be the set of all such leaders. For each $l_j \in S_L$, by Lemma 12 and Markov inequality, l_j, $l_{j+1 \bmod s}$, or $l_{j+2 \bmod s}$ becomes a follower within $O(n^2/s)$ steps with probability $1/2$. Generally, if $X_0, X_1, \ldots, X_i$ are (possibly non-independent) events each of which occurs with probability at least $1/2$, at least half of the events occur with probability at least $1/2$. Thus, with probability $1/2$, for at least half of the leaders l_j in S_L, the event that l_j, $l_{j+1 \bmod s}$, or $l_{j+2 \bmod s}$ becomes a follower occurs within $O(n^2/s)$ steps. Thus, at least $|S_L| \cdot (1/2) \cdot (1/3) = s/12$ leaders become followers within $O(n^2/s)$ steps with probability at least $1/2 = \Omega(1)$. Repeating this analysis, we observe that Ξ reaches a configuration in $\mathcal{L}_{\alpha^k} \cap \mathcal{C}_{\mathrm{NI}}$ within $O(n^2/\alpha^k)$ steps in expectation.   □

By Claim 3, for sufficiently large integer $k = O(1)$, the number of leaders becomes a constant (*i.e.*, $O(\alpha^k) = O(1)$) within $\sum_{i=k}^{\lceil \log_\alpha n \rceil} O(n^2/\alpha^i) = O(n^2)$ steps in expectation in Ξ. Thereafter, by Lemma 12, the number of leaders decreases to one within $O(n^2)$ steps in expectation.   □

Lemmas 8, 11, and 13 give the following main theorem.

Theorem 1. *Given an integer N, P_{RL} is a self-stabilizing leader election protocol for any directed rings of any size $n \leq N$. The convergence time is $O(nN)$. The number of states is $O(N)$.*

5   Conclusion

We have presented a self-stabilizing leader election protocol for directed rings in the population protocol, given the knowledge of an upper bound N of the population size n. Specifically, an execution of the protocol starting from any initial configuration elects a unique leader within $O(nN)$ steps in expectation, by using $O(N)$ states per agent. If a given knowledge N is asymptotically tight, *i.e.*, $N = O(n)$, this protocol is time-optimal.

References

1. Angluin, D., Aspnes, J., Fischer, M.J., Jiang, H.: Self-stabilizing population protocols. ACM Trans. Auton. Adapt. Syst. **3**(4), 1–28 (2008)
2. Angluin, D., Aspnes, J., Diamadi, Z., Fischer, M.J., Peralta, R.: Computation in networks of passively mobile finite-state sensors. Distrib. Comput. **18**(4), 235–253 (2006)
3. Angluin, D., Aspnes, J., Eisenstat, D.: Fast computation by population protocols with a leader. Distrib. Comput. **21**(3), 183–199 (2008)
4. Beauquier, J., Blanchard, P., Burman, J.: Self-stabilizing leader election in population protocols over arbitrary communication graphs. In: International Conference on Principles of Distributed Systems, pp. 38–52 (2013)

5. Burman, J., Doty, D., Nowak, T., Severson, E.E., Xu, C.: Efficient self-stabilizing leader election in population protocols. arXiv preprint arXiv:1907.06068 (2019)
6. Cai, S., Izumi, T., Wada, K.: How to prove impossibility under global fairness: on space complexity of self-stabilizing leader election on a population protocol model. Theory Comput. Syst. **50**(3), 433–445 (2012)
7. Canepa, D., Potop-Butucaru, M.G.: Stabilizing leader election in population protocols (2007). http://hal.inria.fr/inria-00166632
8. Chen, H.P., Chen, H.L.: Self-stabilizing leader election. In: Proceedings of the 38th ACM Symposium on Principles of Distributed Computing, pp. 53–59 (2019)
9. Chen, H.P., Chen, H.L.: Self-stabilizing leader election in regular graphs. In: Proceedings of the 39th Symposium on Principles of Distributed Computing, pp. 210–217 (2020)
10. Cordasco, G., Gargano, L.: Space-optimal proportion consensus with population protocols. In: International Symposium on Stabilization, Safety, and Security of Distributed Systems, pp. 384–398 (2017)
11. Dijkstra, E.: Self-stabilizing systems in spite of distributed control. Commun. ACM **17**(11), 643–644 (1974)
12. Fischer, M.J., Jiang, H.: Self-stabilizing leader election in networks of finite-state anonymous agents. In: International Conference on Principles of Distributed Systems, pp. 395–409 (2006)
13. Izumi, T.: On space and time complexity of loosely-stabilizing leader election. In: International Colloquium on Structural Information and Communication Complexity, pp. 299–312 (2015)
14. Mertzios, G.B., Nikoletseas, S.E., Raptopoulos, C.L., Spirakis, P.G.: Determining majority in networks with local interactions and very small local memory. In: International Colloquium on Automata, Languages, and Programming, pp. 871–882 (2014)
15. Sudo, Y., Ooshita, F., Kakugawa, H., Masuzawa, T.: Loosely-stabilizing leader election on arbitrary graphs in population protocols. In: International Conference on Principles of Distributed Systems, pp. 339–354 (2014)
16. Sudo, Y., Eguchi, R., Izumi, T., Masuzawa, T.: Time-optimal loosely-stabilizing leader election in population protocols. arXiv preprint arXiv:2005.09944 (2020)
17. Sudo, Y., Masuzawa, T., Datta, A.K., Larmore, L.L.: The same speed timer in population protocols. In: The 36th IEEE International Conference on Distributed Computing Systems, pp. 252–261 (2016)
18. Sudo, Y., Nakamura, J., Yamauchi, Y., Ooshita, F., Kakugawa, H., Masuzawa, T.: Loosely-stabilizing leader election in a population protocol model. Theoret. Comput. Sci. **444**, 100–112 (2012)
19. Sudo, Y., Ooshita, F., Kakugawa, H., Masuzawa, T.: Loosely stabilizing leader election on arbitrary graphs in population protocols without identifiers or random numbers. IEICE Trans. Inf. Syst. **103**(3), 489–499 (2020)
20. Sudo, Y., Ooshita, F., Kakugawa, H., Masuzawa, T., Datta, A.K., Larmore, L.L.: Loosely-stabilizing leader election for arbitrary graphs in population protocol model. IEEE Trans. Parallel Distrib. Syst. **30**(6), 1359–1373 (2018)
21. Sudo, Y., Ooshita, F., Kakugawa, H., Masuzawa, T., Datta, A.K., Larmore, L.L.: Loosely-stabilizing leader election with polylogarithmic convergence time. Theoret. Comput. Sci. **806**, 617–631 (2020)

22. Sudo, Y., Shibata, M., Nakamura, J., Kim, Y., Masuzawa, T.: The power of global knowledge on self-stabilizing population protocols. In: International Colloquium on Structural Information and Communication Complexity, pp. 237–254 (2020)
23. Yokota, D., Sudo, Y., Masuzawa, T.: Time-optimal self-stabilizing leader election on rings in population protocols. arXiv preprint arXiv:2009.10926 (2020)

Brief Announcement: Effectiveness of Code Hardening for Fault-Tolerant IoT Software

Igor Zavalyshyn[✉], Thomas Given-Wilson, Axel Legay, and Ramin Sadre

UCLouvain, Louvain-la-Neuve, Belgium
igor.zavalyshyn@uclouvain.be

Abstract. Internet of Things (IoT) device software has to be resistant to faults to ensure data privacy and security. In this work, we examine five common software hardening techniques and study their impact on software fault-tolerance and security. We experimentally show that some of these techniques may improve fault-tolerance, while the others can *reduce* overall security. We offer a guideline for IoT developers seeking to make their software robust, and propose a tool for automatic software fault-tolerance evaluation.

1 Introduction

In order to preserve the end-users privacy IoT devices usually encrypt the sensor data before transmitting it (e.g. to the local hub, mobile phone or cloud server). However, a single fault in the encryption logic, introduced either accidentally (e.g. electromagnetic interference) or intentionally by a malicious attacker, may cause sensitive data leaks.

To make the devices more resistant to faults, IoT manufacturers often harden the software components by adding a safety logic that aims to detect the presence of faults and minimize their impact. While there is a variety of hardening techniques [6], in practice, IoT software developers rarely have a clear understanding of the real impact of a chosen hardening technique on their software's fault-tolerance and performance.

In this paper we present a thorough IoT software hardening analysis. We select five popular and (potentially) automated hardening techniques applied to implementation of PRESENT – a lightweight block cipher targeting IoT devices. We then expose the hardened versions of PRESENT to six kinds of faults, and evaluate the effectiveness of the hardening techniques on three fronts: (1) their ability to prevent sensitive data leaks; then (2) their general fault-tolerance and the impact of each fault type; and, finally, (3) the impact of hardening techniques on software performance and binary size.

We show experimentally that techniques exploring redundancy on a function level provide a good balance between software security and general fault-tolerance, while classic loop hardening techniques in some cases make the hardened software less fault-tolerant. To facilitate the analysis we have developed

© Springer Nature Switzerland AG 2020
S. Devismes and N. Mittal (Eds.): SSS 2020, LNCS 12514, pp. 317–322, 2020.
https://doi.org/10.1007/978-3-030-64348-5_25

Chaos Duck – a tool for automatic software fault-tolerance evaluation. The results in this paper exploit Chaos Duck and demonstrate its utility in detecting sensitive data leaks, program crashes and corruptions in data and control flows caused by injected faults.

2   Case Study and Methodology

In this work we evaluate the efficiency of five common software hardening techniques applied to the implementation of PRESENT [3] – a block cipher specifically designed for low-power resource-constrained IoT devices. Listing 1.1 illustrates an implementation of PRESENT used in a heart rate monitor software.

We consider five popular hardening techniques in this case study: *classic loop hardening* (CLH) [2] (Listing 1.2); *variable duplication* (VD) [4] (Listing 1.3); *function duplication* (FD) [1] (Listing 1.4); *decryption at place* (DaP) [1] (Listing 1.5); and *statement-based counters* (SC) [11] (Listing 1.6).

To evaluate the effectiveness of hardening we consider a set of fault models commonly used to find vulnerabilities in implementations of encryption algorithms [7]: branch instruction faults (both unconditional B and conditional BC) that aim to disrupt the control flow; *FLP* faults flipping instruction bits; *NOP* faults that skip an instruction; and *Z1B* and *Z1W* faults that set a single instruction byte or word (respectively) to zero.

We consider fault injection on a binary level, as it remains one of the main attack vectors in the IoT environment [8,9]. For our experiments we developed Chaos Duck[1] – an automatic tool that given a binary file produces *"faulted"* binaries (one binary per fault injected), and evaluates the impact of each fault.

We compare five implementations of PRESENT hardened with techniques described above with a non-hardened one (*i.e. baseline*)[2]. For each fault model, every possible fault location is considered. We use a set of three encryption keys and three plaintexts resulting in nine executions per binary with a 3 s timeout for each. We differentiate between normally terminated binaries and those that were interrupted by timeout. We measure the total number of faulted binaries and collect statistics on fault types and their success rate. The hardening efficiency is evaluated based on its ability to prevent sensitive data leaks in general and under a differential fault attack (DFA) [5]. Finally, we measure the average execution time across 10000 executions with randomly generated *key,plaintext* pairs, and record the size in bytes for all binaries.

3   Results

With respect to data leaks, none of the hardening techniques were able to prevent plaintext leakage (see Table 1), which is in line with our previous results [10].

---

[1] Available from https://github.com/zavalyshyn/chaosduck.
[2] Available from http://www.lightweightcrypto.org/implementations.php.

```
1  encrypt(state,key) {
2    int round = 0;
3    while(round < 31) {
4      addRoundKey(state,key);
5      sBoxLayer(state);
6      pLayer(state);
7      round++;
8    }
9    addRoundKey(state,key);
10 }
11 reportcycle() {
12   state = sense();
13   key = {0xd3, 0xe4 ... 0xba};
14   encrypt(state,key);
15   transmit(state);
16 }
```

Listing 1.1. A pseudocode of heart rate monitor's software using PRESENT.

```
1  encrypt(state,key) {
2    int round = 0, round_dup = 0;
3    while((round < 31) &&
4    (round_dup < 31)) {
5      addRoundKey(state,key);
6      sBoxLayer(state);
7      pLayer(state);
8      round++; round_dup++;
9    }
10   if (round!=round_dup) error();
11   addRoundKey(state,key);
12 }
```

Listing 1.2. Classic loop hard. (CLH).

```
1  encrypt(state,key) {
2    int round = 0, round_dup = 0;
3    while(round < 31) {
4      addRoundKey(state,key);
5      sBoxLayer(state);
6      pLayer(state);
7      if (round!=round_dup) error();
8      round++; round_dup++;
9    }
10   if (round!=round_dup) error();
11   addRoundKey(state,key);
12 }
```

Listing 1.3. Variable duplication (VD).

```
1  reportcycle() {
2    state = sense();
3    key = {0xd3, 0xe4 ... 0xba};
4    for (int i=0; i<8; i++) {
5      copy[i] = state[i];
6    }
7
8    encrypt(state,key);
9    encrypt_dup(copy,key);
10
11   for (int i=0; i<8; i++) {
12     if (state[i]!=copy[i]) error();
13   }
14   transmit(state);
15 }
```

Listing 1.4. Function duplication (FD).

```
1  reportcycle() {
2    state = sense();
3    key = {0xd3, 0xe4 ... 0xba};
4    for (int i=0; i<8; i++) {
5      copy[i] = state[i];
6    }
7
8    encrypt(state,key);
9    decrypt(state,key);
10
11   for (int i=0; i<8; i++) {
12     if (state[i]!=copy[i]) error
13     ();
14   transmit(state)
15 }
```

Listing 1.5. Decryption at place (DaP).

```
1  #define DECL_INIT(cnt,x) int cnt;
        if((cnt=x)!=x) error();
2  #define CHECK_INC(cnt,x) cnt=(cnt
        ==x ? cnt+1 : error());
3  #define RESET_CNT(cnt_while,val)
        (cnt_while==1||cnt_while==
        val) ? cnt_while=1 : error()
        ;
4  #define CHECK_LOOP_INC(cnt_loop,x
        ) (cnt_loop==x) ? cnt_loop
        +=1 : error();
5  #define CHECK_LOOP_END(cnt_loop,
        val) if (cnt_loop!=val)
        error();
6  encrypt(state,key) {
7    DECL_INIT(enc_cnt,1);
8    CHECK_INCR(enc_cnt,1);
9    int round = 0;
10   CHECK_INC(enc_cnt,2);
11   DECL_INIT(while_cnt,1);
12   CHECK_INC(enc_cnt,3);
13   DECL_INIT(loop_cnt,0);
14   CHECK_INC(enc_cnt,4);
15   while(round < 31) {
16     RESET_CNT(while_cnt,6);
17     CHECK_LOOP_INC(loop_cnt,round)
          ;
18     CHECK_INC(while_cnt,1);
19     addRoundKey(state,key);
20     CHECK_INC(while_cnt,2);
21     sBoxLayer(state);
22     CHECK_INC(while_cnt,3);
23     pLayer(state);
24     CHECK_INC(while_cnt,4);
25     round++;
26     CHECK_INC(while_cnt,5);
27   }
28   CHECK_INC(enc_cnt,5);
29   CHECK_LOOP_END(loop_cnt,31);
30   CHECK_INC(enc_cnt,6);
31   addRoundKey(state,key);
32   CHECK_INC(enc_cnt,7);
33 }
```

Listing 1.6. Statement counters (SC) hardening.

Table 1. Sensitive data leakage across five hardening techniques.

	Baseline	CLH	VD	SC	FD	DaP
Binaries	1314549	4818348	9267993	52341831	3580380	3902400
Leaked key (normal/timeout)	0 / 0	0 / 0	0 / 0	0 / 0	0 / 0	0 / 0
Leaked plaintext (normal/timeout)	1713 / 180	1449 / 0	3559 / 4	567 / 0	0 / 72	108 / 72
DFA vulnerable	2 / 0	9 / 0	2 / 0	0 / 0	0 / 0	0 / 0

Table 2. Fault types statistics for faulted binaries leaking sensitive data (normally terminated vs. terminated by timeout).

	Baseline	CLH	VD	SC	FD	DaP
Binaries	1713 / 180	1449 / 0	3559 / 4	567 / 0	0 / 72	108 / 72
FLP	45 / 108	9 / 0	28 / 0	234 / 0	0 / 0	36 / 0
Z1B/Z1W	0 / 0	0 / 0	0 / 0	0 / 0	0 / 0	0 / 0
NOP	0 / 0	0 / 0	0 / 0	9 / 0	0 / 0	0 / 0
B	852 / 0	396 / 0	341 / 1	324 / 0	0 / 0	0 / 0
BC	816 / 72	1044 / 0	3190 / 3	0 / 0	0 / 72	72 / 72

Table 3. Execution results for five hardening techniques in presence of faults.

	Baseline	CLH	VD	SC	FD	DaP
Binaries	1314549	3836889	5639832	52341831	3580380	3902400
Valid cipher	16.7 %	38.64 %	34.07 %	62.87 %	16.99 %	14.02 %
Invalid cipher	17.3 %	2 %	1.64 %	0.19 %	0.57 %	0.55 %
No output	66 %	59.36 %	64.29 %	36.94 %	82.45 %	85.43 %

FD and DaP techniques were more effective since they check final output values instead of intermediate values, e.g. loop counters.

Branch instruction faults were the most common cause of sensitive data leaks (see Table 2). Hence hardening techniques that add more branches to the original code (e.g. SC technique) should be avoided, as new branches inadvertently increase the attack surface. Alternative techniques, e.g. FD and DaP, proved to be less affected by this type of fault.

Our experiments showed that the majority of faulted binaries simply failed to execute correctly without producing a valid ciphertext (see Table 3). However, the SC technique turned out to be less affected by the fault presence.

Failures were mostly caused by the binaries getting stuck in an infinite loop and then interrupted by timeout (see Table 4). The next most common cause was a segmentation fault, followed by illegal instruction and aborted faults. The fault detection rate is rather low, barely reaching 10% for most of the hardening techniques, with the sole exception of CLH reaching 80% detection rate.

Table 4. Statistics on failed executions.

	Baseline	CLH	VD	SC	FD	DaP
Crashed binaries	**867852**	**1862129**	**4807851**	**19334285**	**3505885**	**3845470**
Seg. fault	56.11 %	35.21 %	20.99 %	17.36 %	45.28 %	60.81 %
Timed out	30.65 %	6.81 %	43.46 %	0.48 %	15.33 %	9.87 %
Illegal instr.	8.83 %	1.82 %	0.85 %	0.65 %	1.83 %	2.25 %
Aborted	0.63 %	1.7 %	0.68 %	0.11 %	0.29 %	0.29 %
Fault detected	n/a	52.4 %	33.2 %	80.22 %	35.3 %	25.03 %
Other	3.78 %	2.06 %	0.82 %	1.18 %	1.97 %	1.75 %

All hardening techniques have no significant impact on execution time (52.4$\pm$ 2 ms), and (except SC) have little impact on binary size (+1KB).

4 Conclusions

Hardening techniques exploring redundancy on a function level strike a good balance between security and performance properties, while some of the classic techniques make the software more vulnerable to faults resulting in a plaintext being leaked. Although the experiments were performed on PRESENT, these results indicate hardening is generally ineffective in preventing faults. Chaos Duck proved to be a powerful tool for evaluating an IoT software security and robustness in the presence of faults.

References

1. Bar-El, H., Choukri, H., Naccache, D., Tunstall, M., Whelan, C.: The sorcerer's apprentice guide to fault attacks. Proc. IEEE **94**(2), 370–382 (2006)
2. Barbu, G., Andouard, P., Giraud, C.: Dynamic fault injection countermeasure. In: Proceedings of CARDIS (2012)
3. Bogdanov, A., et al.: Present: an ultra-lightweight block cipher. In: Proceedings of CHES (2007)
4. Cheynet, P., Nicolescu, B., Velazco, R., Rebaudengo, M., Reorda, M.S., Violante, M.: Experimentally evaluating an automatic approach for generating safety-critical software with respect to transient errors. IEEE Trans. Nucl. Sci. **47**(6), 2231–2236 (2000)
5. Ghalaty, N.F., Yuce, B., Schaumont, P.: Differential fault intensity analysis on present and led block ciphers. In: Proceedings of COSADE (2015)
6. Gilley, G.C., et al.: Ftcs, digest of papers (1971)
7. Given-Wilson, T., Heuser, A., Jafri, N., Legay, A.: An automated and scalable formal process for detecting fault injection vulnerabilities in binaries. Concurr. Comput.: Pract. Exp. **31**(23), e4794 (2019)
8. Given-Wilson, T., Jafri, N., Legay, A.: The state of fault injection vulnerability detection. In: Proceedings of VECoS (2018)

9. Given-Wilson, T., Jafri, N., Legay, A.: Combined software and hardware fault injection vulnerability detection. Innovations Syst. Softw. Eng. **16**(2), 101–120 (2020)
10. Given-Wilson, T., Legay, A.: Formalising fault injection and countermeasures. In: Proceedings of ARES (2020)
11. Lalande, J.F., Heydemann, K., Berthomé, P.: Software countermeasures for control flow integrity of smart card C codes. In: Proceedings of ESORICS (2020)

Author Index

Printed in the United States
By Bookmasters